T0314492

Data Science with Semantic Technologies

As data is an important asset for any organization, it is essential to apply semantic technologies in data science to fulfill the need of any organization. This first volume of a two-volume handbook set provides a roadmap for new trends and future developments of data science with semantic technologies.

Data Science with Semantic Technologies: New Trends and Future Developments highlights how data science enables the user to create intelligence through these technologies. In addition, this book offers the answers to various questions such as: Can semantic technologies facilitate data science? Which types of data science problems can be tackled by semantic technologies? How can data scientists benefit from these technologies? What is the role of semantic technologies in data science? What is the current progress and future of data science with semantic technologies? Which types of problems require the immediate attention of the researchers? What should be the 2030 vision for data science?

This volume can serve as an important guide toward applications of data science with semantic technologies for the upcoming generation and, thus, it is a unique resource for scholars, researchers, professionals, and practitioners in this field.

Data Science with Semantic Technologies

New Trends and Future Developments

Edited by

Archana Patel and Narayan C. Debnath

CRC Press is an imprint of the
Taylor & Francis Group, an **informa** business

Designed cover image: Shutterstock

First edition published 2023
by CRC Press
6000 Broken Sound Parkway NW, Suite 300, Boca Raton, FL 33487–2742

and by CRC Press
4 Park Square, Milton Park, Abingdon, Oxon, OX14 4RN

CRC Press is an imprint of Taylor & Francis Group, LLC

Library of Congress Cataloging-in-Publication Data
Names: Patel, Archana (Lecturer in software engineering), editor. | Debnath, N. C. (Narayan C.), editor.
Title: Data science with semantic technologies / edited by Archana Patel, Narayan C. Debnath.
Description: First edition. | Boca Raton : CRC Press, [2023] | Includes bibliographical references and index. | Contents: volume 1. New trends and future developments — volume 2. Deployment and exploration.
Identifiers: LCCN 2022056304 (print) | LCCN 2022056305 (ebook) | ISBN 9781032316666 (hardback; volume 1) | ISBN 9781032316673 (paperback; volume 1) | ISBN 9781032316680 (hardback; volume 2) | ISBN 9781032316697 (paperback; volume 2) | ISBN 9781003310785 (ebook; volume 1) | ISBN 9781003310792 (ebook; volume 2)
Subjects: LCSH: Big data. | Semantic computing. | Artificial intelligence. | Information science. | Information technology.
Classification: LCC QA76.9.B45 D397 2023 (print) | LCC QA76.9.B45 (ebook) | DDC 005.7—dc23/eng/20230109
LC record available at https://lccn.loc.gov/2022056304
LC ebook record available at https://lccn.loc.gov/2022056305

ISBN: 978-1-032-31666-6 (hbk)
ISBN: 978-1-032-31667-3 (pbk)
ISBN: 978-1-003-31078-5 (ebk)

DOI: 10.1201/9781003310785

Typeset in Times
by Apex CoVantage, LLC

Contents

Preface

As the world enters the era of big data, there is a serious need to give a semantic perspective to the data. At the end of 2020, more than 80% of the data collected by organizations was unstructured, and such enormous data required advanced techniques to discover meaningful insights from large and complex datasets. This requirement is being fulfilled by data science. Data Science provides an invaluable resource that deals with vast volumes of data using modern tools and techniques to find unseen patterns, derive meaningful information, and make business decisions. However, there are some major challenges of data science that include imbalanced data, data preparation, identifying patterns and anomalies, multiple data sources, data security and privacy, interoperability, prediction of accurate outcomes, visualization, effective data analysis, measuring the accuracy of results, and many more. These existing challenges create big obstacles for the data scientists in decision making. This is the place where semantic technologies play a vital role in eradicating these problems and offering accurate decisions for the organization by creating intelligence in data science.

The semantic technologies pillars are: Internationalized Resource Identifiers for indexing resources, the Resource Description Framework (RDF) and Resource Description Framework Schema (RDFS) for describing graphs of resources in terms of triples, the Web Ontology Language (OWL) for defining sophisticated logical models that describe resources, the SPARQL query language for querying graphs, and the Shapes Constraint Language (SHACL) for defining data integrity constraints on resources. Together these pillars provide the tools for a new kind of Internet and new kinds of systems that can leverage big data and provide synergy with machine learning algorithms.

As data is an important asset for any organization, it is essential to apply semantic technologies in data science to fulfil the need of any organization. This book aims to provide a roadmap for a new trend and future development of data science with semantic technologies. Moreover, it highlights how data science enables the user to create intelligence through these technologies. In addition, this book intends to provide the answers to various questions like: Can semantic technologies facilitate data science? Which types of data science problems can be tackled by semantic technologies? How can data scientists benefit from these technologies? What is the role of semantic technologies in data science? What is the current progress and future of data science with semantic technologies? Which types of problems require the immediate attention of the researchers? This book will potentially serve as an important guide toward applications of data science with semantic technologies for the upcoming generation and, thus, it is a unique resource for scholars, researchers, professionals, and practitioners in this field.

Preface

Contributors

Prakash Andugula
Independent Researcher
Maharashtra, India

Ahsar Ansari
NED University
Karachi, Pakistan

Y. Suresh Babu
Jagarlamudi Kuppuswamy Choudary College
Guntur, Andhra Pradesh, India

Ujwala Bharambe
Thadomal Shahani Engineering College
Mumbai, India

D.V. Chandrashekar
PG Department of Computer Science
Tellakula Jalayya Polisetty Somasundaram College
Guntur, India

Meghana Nagaraj Cilagani
Vellore Institute of Technology
AP University
Amaravathi, India

Christopher Connor
Independent Consultant
Columbus, Ohio, USA

Michael DeBellis
Independent Researcher
USA

Narayan C. Debnath
Department of Software Engineering
Eastern International University
Binh Duong Province, Vietnam

Houda El Bouhissi
University of Bejaia
Bejaia, Algeria

Marischa Elveny
Data Science & Computational Intelligence Research Group
Excellent Centre of Innovation and New Science Medan
Sumatera Utara, Indonesia

Syed Md Fazal
MANUU – Central University
Hyderabad, Telangana, India

Anjali Gautam
Vellore Institute of Technology
AP University
Amaravathi, India

Noman Islam
Karachi Institute of Economics and Technology
Karachi, Pakistan

Manh-Kha Kieu
Eastern International University
Binh Duong Province, Vietnam
and
RMIT University
Ho Chi Minh City, Vietnam

Nouluri Vamsi Krishna
Vellore Institute of Technology
AP University
Amaravathi, India

A.V. Senthil Kumar
Hindusthan College of Arts and Science (A)
Coimbatore-Tamilnadu,
India

Ngoc-Bich Le
International University
Ho Chi Minh City, Vietnam
and
Vietnam National University
Ho Chi Minh City, Vietnam

Ngoc-Huan Le
Eastern International University
Binh Duong Province, Vietnam

Paola Di Maio
CSKRNS (Center for Systems Knowledge Representation and Neuroscience)
Edinburgh, UK

Koteswara Rao Makkena
School of Computer Science and Engineering
Vellore Institute of Technology
AP University
Amaravathi, India

S. Divya Meena
Vellore Institute of Technology
AP University
Amaravathi, India

Mehmet Milli
Bolu Abant İzzet Baysal University
Bolu, Turkey

Musa Milli
Computer Engineering Department
Turkish Naval Academy
National Defense University
Istanbul, Turkey

Sheraz Mohsin
NED University
Karachi, Pakistan

Chhaya Narvekar
Xavier Institute of Engineering
Mumbai, Maharashtra, India

Mahyuddin K. M. Nasution
Data Science & Computational Intelligence Research Group
Excellent Centre of Innovation and New Science Medan
Sumatera Utara, Indonesia

Karthika Nataranjan
Assistant Professor Sr. Grade,
School of Computer Science and Engineering
Vellore Institute of Technology
AP University
Amaravati, India

Duc-Canh Nguyen
Eastern International University
Binh Duong Province, Vietnam

Vu-Anh-Tram Nguyen
Eastern International University
Binh Duong Province, Vietnam

Xuan-Hung Nguyen
Eastern International University
Binh Duong Province, Vietnam

Tran-Thuy-Duong Ninh
Eastern International University
Binh Duong Province, Vietnam

Archana Patel
National Forensic Sciences University
Gandhinagar, Gujarat, India

Amit Prakashrao Patil
RCPET Institute of Management Research and Development
Shirpur, Maharashtra, India

Chhaya Suhas Patil
RCPET Institute of Management Research and Development
Shirpur, Maharashtra, India

Livia Pinera
Reddit.com
San Francisco, California, USA

Pedaballi Rajeswari
Vellore Institute of Technology
AP University
Amaravathi, India

Thoom Purna Chander Rao
Vellore Institute of Technology
AP University
Amaravathi, India

Enayat Raza
NED University
Karachi, Pakistan

Fatmana Şentürk
Computer Engineering Department
Pamukkale University
Denizli, Turkey

J. Sheela
Vellore Institute of Technology
AP University
Amaravathi, India

Razeen Shuja
Institute of Business Administration
Karachi, Pakistan

P. Anushri Sowmya
Vellore Institute of Technology
AP University
Amaravathi, India

K. Suneetha
PG Department of Computer Science
Tellakula Jalayya Polisetty Somasundaram College
Guntur, India

Rahmad Syah
Data Science & Computational Intelligence Research Group
Excellent Centre of Innovation and New Science Medan
Sumatera Utara, Indonesia

Darakhshan Syed
Iqra University
Karachi, Pakistan

1 What is Data Science

Mahyuddin K. M. Nasution, Rahmad Syah and Marischa Elveny

CONTENTS

1.1 INTRODUCTION

From ancient times to modern times, whether humans realized it or not, data has always been a part of the lives of humans, even for simple things like the need for daily food.[1] In this case, eating becomes the subject of human thought, and the remaining internal data resides in that subject. Psychologically, reasoning will only find its essence when data becomes fact in the logic of language.[2] Indirectly, words are created in every language not just by assembling sounds but by stringing sound frequencies together according to the laws of nature, physics, or non-physics[3] and then by a culture creating symbols in the form of characters. Words become a series of characters generally recognized as alphabetic. The encoding has gone from frequency to character. Thus, data now consists of descriptions that flow into a recording device in a storage area in any computer machine or server.[4] The Internet and the web have made it easy for the encoding to go continuously through several built-in sensory-electronic devices, and this automatically serves the needs of humans.[5] So from the beginning, data was through the activities of humans until the presence of technology—computers, computer networks, the Internet, and the web—in addition to the basic sciences such as mathematics and statistics, provided an abundance of data that was only measurable through a scientific framework.[6]

DOI: 10.1201/9781003310785-1

Each science provides a scientific framework through its object of study and then organizes it through scientific methodology.[7] Organization based on the interaction between its components is a systematic system. That is, there are relationships between the objects themselves.[8] The essence of the human senses is generally in capturing the image of an object in accordance with what the object represents. Of course, conversely, data is not the object itself but the nature, characteristic, or description of that object. Ontologically, data is the existence of objects that can be certain or uncertain.[9, 10] Thus, the framework for data will differ from other sciences, and that scientific framework is introduced as data science. This chapter starts by introducing terms, concepts, and paradigms, and then it explains things that are necessary to recognize what data science is.

1.2 RELATED WORK

Data, is a subject that has always accompanied the thinking of humans since they appeared on earth.[11, 12–14] The subject—everything that affects the human mind—is indirectly attached to the spoken narrative, implicitly in the reasoning of or for reasoning it in any natural language, or explicitly being a fact that explains objects, whether tangible or intangible. From a language framework, and it is almost impossible to introduce it from another framework, *data* [15, 16] is an affirmation of the subject of thought and a term derived from the Latin plural form of *datum*,[17] which means "something which is given".[18, 19] For example, the mind of Thales of Miletus pioneered a science of measurement and created geometry based on a series of ground measuring instructions that the Babylonians and Egyptians had practiced for centuries.[20–22] However, rectangular geometric shapes have length and width. Of course, two such lines express the area between the units of measurement. Language reasoning of the area is the existence of length and width measurements in units with their respective values, and it produces what is called data. Simultaneously, in the same reasoning, this information describes a geometric shape—a rectangle—and geometry is a branch of mathematics. There is knowledge inherent in the minds of humans in the form of geometry and mathematics.[23]

Knowledge comes from data, especially during decision-making or for reaching wisdom.[24] Every decision determines the continuity and direction of a human's life journey. Wisdom aims to promote well-being. Therefore, languages reveal knowledge. However, the word *science* becomes the anchor for all terms about knowledge that come from different languages while insisting that there is a difference between each knowledge—a framework that distinctly organizes the subjects of thoughts. Science comes from the Latin word, *scientia*.[25] The terminology means a systematic effort [26] to build and organize [27] human familiarity, [28] awareness, [29] understanding, [30] or something in the form of facts, skill, entities, or objects [31] where explanations and predictions [32] make it possible to reveal [33] the universe of discourse [34] naturally within a framework.[35] Thus, science is not just knowledge; science is also an ability to organize knowledge based on agreed theories and is systematically tested with a set of methods recognized by that science.

On the one hand, the organization of knowledge makes sense of a complex system consisting of interacting components which reasonably express the characteristics of the cause and effect of the embodiment of that knowledge.[36] The organization converges all characteristics into an abstraction that reduces complexity. Abstraction of the basic story of the universe as the organization of the structured items of atoms is called physics;[37] accumulation of interactions between atoms into their natural compound is a science called chemistry;[38] compounds based on the atoms' interactions that form a large structure in an organism is summarized in the science name of biology.[39] Physics, chemistry, and biology are called natural sciences.[40] Naturally, human live on earth. Humans, one of the organisms that interact with the earth and other organisms as life support, call it the living environment.[41] The earth also depends on other planets and stars, such as the moon and sun in the sky. The latter part of the previous sentence refers to earth science,[42] while the former part of the sentence refers to space science.[43] It is strange to explore the potential of all the sciences in one natural framework. It naturally allows the study within a framework based on its components. Based on each framework, physics, chemistry, biology, earth science, and space science are one collection of sciences called basic natural science.[44] However, the different sciences come from the same parent; therefore the sciences are one unit.

On the other hand, objects and their characteristics inter-influence each other; animals, plants, earth, and other entities provide life challenges for every organism. The challenges may call for behavioural changes or adaptations of organisms, and solutions to these accumulate into applied natural sciences such as engineering, agricultural science, medicine, materials science, and others.[45] Adaptation of organisms requires ever-growing data about objects, their interactions, and the relationships between them. The data is attached to ideas and thoughts, and that prove that existence of the formation of ideas in the form of natural laws.[46] Natural law directly degrades ethics.[47] Meanwhile, the adaptation of organisms is an input to every basic natural science, especially applied science, which is guided by scientific ethics where, case by case, the events of the struggles of humans with their environment spill over into ethics.[48] The story of the human struggle between the life interests their society typically organize into the social sciences.[49] This is a science that determines the rules that apply to individuals and the social status of humans. It is a science that also includes relationships between humans. Although social science involves data in its unstructured characteristics differently, it still involves a series of reasoning in which data is inherent in social interpretations.[50] Thus, like the natural sciences, the social sciences are also empirical.

Empirical science[51] is the science that theoretically explains that knowledge comes from sensory experience. The sensory experience is recorded into memory (memory brain) as a unity for the sake of a meaningful entity, namely data.[52] Organizing these unities requires a new science: data science, a framework for talking about data. On that basis, Thales is not a data scientist, and mathematics is not a data science.

1.2.1 The Term "Data Science"

Data science[53] consists of two words to form a term that refers to the science of data, namely the knowledge related to the intricacies, scientific activities, or what is known as data;[54] it has been stated that data science is "an interdisciplinary field that extracts insights from data through a multi-stage process of data collection, analysis and use". This statement is the base expression to briefly integrating the meanings of the words—data and science—from their respective terms. However, differing interests in data have shaped the conceptual framework for thinking.[55] This breakthrough causes the concept of data science to be incomplete. A fundamental concept becomes the scientific basis of a science. As revealed from both basic natural science and applied natural science, the formulation of concepts based on terminology will continue to shift along with the challenges of change according to the interests of stakeholders. Of course, the reduction of these interests primarily requires introducing the term data science.[56]

The first challenge was not long ago from the advent of computer science. In terminology, the concept that data science is a "substitute for computer science" framing the scientific framework of data does not come out of computer science. At the same time, it monopolizes the data part by part based on studies with the principle—input, process, and output—in which algorithms play a role and are the focus of the study.[15] Because of this condition, computer science was not well developed until the early 2000s.[57] This framework suggests the science of data needs a different name. When adding the *-logy* suffix to the data, *datalogy* sounds like a word that expresses the purpose of the idea.[58] The word suffix in *-logy* comes from the Greek loan word λόγος (read *logos*), which means "account, explanation, and narrative". The intent of the words directly reveal that the term datalogy is only a scientific activity about data around computing and data protection, which is an explanation of the results about processing and gives a little encouragement to the existence of studies around data.[59] It also appears that the social sciences, such as anthropology, psychology, and sociology, for example, depend heavily on statistics or other basic natural sciences.[60] The use of the term datalogy will allow the scope of studies on data depending on other scientific field according to the interests of that field. So, the term datalogy is not considered to spark the scientific potential of the science of data.

The lack of a concentration of ideas has led to the information concept of data science. It is a result of the lack of information about data. A scientific event, "The Japanese-French Statistics Symposium at French", has tried to sketch a modern definition of data science, namely "a new field that integrates aspects of computer science, statistics and information management".[61] It is a conceptual description framework through terminology that strongly binds knowledge about data with statistics through computer as technology.[62, 63] Then, on a different occasion, a scientific meeting formulated a connection between science about data and various methods for classifications and clustering.[64] However, the insistence on data science as statistics is straightforwardly enshrined in the title of the lecture, "*Statistics = Data Science*",[65] which conceptually states that "The characteristics of statistical work are as a trilogy: data collection, data modelling and analysis, and decision making".[66] Furthermore, these claims are then softened into a definition that states "Data science as an independent discipline, but an extension of the field of statistics to include advances in data computing".[67] Therefore, referring to the term is not enough to express the meaning and purpose of the existence of data science. It is necessary to introduce data science in a framework and concept: the definition.

1.2.2 Concept of Data Science

Recently, the term data science has been discussed in several scientific meetings and used in the health sector for the first time.[68] It was embraced as an academic discipline in the mid-1990s,[69] and the concept's ideas have featured widely in the literature. However, data science is latently evolving through study trait tracing to feature in studies on data throughout the range of scientific publications. It has given birth to a new science,[70] and there is a concentration of studies in directions such as data capture, data analysis, metadata, rapid retrieval, archiving, exchange, data mining, data relational, visualization data, data management, and more. The study is generally to find knowledge and make intellectual property that is beneficial to the welfare of humans.

Humans are interested in many types knowledge. Chemistry as a science is popular. Likewise, computer science has become fashionable for its ability to overcome the complex human mindlessness in the face of speed. The role of data itself in this achievement has not become the primary focus. Thus, computer science has only become popular because of the limitations of the human brain in processing data, not because it is directly related to the data itself. As stated, data science is a new study that combines computer science, data mining, data engineering, and software development.[71] However, is it because of the last reason that data science has become a exciting field of science, or is it a buzz? Like the complexity of a science, any knowledge is always based on concepts. The concept of data science is the basis for its development. Of course, it should not divert to other scientific fields as a result of misunderstanding, which can result in overlapping studies and suboptimal knowledge development. That is, computer science will not become data science, and data science will not become part of computer science. At the outset, it may seem that data science is a growing field at the intersection of mathematics, computer science, and the domain expertise.[72] Furthermore, when optimal knowledge development is carried out, the learning outcomes of any science produce educational achievement targets. The concept of data science does not just introduce data science but especially targets what human wants to achieve about data science in the three pillars of learning: education, research, and community service. It conceptually describes data science as "a set of fundamental principles that support and guide the principled extraction of information and knowledge from data".[73]

The concept of data science in the previous statement is an effort to state the scope that provides an opportunity to develop a useful model to support data collection, integration, and analysis, and then make data science the leading science for determining database resources in terms of open access or deal with login.[39] For this reason, data science is a science that involves the ability to find relevant relationships in data to determine the quality of a method or the performance of software as a tool.[74] Quality is the value of a method's ability to deal with data and map it without changing its treatment, but it is not related to the intricacies of method performance where performance is related to the possibility of changes from the formulation in the method. The treatment of methods provides the appropriate technology so that there is no need to change the behaviour of the original data. Meanwhile, performance, in this case, becomes the value of the software when carrying out all activities that result in data turning into knowledge.[75]

Data science is a science that includes new studies ranging from concepts to theories and then gets feedback from its implementation in a framework, namely innovation around data exploration that provides scientific competence as a goal. It is a framework that allows studying or reviewing data sources for achieving uniformity in point of view to data, even if the sources come from different empirical sciences. This concept does not avoid software development as a tool, but it puts scientific organization into a language that clarifies its reasoning. Invariably, an emerging science is uniquely different from existing sciences, and it begins with both the discussion that establishes its concepts and the reasoning that operates within a framework. Thus, data science differentiates itself by involving the word "extraction" and its derivatives with data subjects and knowledge objects. One word in the definition of data science intent to eliminates the overlapping of business purposes and functions, i.e., by assigning one word, extraction.[76] A task takes over the extraction in the optimization of data mining.

The involvement of new concepts provides the scientific independence of data science. At first glance, that is in different disciplines; however, new sciences such as data science are always at the crossroads between basic natural science, applied natural science, and formal science. Data science is an emerging interdisciplinary field that combines elements of mathematics, statistics, computer science, and knowledge in application domains for the purpose of extracting meaningful information from the increasingly sophisticated datasets available in many settings.[77] Mathematics is the scientific basis of all sciences generally and with computer science to service it, together they are a formal science involving set theory rules for generating any knowledge. Statistics, data engineering, data analysis, optimization, and decision making need a source of knowledge: data. Thus, data science differs from statistics and other disciplines in several important ways[78] that can be expressed in the following concept: data science is the process of extracting knowledge from data using base principles of analytical techniques such as statistics to achieve business goals.[79] Once again, data science connects with statistics!

With the unique concept of data science, this science is not a craze that erodes its own competitiveness, but it is something phenomenal. However, a phenomenon requires a partner, namely a paradigm, and it is all related to the laws of nature.

1.2.3 Paradigms

Assessing the natural laws of life itself, as a science subject, always start with modelling. It's a solution to the problem of explaining the subjects of thoughts (what observation and investigation are for). Proof is logically understanding the reasons (why) for the issue and the answers (to the types of questions). It's a description that systematically composes a sequence of reasoning for a series of opinions about a subject where there is a way to interpret the results of the study (how and methods), implicitly perpetuate science (with the possibility of prediction), and crawl in-depth on the subject—the environment—for the welfare of humans through simulation and experimentation.[80–83] The output of all problems concerning the subject of thought becomes a universally recognized scientific achievement, namely a paradigm.[84] With that enveloping, data as a subject of thought together with the methodology of building science is called data science.[85] A way to recognize a science of data is to confirm it is from data, about data, and for data. In that way, data science becomes a collection of principles and techniques used to measure and improve the correctness, completeness, and efficiency of intensive data analysis.[86] As stated by the U.S. National Institute of Standards and Technology, the data science paradigm is the [87] "extraction of actionable knowledge directly data through a process of discovery, hypothesis, and hypothesis testing".[88] In general, a hypothesis is an argument put forward to explain one or more phenomena, but the best hypothesis is the useful assumption. In this case, the assumption is something that is taken for granted or believed to be true.

Like other sciences, data science is identified by a set of assumptions that justify its scientific method. In the scientific world, all methods deal directly or indirectly with data, but in particular, some scientific methods deal only with data.[89] With the basic concepts of set theory in mathematics, probability theory expresses the behaviour of data from a set of data. In theory, the distribution (probability distribution) describes the behaviour that data must meet to qualify to explain the meaning of the data, where the data relate, mainly for generating information. By that, data science is the art of producing information and knowledge.[90] The randomness test, for example, determines whether the sample is representative of the population. Random or randomness is the behaviour of data such as discrete, continuous, or fuzzy. A set of data may or may not come from the same population where the homogeneity test answers it and based on normal distribution test.[91] On the other hand, when the data tests and is not normal, the data behave heterogeneously. Statistics describe the behaviour by formulas involving the calculated average, median, modus, or other measures. For example, data homogeneity states the closeness of the difference between all data and the mean. So indirectly, it expresses the size of the data that meet the meaning of something or is for producing information. Size is one of the characteristics of the data. Data science is closely related to the assumption that it will produce information. In other words, the output of data science is the input to the process of producing information, if knowledge is assumed to be derived from information.[92]

There is an assumption that there is an objective reality that is equally present with the subject of reasoning. The assumption is a requirement in the paradigm to justify scientific methods of data. The subject of reasoning is the object that essentially realizes the subject of thought. It is in the form of the characteristics of the object recognized through the experience of humans. Each characteristic becomes a unity—that is datum—and those unities are bound in a bond by objects, but they have no structure.[93, 94] Through this assumption, the data have a description name known as metadata, sometimes called a variable, that serves to accommodate the existence of the data. For the same reasons treat the naming any object of human thought, unities bring up an indirect bond that embodies the relationship between the data. In data science, the unity is datum and indirectly causes data to be tied to one another. Engagement is the study objective of data science. Although the binding between data is possible in statistical studies externally, internal data binding is an obligation of the integration function called convergence [95]: a space in the universe of discourse that optimally exists at one point. The opposite is called divergence.

However, by the framework of each science, something that is the general becomes part. It adapts to the methodology of each science. Assuming that framework, data science is a growing field with significant research focus on improving the techniques available to analyse data.[96]

Furthermore, some rules naturally govern objective reality. The rules are known as natural laws. Every law of nature is based on logical assumptions through experience that propagates into data.[97] Any object always rests on another object at least on thought. Data as objects depend on thought. Objects, or entities such as humans, are on the earth. Thought is in the system that organizes data on earth. When any object cannot stand alone, data as a representation of objects in thought also has a relationship with one another, but its dependence or independence in size may change according to scale.[98] The change corresponds to a condition that satisfies the scientific will.

By exploring the potential meaning of the data, interpretations continue by revealing the rules naturally. The laws of nature squeeze the essence of data to produce something as information, which becomes knowledge or helpful information. In the paradigm, assumptions encourage observation and experiments to realize the laws of nature.[99] Internally, the natural law about data is a description of the behaviour of an object, while externally it is its environmental behaviour.[100] The natural law of data is quantitatively a concise description with a number, a little or a lot, about the object, while qualitatively a concise description involves ranking, good or not. The natural law of historical data reveals that data not only grow in size but also that data is dynamic. Meanwhile, the natural law of data is predictive in response to its existence in the future.[101]

Through the existing science, the scientific will always ask for evidence in the form of data. In the universe of discourse, data do not only exist in dimensions and therefore has dimensions but also has a fundamental nature independent of that dimension.[102] This philosophical essence is to provide an understanding of logically valid assumptions. Thus, the assumptions in the paradigm about data science are provided as a scientific framework that philosophically follows the sequence of reasoning, which systematically flows descriptions in language logically, expresses variables and functions as mathematical proofs, and presents implemented unities in computations.[103]

1.3 AN EXPLANATION FOR UNDERSTANDING

Science is a conscious effort of humans to investigate, discover, and improve their understanding of various aspects of themselves and their environment. Science provides a framework involving language as a description, reason as a tool for analysing that description, and logic as the philosophical basis for a coherent explanation of the subject of science.[104] Every science depends on the description of the framework, which then abstracts into its foundation, namely mathematics. Data science is the science that describes data in order to achieve knowledge.[105] A description involves data for data, meaning that gives validity to the description, reality that explains the reason, and facts that reveal the truth of something.

1.3.1 Data Encoding

Word for word has compiled a description stating that data science (p) is "the extraction of knowledge from high-volume data, using the computational intelligence and the specialist domain knowledge of experts" (q).[54] A statement logically contains bi-implication between data science p and its description q. The statement forms a relationship in the composition of $p \leftrightarrow q$. That is, there are two implications, namely if p then q ($p \rightarrow q$) and if q then p ($q \rightarrow p$), which means that data science exists (with a truth value–True) if the description becomes a fact (with a truth value–True), or vice versa there is an attempt–effort as described in the description (with truth value–True) and the effort as data science (with truth value–True). In technology, it is codification.[106]

In logic, every statement p is a sentence that evaluates to true (B) or false (S), $\rho(p) = \text{B}$ or $\rho(p) = \text{S}$. One statement has two possible truth values (as quantitative description) or with numbers expressed as 2^1 units of description. Two statements generate four possible truth values, i.e., 2^2 descriptions. Furthermore, a logical description with three or more statements is 2^n description, $n = 1, 2, 3, \ldots, N$. $\rho(p)$ is the description of the statement. Any statement in logic is symbolically represented by letters, while true or false is the content of the statement value, p. $\rho(p)$ are logical data; data come from logic. The description contained in each statement consists of words. Each word has its categories, such as nouns, verbs, etc., which are also data. The words specifically exist in dictionaries. At different positions, each word consists of an alphabetical character, like genes.[107] The alphabet consists of 26 characters. In all languages, permutations of one to 26 characters build words (strings), a sequence of symbols, and words to refer to the meaning or naming of any object according to the attention of humans. Character permutations can compute directly by

$$\text{P}^{\text{r}}_{\text{s}} = \text{s}!/(\text{s}-\text{r})! \quad (1.1)$$

where r represents several characters in a subset of the alphabet set, the alphabet set contains s elements (characters). So it is possible for a single character to be a word, and there are 26 possible words. Additionally, in a collection of words with two characters of the alphabet, there will be as many as 650 possible words. This continues on so that cumulatively the number of permutations jp is the sum of all permutations, that is

$$jp = \sum_{r=0\ldots s} s!/(s-r)! \quad (1.2)$$

Not all permutations from r characters to s alphabetical, $r \leq s$, become words that have meaning and are used by humans. For example, for $r = 4$, the result of a permutation such as "czlb" may be a word that possibly does not exist in any of the languages of the world. However, not all words that have been used by humans are in the set of words that come from a permutation of r characters to s alphabetically. For example, for $r = 4$ characters from $s = 26$ alphabetical characters, there will not be the word "coco" (in English), except in a permutation environment with $(r = 2)^2$

characters or $r^2 = 2^2$ characters with $(s = 26)^2$ or $s^2 = 26^2$, where there are repeated permutations of the same space, and they may act like logical descriptions where r^n or s^n, and consequently the number of words jk that is derived based on permutation, is

$$jk = \sum_{n=1...N} \sum_{r=0...s} s^n!/\left(s^n - r^n\right)! \tag{1.3}$$

where $r^n \leq s^n$. While the combination of r characters from the entire s alphabetic character yields a normal distribution of possible combinations of words, the same probability distribution holds for r^n with s^n.[61]

A character is a symbol of the sound produced by the articulation tools. However, the sound from one community of humans to another may differ, as is the case in languages like German, Arabic, French, and others. Additional characters appear in some of the alphabets. Every new character accord with the sound for representations of the word in that language. In addition, numeracy in language has words that represent numbers. Computationally, the count involves a set of stones (*calculi*) replaced with numeric symbols, such as 0, 1, 2, 3, 4, 5, 6, 7, 8, 9.[108]

In human thought, the reality is not only an object that is recognized and has a name but reality in the form of quantity or number. Thus, the set of numbers is formed one by one according to the importance and type of quantity. Integers consist of the set {. . ., −3, −2, −1, 0, 1, 2, 3, . . .}, fractional numbers of the form a/b, and together with numbers have a floating-point or of the form $\cdots aaa.aaa\cdots$ if each a is a number (different or the same), for example, 3.14, are called a real number. In data science, symbols and symbol arrangements are descriptions that mean something. The unity of one or more symbols describes a reality that is understood by humans as data. In other words, the transformation of one or more symbols into other symbols as words, already known as encoding, where word one has character "1" and vice versa, are also data. Similar to logical values, electronically turning an object or a unit on and off transforms into symbolic meaning 0 and 1 as clues, on cue, or command. These two symbols are known as binary numbers (bits), and their arrangement depends on the scope of technology, namely computer microprocessors (machines). The standard encoding or symbolization of the bit string consists of 8 bits or is called 1 byte. It causes the number 0 to have an arrangement of 00000000 or in hexadecimal 00h, the number 1 to have a composition of 00000001 or 01h. The results digitally for 1 byte are $2^8 = 256$. Therefore, the standard machine-recognized character set is the American Standard Code for Information Interchange (ASCII) with as many as 256 characters. ASCII consists of various additional symbols in alphabetic and numeric (alphanumeric). For example, character A has a machine language code or binary 01000001 = 41h (h = hexadecimal). The number "one" has coding in machine language or binary 00110001 = 31h, but the numeric one has coding in the machine or binary 00000001 = 01h, etc. Thus the numeric 1 and the number symbol "1" in bits are different encodings.[109]

Starting from symbols, words, strings, and binary digits take different forms, and along with their descriptions, as distinct entities or as an encoding. All or a part of them are data. It is not only a subject of thought but also an object of data science. Based on the accompanying technology, their categorization becomes types of data such as characters, integers, floats, strings, etc. So it is not wrong to say that data are the smallest unit that has meaning.[110] However, shifting characteristics of data have changed the landscape of many methodologies dealing with data, and that is the study of data science framework.

1.3.2 Concept Description

As a subject of thought, data are something that is accepted as it is.[111, 112] Data science captures it as "an essential concept for . . . workforce and as a result the need to help . . . acquire such skill".[113] A concept of data science is useful when asserting that data are facts or figures from which conclusions can be drawn.[114] This concept makes it easier for humans to understand their world through data, and data require processing to have added value or be useful. [115] Each processor has a function that is employed to reveal what is carried with the data both internally and externally,[116] a result called information[117] that will then provide something behind it. That is the knowledge that is useful in decision-making.[118] Based on this, the first definition given by the American National Standards Institute (ANSI) of data is as follows: "A representation of facts, concepts, or instructions in a formalized manner suitable for communication, interpretation, or processing by humans or by automatic means".[119]

Any data processing to produce information requires integrated support from mathematics, statistics, and computer science.[120] Linearly, referring to computational principles, data science is a contemporary discipline based on a computational approach to mathematics to gain meaningful knowledge.[121] Data science reasoning on the subject of thought through statistical inference—makes propositions about universe of discourse—can provide a more focused description of conclusions,[122] and data science systematizes the structure of the subject of thought through the development of algorithms from computer science to solve more complex problems.[102]

In the implementation of science, data science does not take over the roles of mathematics, statistics, and computer science. However, data science follows up on all those roles and then provides feedback. Data science is not about mathematically making derivative of the function $y = x^2+x$, for example, and proving the existence of an optimal point. While it is possible to explore statistically that any mathematical function has an average of its points of intersection with the x-axis so there is a line of approximation, data science does not replace the role of statistics. Likewise, giving meaning to the points as a whole through their values, respectively, by placing them in a one-dimensional space requires an algorithm that involves one times of iteration. Data science may require the role of algorithms, but only computer science deals with the complexity of algorithms. Furthermore, by involving other fields—information science, information management, data mining, and others that have the tasks of framing data according to their interests and methodologies—data science has the ability (method) to understand and analyse actual phenomena with data to gain knowledge. On one hand, data science is relatively new. Therefore, data science requires an ever-evolving framework along with the importance of conducting data analysis effectively.[123]

On the other hand, the formation of data science with several trans-disciplinary fields, such as computer science, information systems, information science, bioinformatics, ethnography, and others, can produce data scientists, particularly academic alumni who understand data well and, with the help of experts, can become specialists in domain knowledge. Thus, data science education is a pedagogical pyramid based on data, information, and knowledge to produce scientific insights about data.[124]

1.3.3 Description of Change

The subjects of thoughts, or objects and their characteristics, together with descriptions constantly build content of the learning and the study that is increasingly burdensome for humans.[125] Description after description grows, changes, and develops dynamically as a cultural product.[126] That is inherent wherever humans go. Over time, the collective agreements of humans became binding as laws of nature and ethics in communities were in a position to reproduce that description. Agreements passed down from generation to generation and spanned from the era of nomadic life to the present.[127] The shift to a sedentary lifestyle resulted in humans living in agricultural pockets, which resulted in the struggle for land commodities for agriculture production. At its peaks, there were wars to defend or seize agricultural products.[128] The pauses between fighting fellow humans due to exhaustion led to agreements to build administrative centres in the form of cities to serve the requirements for society and power.[129] These cultural products changed the human lifestyles. Several technologies are considered tools used to improve welfare.[130] These changes shaped the industrial era that required energy sources. Once again, the struggle for energy commodities has led to war between communities.[131]. These quarrels have crossed country borders. However, now computer technology adds to the complexity of the energy struggle. At first, the influence of computers was considered underestimated. Now, computers have shifted principles from the academic to the social. The abacus has been replaced with a calculator. Writing something on stacks of paper has shifted to a word processing application that is always ready to correct a word or typing error. The pen has become a table decoration display. The accountant's worksheet has been replaced with a spreadsheet, and their books sit nicely on the shelf. Yet the attention to data is still neglected even though everyone from the academic to the accountant performs activities that process data.[57]

Computers have changed the fabric of human life, especially in academia. The battle to seize key technologies from computers, such as algorithms, has become a battleground for humans. The birth of computer networks narrowed the world by shortening the distance of communications between humans by connecting them from everywhere. However, there are issues with the transfer of computer technology from one place to another, either through learning or through technological invasion. In the decades since World War II, programs and programming have been at the heart of the academic discussion.[132]

Intrigues to undermine information systems and software application systems, and programs are continuously developed. Then, cyber warfare potentially emerged with the intention that people who have an interest in social change could impact decision by each people. Data have become a bone of contention, and a new battleground has emerged. Any system can be reinstalled, but if there is no backup for a data loss, then the data will be lost forever! Data can be deterministic or nondeterministic depending on human interests. The interests determine where data need to be present, and therefore data science frames data into phenomena. Data science is increasingly important and challenging,[133] and data science is an old disease and turned into a gold rush.[134] Thus, "data science is an evolving field, existing definitions reflect this uncertainty with excessive terms and inconsistencies".[135] This definition gives the basic concept of data science. Definitions are initial data to present the framework of science.

1.3.4 Description of Complexity

The subject of thought, namely data, has the descriptions in the flow of the words and coding symbols. Encoding, with the characters in the alphabet, represents conversations between humans. Descriptions in language have explained all aspects of humankind; many have been recorded in a written form called documents. In addition, technologies such as radio, tape recorder, telephone, and television emerged to facilitate descriptions in the form of sounds, movements, encounters, and other activities. Not long ago, after the advent of computer networks and the Internet, web technology brought together all the possible humankind descriptions. Classic data types like characters, numbers, and strings and traditional data types like those found in database management systems, text and date, for example, expanded to data types such as hypertext, image, audio, and video.[136] Text as a description of any object or entity flows continuously to and changes dynamically in the information space: the Internet and the web. One or more images are inserted among the text. The image describes an object visually in a two-dimensional or area space. Conversations, dialogues, or something in the form of sound usually change into the text, which becomes a stream of sound waves or audio. Moving images are used as a description of an activity, and that movement in the frame by frame flows as the form of video. Several data types may exist in a single document or application system on the web. They are presented by the tags that define them as information. Some of the types of data in the document establish a connection via hyperlinks to other documents as descriptions. It agrees with the second ANSI definition of data, which is "Any representation such as characters or analogue quantities to which meaning is or might be assigned. Generally, we perform operations on data or data items to supply some information about an entity".[137] Therefore, document-by-document collaboration builds on the ever-expanding potential of data, and those data sets accumulate in information subspaces that become the platform of big data. Examples of big data platforms include Google, Scopus, Facebook, ResearchGate, etc. (Figure 1.1). Thus, big data indirectly assigns a task to science to build an organizational framework of data.[138]

Separately, through electronic devices that serve the human interests, descriptions flow into the storage space of a computer or other machine. Likewise, the Internet of Things (IoT) integrates sensor devices for the same purposes and then records explanations of the nature of the related objects description by description. Learning benchmarks use original descriptions for giving standards to the derived descriptions. That is, information comes from automation in industry, trade, government,

FIGURE 1.1 Multiple platforms of big data.

industry, and academia. Of course, in the industrial era 4.0,[139] data have a decisive role in transforming changes in the industrial world that produce products or services to more effectively and efficiently serve welfare demands.[140] In this case, data science has the potential to leverage leaner production tools and workers with the capacity to match that demand.[141] For that reason, data science serves interest through the knowledge that results from those descriptions.

Description after description as a result of human recognition is recorded in the information space both quantitatively and qualitatively. The description is an explanation: information that is then turned into data. "Data is the encoded representation of information".[142] Then, "Information is derived from data and useful in solving problems".[142] Data turn into information or vice versa according to the processing procedure as expressed by the definition of information from ANSI and ISO, "Information is the meaning that a human assigns to data by means of the known conventions used in their representation".[143] Recognition results, derived from responses to environmental stimuli or derived from human interactions with other entities, provide a sensation of complexity to human estimation, memory, and reasoning. That sensation is related to volume, variety, and velocity (3V), which then states it as characteristics of big data.[144] Thus, data science is an attempt to describe the problem of analysis constrained by the characteristics of big data.[145] *Data science is another term for big data*.[146] When mathematics formulates responses to data size in various formulas, it may be just a dream for computer science to implement them without dealing with the complexity of such data. Likewise, the spirit of statistics with the heroic variance of data randomness is just a knife trying to chop down a big heterogeneous tree. After all, big data are growing continuously from a wide variety of sources and at an astonishing rate. It encourages new theories of concepts that redefine old concepts to adapt to big data.[147] Therefore, data science is a science that can overcome the challenges of big data,[148] specifically scientific disciplines that restate themself with strong links to big data.[149] Therefore, when using word extraction in an effort to frame the data analysis,[150] for example, in applied domains such as bioengineering, the source becomes *high-volume data*.

1.3.5 Framework Description

A set of designs for understanding data, including big data, derives common essences and patterns that reveal the prescription of the framework for optimally handling hyper-structured data. Hyper-structure is a mathematical interpretation of the modelling side of the data. Data modelling, such as relational databases, is an interpretation of the object and the relationship between data in a linear fashion in tabular form. However, mathematics also provides other interpretations of different models, such as tiers and networks, for example. In contrast, statistics provide the basis for extraction to frame the data according to the behaviour of the data. In this regard, researchers introduce data science as "the interdisciplinary field responsible for extracting knowledge from data".[151]

Complexity occurs when the ability of science with a broad framework is limited by technological capabilities. Thus, computing borders two disciplines: mathematics and computer science require another spark to improve their capabilities, such as the intelligence inherent in computing like artificial intelligence or computational intelligence. On one hand, the human ability to handle the details of something in complexity is better than a computer machine. On the other hand, the computer's ability to solve the complexity of speed is better than humans' abilities. These two issues gave birth to computational intelligence. So the researcher introduces data science to emphasize its importance, as "an umbrella term used for referring to concepts and practices of subset of topics under artificial intelligence (AI) methodologies".[152] Learning is the key to implementing computational intelligence. The descriptions in the big data platform always have patterns according to the templates prepared by technology. Learning from all technologies attached to extraction produces a general paradigm that functions effectively and efficiently. It fits with the researchers' introduction of data science as a "promising area for processing data and extracting hidden patterns using machine learning techniques".[153]

Learning is important for artificial intelligence technology to ensure that the output of extraction has quality, but learning is also improves the quality of humans. Learning produces specialists. When data science provides a framework for data for all scientific activities, it's important to establish data science as a science that has scientific autonomy and a framework for study programs. Education experts introduce data science in education as educational data science by claiming it "is an emerging methodological field which possesses the algorithm-driven technologies required to generate insights and knowledge from educational big data".[155] The experts reiterate data science by introducing it as "a tool to be taught to science and engineering students in addition to knowledge of their chosen domain".[154] Data science is "not just rote application of trainable skill sets; it requires the ability to think flexibly about all these areas and understand the connections between them".[155–156]

Study programs should include information about education, research, and community service from a scientific science. In this case, a curriculum guide for systematizing scientific subjects from the beginning to the end of the lecture structures the discussions. The subjects should be arranged according to their level to develop competent graduates. This book not only provides guidance for uncovering data science but also suggests that "Data science is a form of data-oriented science, which serves as a set of theories, methodologies, and technologies for data exploration and analysis".[157] Therefore, this chapter introduces data science is as "the extraction of knowledge from high-volume data, using the computational intelligence and the specialist domain knowledge of experts".[158]

1.4 CONCLUSION

Although every science pays attention to the treatment of its framework on data, data science frames data as the foundation of its framework, and it differs both scientifically and theoretically. Conceptually, data science gives cues through the word "extraction" with equipment; origin sources that have the high-volume capacity, output as knowledge, and scientific tools in the form of computational intelligence or something that comes from specialist domains. Therefore, mathematics does not become data science but underlies the scientific principles of data science. Statistics cannot be confused with data science, but data science involves statistics to reveal the behaviour of data. Data science is not a shadow of computer science, but it poses a different challenge.

REFERENCES

1. Andrienko, N., Andrienko, G., Pelekis, N., & Spaccapietra, S. (2008). Basic concepts of movement data. *Mobility, Data Mining and Privacy: Geographic Knowledge Discovery*, 15–38. https://doi.org/10.1007/978-3-540-75177-9_2
2. Sun, H. (2020). Psychology of adolescents' preference for E&A vocal music and its influencing factors. *Revista Argentina de Clinica Psicologica*, 29(2), 635–641. https://doi.org/10.24205/03276716.2020.289
3. Levy, J., & Démonet, J.-F. (2020). MEG data representing a gamma oscillatory response during the hold/release paradigm. *Data in Brief*, 23. https://doi.org/10.1016/j.dib.2019.103787
4. Ahuja, S., & Krunz, M. (2008). Server placement in multiple-description-based media streaming. *Data Compression Conference Proceedings*, 372–381. https://doi.org/10.1109/DCC.2008.63
5. Nakano, M., Sagane, Y., Koizumi, R., Nakazawa, Y., Yamazaki, M., Ikehama, K., Yoshida, K., Watanabe, T., Takano, K., & Sato, H. (2018). Clustering of commercial fish sauce products based on an e-panel technique. *Data in Brief*, 16, 515–520. https://doi.org/10.1016/j.dib.2017.11.083
6. Lima, V. M. A., Dos Santos, C. A. C. M., & Rozestraten, A. S. (2020). The ARQUIGRAFIA project: A web collaborative environment for architecture and urban heritage image. *Journal of Data and Information Science*, 5(1), 51–67. https://doi.org/10.2478/jdis-2020-0005
7. Nasution, M. K. M. (2019). Methodology. *Journal of Physics: Conference Series*, 1566(1). https://doi.org/10.1088/1742-6596/1566/1/012031
8. Vakouftsis, C., Mavridis-Tourgelis, A., Kaisarlis, G., Provatidis, C. G., & Spitas, V. (2020). Effect of datum system and datum hierarchy on the design of functional components produced by additive manufacturing: A systematic review and analysis. *International Journal of Advanced Manufacturing Technology*, 111(3–4), 817–828. https://doi.org/10.1007/s00170-020-06152-6

9. Nasution, M. K. M. (2018). Ontology. *Journal of Physics: Conference Series*, 1116(2). https://doi.org/10.1088/1742-6596/1116/2/022030
10. Nasution, M. K. M. (2018). The uncertainty: A history in mathematics. *Journal of Physics: Conference Series*, 1116(2). https://doi.org/10.1088/1742-6596/1116/2/022031
11. Moore, H. T. (1922). Further data concerning sex differences. *Journal of Abnormal Psychology and Social Psychology*, 17(2), 210–214. https://doi.org/10.1037/h0064645
12. Whelden Jr, C. H. (1933). Training in Latin and the quality of other academic work. *Journal of Educational Psychology*, 24(7), 481–497. https://doi.org/10.1037/h0074703
13. Yela, M. (1949). Application of the concept of simple structure to Alexander's data. *Psychometrika*, 14(2), 121–135. https://doi.org/10.1007/BF02289148
14. Mali, R., Sipal, S., Mali, D., & Shakya, S. (2022). Parkinson's disease data analysis and prediction using ensemble machine learning techniques. *Lecture Notes on Data Engineering and Communications Technologies*, 68, 327–339. https://doi.org/10.1007/978-981-16-1866-6_24
15. Nasution, M. K. M., Aulia, I., & Elveny, M. (2019). Data. *Journal of Physics: Conference Series*, 1235(1). https://doi.org/10.1088/1742-6596/1235/1/012110
16. Nasution, M. K. M. (2022). Understanding data toward going to data science. *Cybernetics Perspectives in Systems* LNNS 503, 478–489. https://doi.org/10.1007/978-3-031-09073-8_42
17. Buchner, E. F. (1900). Volition as a scientific datum. *Psychological Review*, 7(5). https://doi.org/10.1037/h0067763
18. Armstrong, J., & Stanley, J. (2011). Singular thoughts and singular propositions. *Philosophical Studies*, 154(2), 205–222. https://doi.org/10.1007/s11098-010-9532-1
19. Vallee, M. (2020). Doing nothing does something: Embodiment and data in the COVID-19 pandemic. *Big Data and Society*, 7(1). https://doi.org/10.1177/2053951720933930
20. Huestis, R. R. (1934). The borderlands of science. *Journal of Heredity*, 25(4), 143–144. https://doi.org/10.1093/oxfordjournals.jhered.a103901
21. Seidenberg, A. (1975). The ritual origin of geometry. *Archive for History of Exact Sciences*, 1(5), 488–527. https://doi.org/10.1007/BF00327767
22. Forbes, E. G. (1977). Descartes and the birth of analytic geometry. *Historia Mathematica*, 4(2), 141–151. https://doi.org/10.1016/0315-0860(77)90105-7
23. Nasution, M. K. M. (2020). Mathematical philosophy. *Journal of Research in Mathematics Trends and Technology* (JoRMTT), 2(2), 45–60. https://doi.org/10.32734/jormtt.v2i2.4678
24. Zins, C. (2007). Conceptual approaches for defining data, information, and knowledge. *Journal of the American Society for Information Science and Technology*, 58(4), 479–493. https://doi.org/10.1002/asi.20508
25. Niiniluoto, I. (2001). Futures studies: Science or art? *Futures*, 33(5), 371–377. https://doi.org/10.1016/S0016-3287(00)00080-X
26. Bergin, A. E., & Strupp, H. H. (2017). Changing frontiers in the science of psychotherapy. *Changing Frontiers in the Science of Psychotherapy*, 1–468. https://doi.org/10.4324/9781315081564
27. Sun, R.-T., Hu, H.-C., & Chang, H.-L. (2016). A design science approach to improve adherence on exercise plan via mobile application built by researchkit framework. In *Proceedings of the international conference on electronic business (ICEB)*. Xiamen, CEUR-WS, pp. 269–272.
28. Cioffi, F. (2000). The propaedeutic delusion: What can 'ethogenic science' add to our pre-theoretic understanding of 'loss of dignity, humiliation and expressive failure'? *History of the Human Sciences*, 13(1), 108–123. https://doi.org/10.1177/09526950022120638
29. Nursall, A. (2003). Building public knowledge: Collaborations between science centres, universities and industry. *International Journal of Technology Management*, 25(5), 381–389. https://doi.org/10.1504/IJTM.2003.003107

30. Goldsmith, M. (1974). 'Popularisation' of science. *Nature*, 250(5469), 752–754. https://doi.org/10.1038/250752a0
31. Teller, P. (2013). Representation in science. *The Routledge Companion to Philosophy of Science*, 490–496. https://doi.org/10.4324/9780203744857-56
32. Janda, K. (1967). Computer applications in political science. In *AFIPS conference proceedings—1967 fall joint computer conference, AFIPS*, pp. 339–345. https://doi.org/10.1145/1465611.1465655
33. Abell, S. K., & Smith, D. C. (1994). What is science? Preservice elementary teachers' conceptions of the nature of science. *International Journal of Science Education*, 16(4), 475–487. https://doi.org/10.1080/0950069940160407
34. Guo, H., Nativi, S., Liang, D., Craglia, M., Wang, L., Schade, S., Corban, C., He, G., Pesaresi, M., Li, J., Shirazi, Z., & Liu, J. (2020). Big earth data science: An information framework for a sustainable planet. *International Journal of Digital Earth*, 13(7), 743–767. https://doi.org/10.1080/17538947.2020.1743785
35. Rankin, Y., Thomas, J. O., Brown, Q., & Hatley, L. (2013). Shifting the paradigm of African-American students from consumers of computer science to producers of computer science. In *SIGCSE 2013—proceedings of the 44th ACM technical symposium on computer science education*, pp. 11–12. https://doi.org/10.1145/2445196.2445204
36. Thagard, P. (2013). Cognitive science. *The Routledge Companion to Philosophy of Science*, 597–608. https://doi.org/10.4324/9780203744857-67
37. Medel-Esquivel, R., Gómez-Vargas, I., García-Salcedo, R., & Vázquez, J. A. (2021). A Simple estimation of the size of carbon atoms using a pencil lead. *Physics Teacher*, 59(6), 480–481. https://doi.org/10.1119/10.0006135
38. Pierce, O. R., & Lovelace, A. M. (1962). New and varied paths for fluorine chemistry. *Chemical and Engineering News*, 40(28), 72–80. https://doi.org/10.1021/cen-v040n028.p072
39. Swan, M. (2013). The quantified self: Fundamental disruption in big data science and biological discovery. *Big Data*, 1(2), 85–99. https://doi.org/10.1089/big.2012.0002
40. Lisichkin, G. V., & Leenson, I. A. (2013). Natural-sciences education in secondary school in the USSR and Russia: History, trends, and challenges of modernization. *Russian Journal of General Chemistry*, 83(6), 1185–1203. https://doi.org/10.1134/S1070363213060388
41. Berry, R. S. (1971). Environmental problems and the basic natural sciences. *Growth and Change*, 2(2), 25–29. https://doi.org/10.1111/j.1468-2257.1971.tb00208.x
42. Iida, K. (1965). On the study of mechanical properties of rocks in earth sciences. *Journal of the Society of Materials Science, Japan*, 14, 455–463. https://doi.org/10.2472/jsms.14.141Appendix_455
43. Dessler, A. J. (1968). The role of space science in graduate education. *Eos, Transactions American Geophysical Union*, 49(3), 549–554. https://doi.org/10.1029/TR049i003p00549
44. Mardiana, D., & Cahyani, R. (2018). The development of basic natural science learning materials to improve students' competence. *Journal of Physics: Conference Series*, 1028(1). https://doi.org/10.1088/1742-6596/1028/1/012206
45. Goldfein, M. D., & Ivanov, A. V. (2016). *Applied natural science: Environmental issues and global perspectives*. Apple Academic Press, pp. 1–458. https://doi.org/10.4324/9781315366555
46. Bowden, W. (1929). Are social studies sciences? *Social Forces*, 7(3), 367–378. https://doi.org/10.2307/2569805
47. Snow, N. E. (2003). *Stem cell research: New frontiers in science and ethics*. University of Notre Dame Press, Notre Dame. pp. 1–222. https://doi.org/10.2307/j.ctvpj7b9x

48. Graham, L. R. (1984). Comparing U.S. and soviet experiences: Science, citizens, and the policy-making process. *Environment*, 26(7), 6–37. https://doi.org/10.1080/00139157.1984.9932510
49. Albornoz, O. (2001). The production of knowledge in social sciences. *Acta Cientifica Venezolana*, 52(2), 83–95.
50. Zajdlic, W. (1956). The limitations of social sciences. *Kyklos*, 9(1), 65–76. https://doi.org/10.1111/j.1467-6435.1956.tb02682.x
51. Balzer, W., Moulines, C. U., & Sneed, J. D. (1986). The structure of empirical science: Local and global. *Studies in Logic and the Foundations of Mathematics*, 114(C), 291–306. https://doi.org/10.1016/S0049-237X(09)70697-4
52. Groten, J. P., Heijne, W. H. M., Stierum, R. H., Freidig, A. P., & Feron, V. J. (2004). Toxicology of chemical mixtures: A challenging quest along empirical sciences. *Environmental Toxicology and Pharmacology*, 18(3 SPEC.ISS.), 185–192. https://doi.org/10.1016/j.etap.2004.07.005
53. Earnshaw, R. (2019). Data science. *Advanced Information and Knowledge Processing*, 1–10. https://doi.org/10.1007/978-3-030-24367-8_1
54. Nasution, M. K. M., Salim Sitompul, O., & Budhiarti Nababan, E. (2020). Data science. *Journal of Physics: Conference Series*, 1566(1). https://doi.org/10.1088/1742-6596/1566/1/012034
55. Zegura, E., DiSalvo, C., & Meng, A. (2018). Care and the practice of data science for social good. In *Proceedings of the 1st ACM SIGCAS conference on computing and sustainable societies*. COMPASS. 2018, Menlo Park and San Jose, 20 June 2018. https://doi.org/10.1145/3209811.3209877
56. Jácome de Moura, P., Jr. (2021). Is data science a science? The essence of phenomenon and the role of theory in the emerging field. *Kybernetes*, 51(7), 2416–2434. https://doi.org/10.1108/K-03-2021-0205
57. Nasution, M. K. M., Hidayat, R., & Syah, R. (2022). Computer science. *International Journal on Advanced Science, Engineering and Information Technology*, 12(3), 1142–1159. https://doi.org/10.18517/ijaseit.12.3.14832
58. Sveinsdottir, E., & Frøkjær, E. (1988). Datalogy—The Copenhagen tradition of computer science. *BIT*, 28(3), 450–472. https://doi.org/10.1007/BF01941128
59. Naur, P. (1966). The science of datalogy. *Communications of the ACM*, 9(7), 485. https://doi.org/10.1145/365719.366510
60. Bruhn, J. G. (1974). Human ecology: A unifying science? *Human Ecology*, 2(2), 105–125. https://doi.org/10.1007/BF01558116
61. Saltz, J., Armour, F., & Sharda, R. (2018). Data science roles and the types of data science programs. *Communications of the Association for Information Systems*, 43(1). https://doi.org/10.17705/1CAIS.04333
62. Albright, S. D., Klinge, T. H., & Rebelsky, S. A. (2018, January). A functional approach to data science in CS1. In *SIGCSE 2018—proceedings of the 49th ACM technical symposium on computer science education*, pp. 1035–1040. https://doi.org/10.1145/3159450.3159550
63. Prashantgokul, K., Sundararajan, M., & Paul, P. K. (2019). Big data management, data science and data analytics: What is it and where—an educational in Indian perspective. *International Journal of Innovative Technology and Exploring Engineering*, 8(12), 1231–1236. https://doi.org/10.35940/ijitee.L3978.1081219
64. Chen, L. M. (2015). Machine learning for data science: Mathematical or computational. *Mathematical Problems in Data Science: Theoretical and Practical Methods*, 63–74. https://doi.org/10.1007/978-3-319-25127-1_4
65. He, X., & Lin, X. (2020). Challenges and opportunities in statistics and data science: Ten research areas. *Harvard Data Science Review* (2.3), Summer 2020. https://doi.org/10.1162/99608f92.95388fcb

66. Rajagopal, S. (2018). Data science: Recent developments and future insights. *Deep Learning Innovations and Their Convergence with Big Data*. https://doi.org/10.4018/978-1-5225-3015-2.ch008
67. Danoho, D. (2017). 50 years of data science. *Journal of Computational and Graphical Statistics*, 26(4), 745–766. https://doi.org/10.1080/10618600.2017.1384734
68. Asparouhov, T. (2005). Sampling weights in latent variable modeling. *Structural Equation Modeling*, 12(3), 411–434. https://doi.org/10.1207/s15328007sem1203_4
69. Jack Smith, F. (2006). Data science as an academic discipline. *Data Science Journal*, 5, 163–164. https://doi.org/10.2481/dsj.5.163
70. Nasution, M. K. M. (2020). The birth of a science. *History of Science and Technology*, 10(2), 315–338. https://doi.org/10.32703/2415-7422-2020-10-2-315-338
71. Shcherbakov, M., Shcherbakova, N., Brebels, A., Janovsky, T., & Kamaev, V. (2014). Lean data science research life cycle: A concept for data analysis software development. *Communications in Computer and Information Science*, 466 CCIS, 708–716. https://doi.org/10.1007/978-3-319-11854-3_61
72. Oliver, J. C., Kollen, C., Hickson, B., & Rios, F. (2019). Data science support at the academic library. *Journal of Library Administration*, 59(3), 241–257. https://doi.org/10.1080/01930826.2019.1583015
73. Provost, F., & Fawcett, T. (2013). Data science and its relationship to big data and data-driven decision making. *Big Data*, 1(1), 51–59. https://doi.org/10.1089/big.2013.1508
74. Musson, R., & Smith, R. (2013). Data science in the cloud: Analysis of data from testing in production. In *2013 international workshop on testing the cloud, TTC 2013—proceedings*, pp. 18–20. https://doi.org/10.1145/2489295.2493955
75. Shah, C., Anderson, T., Hagen, L., & Zhang, Y. (2021). An iSchool approach to data science: Human-centered, socially responsible, and context-driven. *Journal of the Association for Information Science and Technology*, 72(6), 793–796. https://doi.org/10.1002/asi.24444
76. Barlas, P., Lanning, I., & Heavey, C. (2015). A survey of open source data science tools. *International Journal of Intelligent Computing and Cybernetics*, 8(3), 232–261. https://doi.org/10.1108/IJICC-07-2014-0031
77. Baumer, B. (2015). A data science course for undergraduates: Thinking with data. *American Statistician*, 69(4), 334–342. https://doi.org/10.1080/00031305.2015.1081105
78. Dhar, V. (2013). Data science and prediction. *Communications of the ACM*, 56(12), 64–73. https://doi.org/10.1145/2500499
79. Asniara, S. K. (2014). Using data science for detecting outliers with k nearest neighbors graph. In *Proceedings—2014 international conference on ICT for smart society: "Smart system platform development for city and society, GoeSmart 2014", ICISS 2014*, pp. 300–304. https://doi.org/10.1109/ICTSS.2014.7013191
80. Lucke, C. E. (1914). Research what—who—where—why. *Industrial and Engineering Chemistry*, 6(11), 880–882. https://doi.org/10.1021/ie50071a001
81. Blei, D. M., & Smyth, P. (2017). Science and data science. *Proceedings of the National Academy of Sciences of the United States of America*, 114(33), 8689–8692. https://doi.org/10.1073/pnas.1702076114
82. Cao, L. (2017). Data science: Challenges and directions. *Communications of the ACM*, 60(8), 59–68. https://doi.org/10.1145/3015456
83. Aerts, M., Molenberghs, G., & Thas, O. (2021). Graduate education in statistics and data science: The why, when, where, who, and what. *Annual Review of Statistics and Its Application*, 8, 25–39. https://doi.org/10.1146/annurev-statistics-040620-032820
84. Freericks, J. K., Nikolic, B. K., & Frieder, O. (2014). The nonequilibrium quantum many-body problem as a paradigm for extreme data science. *International Journal of Modern Physics B*, 28(31). https://doi.org/10.1142/S0217979214300217

85. Priebe, T., & Markus, S. (2015). Business information modeling: A methodology for data-intensive projects, data science and big data governance. In *Proceedings—2015 IEEE international conference on big data, IEEE big data 2015*, pp. 2056–2065. https://doi.org/10.1109/BigData.2015.7363987
86. Brodie, M. L. (2015). Understanding data Science: An emerging discipline for data-intensive discovery. *CEUR Workshop Proceedings*, 1536, 238–245.
87. Yue, P., Ramachandran, R., Baumann, P., Khalsa, S. J. S., Deng, M., & Jiang, L. (2016). Recent activities in earth data science [technical committees]. *IEEE Geoscience and Remote Sensing Magazine*, 4(4), 84–89. https://doi.org/10.1109/MGRS.2016.2600528
88. NIST2015. *National Institute of Standards and Technology*. (2015). NIST big data interoperability framework Volume 1, Definitions. http://nvlpubs.nist.gov/nistpubs/SpecialPublications/NIST.SP.1500-1.pdf
89. Chinn, C. A., & Brewer, W. F. (1993). The role of anomalous data in knowledge acquisition: A theoretical framework and implications for science instruction. *Review of Educational Research*, 63(1), 1–49. https://doi.org/10.3102/00346543063001001
90. Heinemann, B., Opel, S., Budde, L., Schulte, C., Frischemeier, D., Biehler, R., Podworny, S., & Wassong, T. (2018). Drafting a data science curriculum for secondary schools. In *ACM international conference proceeding*. https://doi.org/10.1145/3279720.3279737
91. Ultsch, A., & Lötsch, J. (2017). A data science based standardized Gini index as a Lorenz dominance preserving measure of the inequality of distributions. *PLoS One*, 12(8). https://doi.org/10.1371/journal.pone.0181572
92. Zervos, C. (1985). From data to information to knowledge. The problems of metamorphosis. *ACS Symposium Series*, 235–252.
93. Ewing, A. C. (1925). I.-The relation between knowing and its object. (I.) *Mind*, 34(134), 137–153. https://doi.org/10.1093/mind/XXXIV.134.137
94. Ewing, A. C. (1925). III.-The relation between knowing and its object (II.) *Mind*, 34(135), 300–310. https://doi.org/10.1093/mind/XXXIV.135.300
95. Ersoz, E. S., Martin, N. F., & Stapleton, A. E. (2020). On to the next chapter for crop breeding: Convergence with data science. *Crop Science*, 60(2), 639–655. https://doi.org/10.1002/csc2.20054
96. Saltz, J. S., Shamshurin, I., & Crowston, K. (2017, January). Comparing data science project management methodologies via a controlled experiment. In *Proceedings of the annual Hawaii international conference on system sciences*. IEEE Computer Society, 3 January HICSS 2017. pp. 1013–1022.
97. Abu-el-zeet, Z. H., Becerra, V. M., & Roberts, P. D. (2002). Combined bias and outlier identification in dynamic data reconciliation. *Computers and Chemical Engineering*, 26(6), 921–935. https://doi.org/10.1016/S0098-1354(02)00018-2
98. Kobylinski, P., Pochwatko, G., & Biele, C. (2019). VR experience from data science point of view: How to measure inter-subject dependence in visual attention and spatial behavior. *Advances in Intelligent Systems and Computing*, 903, 393–399. https://doi.org/10.1007/978-3-030-11051-2_60
99. Brooks, C., Hoepner, A. G. F., McMillan, D., Vivian, A., & Wese Simen, C. (2019). Financial data science: The birth of a new financial research paradigm complementing econometrics? *European Journal of Finance*, 25(17), 1627–1636. https://doi.org/10.1080/1351847X.2019.1662822
100. Swan, M. (2015). Philosophy of big data: Expanding the human-data relation with big data science services. In *Proceedings—2015 IEEE 1st international conference on big data computing service and applications, bigdataservice 2015*, pp. 468–477. https://doi.org/10.1109/BigDataService.2015.29
101. Trung, H. D. (2021). Database performance evaluation and applications of data science for IoT platform analysis. *International Journal of Computer Information Systems and Industrial Management Applications*, 13, 124–135.

102. Kenett, R. S., Zonnenshain, A., & Fortuna, G. (2017). A road map for applied data sciences supporting sustainability in advanced manufacturing: The information quality dimensions. *Procedia Manufacturing*, 21, 141–148. https://doi.org/10.1016/j.promfg.2018.02.104
103. Anton Feenstra, K., Abeln, S., Westerhuis, J. A., Brancos Dos Santos, F., Molenaar, D., Teusink, B., Hoefsloot, H. C. J., & Heringa, J. (2018). Training for translation between disciplines: A philosophy for life and data sciences curricula. *Bioinformatics*, 34(13). https://doi.org/10.1093/bioinformatics/bty233
104. Overton, M., & Kleinschmit, S. (2021). Data science literacy: Toward a philosophy of accessible and adaptable data science skill development in public administration programs. *Teaching Public Administration*, 2021. https://doi.org/10.1177/01447394211004990
105. van der Voort, H., van Bulderen, S., Cunningham, S., & Janssen, M. (2021). Data science as knowledge creation a framework for synergies between data analysts and domain professionals. *Technological Forecasting and Social Change*, 173. https://doi.org/10.1016/j.techfore.2021.121160
106. Soman, R. K., & Whyte, J. K. (2020). Codification challenges for data science in construction. *Journal of Construction Engineering and Management*, 146(7). https://doi.org/10.1061/(ASCE)CO.1943-7862.0001846
107. Ultsch, A., Kringel, D., Kalso, E., Mogil, J. S., & Lötsch, J. (2016). A data science approach to candidate gene selection of pain regarded as a process of learning and neural plasticity. *Pain*, 157(12), 2747–2757. https://doi.org/10.1097/j.pain.0000000000000694
108. Huai, J. (2015). Computing paradigms: Transformation and opportunities: Thinking on data science and machine intelligence. In *Lecture notes in computer science (including subseries lecture notes in artificial intelligence and lecture notes in bioinformatics)*, vol. 9219. Springer-Verlag. ISSN 03029743, ISBN 978–331921968–4
109. Albuerne, A., Grau-Bove, J., & Strlic, M. (2018). The role of heritage data science in digital heritage. *Lecture Notes in Computer Science (including subseries Lecture Notes in Artificial Intelligence and Lecture Notes in Bioinformatics)*, 11196 LNCS, 616–622. https://doi.org/10.1007/978-3-030-01762-0_54
110. Schweighofer, E. (2015). The role of AI & law in legal data science. *Frontiers in Artificial Intelligence and Applications*, 279, 191–192. https://doi.org/10.3233/978-1-61499-609-5-191
111. Evans, M., & Blythe, J. (1994). Fashion: A new paradigm of consumer behaviour. *Journal of Consumer Studies & Home Economics*, 18(3), 229–237. https://doi.org/10.1111/j.1470-6431.1994.tb00696.x
112. Li, F., Xu, J., Dou, Z.-T., & Huang, Y.-L. (2004). Data mining-based credit evaluation for users of credit card. *Proceedings of 2004 International Conference on Machine Learning and Cybernetics*, 4, 2586–2591.
113. Deb, D., & Jones, E. (2020). University-wide adoption of data science. In *Annual conference on innovation and technology in computer science education, ITiCSE, 1300*. https://doi.org/10.1145/3328778.3372657
114. Gregory, A. (2011). Data governance protecting and unleashing the value of your customer data assets: Stage 1: Understanding data governance and your current data management capability. *Journal of Direct, Data and Digital Marketing Practice*, 12(3), 230–248. https://doi.org/10.1057/dddmp.2010.41
115. Ziegler, Jr. Kurt. (1978). Distribution: A new impetus toward understanding data. In *Proceedings of the Jerusalem conference on information technology, 3rd(JCIT3)*. North-Holland Publ. Co., August 1978 through 9. pp. 311–318.
116. Power, D. J. (2008). Understanding data-driven decision support systems. *Information Systems Management*, 25(2), 149–154. https://doi.org/10.1080/10580530801941124
117. Okunade, K., Bashan Nkhoma, K., Salako, O., Akeju, D., Ebenso, B., Namisango, E., Soyannwo, O., Namukwaya, E., Dandadzi, A., Nabirye, E., Mupaza, L., & Luyirika, E. (2019). Understanding data and information needs for palliative cancer care

to inform digital health intervention development in Nigeria, Uganda and Zimbabwe: Protocol for a multicountry qualitative study. *BMJ Open*, 9(10). https://doi.org/10.1136/bmjopen-2019-032166

118. Valdés, J. J. (2003). Virtual reality representation of information systems and decision rules: An exploratory technique for understanding data and knowledge structure. *Lecture Notes in Computer Science (Including Subseries Lecture Notes in Artificial Intelligence and Lecture Notes in Bioinformatics)*, 2639 LNCS, 615–618. https://doi.org/10.1007/3-540-39205-x101
119. Finne, T. (2000). Information systems risk management: Key concepts and business processes. *Computers and Security*, 19(3), 234–242. https://doi.org/10.1016/S0167-4048(00)88612-5
120. Saltz, J. S., Dewar, N. I., & Heckman, R. (2008, January). Key concepts for a data science ethics curriculum. In *SIGCSE 2018—proceedings of the 49th ACM technical symposium on computer science education*, 2018, pp. 952–957. https://doi.org/10.1145/3159450.3159483
121. Oudshoorn, M. J., Titus, K. J., & Suchan, W. K. (2020, October). Building a new data science program based on an existing computer science program. In *Proceedings—frontiers in education conference, FIE volume 2020*. https://doi.org/10.1109/FIE44824.2020.9273934
122. Adeboye, N. O., Popoola, P. O., & Ogunnusi, O. N. (2020). Data science skills: Building partnership for efficient school curriculum delivery in Africa. *Statistical Journal of the IAOS*, 36(S1), S49–S62. https://doi.org/10.3233/SJI-200693
123. Anslow, C., Brosz, J., Maurer, F., & Boyes, M. (2016). Datathons: An experience report of data hackathons for data science education. In *SIGCSE 2016—proceedings of the 47th ACM technical symposium on computing science education*, pp. 615–620. https://doi.org/10.1145/2839509.2844568
124. Medeiros, M. M., Hoppen, N., & Maçada, A. C. G. (2020). Data science for business: Benefits, challenges and opportunities. *Bottom Line*, 33(2), 149–163. https://doi.org/10.1108/BL-12-2019-0132
125. Williamson, B. (2017). Who owns educational theory? Big data, algorithms and the expert power of education data science. *E-Learning and Digital Media*, 14(3), 105–122. https://doi.org/10.1177/2042753017731238
126. Klenk, B., & Dennison, L. (2020). Why data science and machine learning need silicon photonics. *Optics InfoBase Conference Papers*, Part F174-OFC 2020.
127. Goodman, A. (2014). Evolution of symposia on the interface of computing and statistics defines data science to be the interface. *Wiley Interdisciplinary Reviews: Computational Statistics*, 6(5), 367–377. https://doi.org/10.1002/wics.1316
128. Devalkar, S. K., Seshadri, S., Ghosh, C., & Mathias, A. (2018). Data science applications in Indian agriculture. *Production and Operations Management*, 27(9), 1701–1708. https://doi.org/10.1111/poms.12834
129. Brady, H. E. (2019). The challenge of big data and data science. *Annual Review of Political Science*, 22, 297–323. https://doi.org/10.1146/annurev-polisci-090216-023229
130. Lyubchich, V., Newlands, N. K., Ghahari, A., Mahdi, T., & Gel, Y. R. (2019). Insurance risk assessment in the face of climate change: Integrating data science and statistics. *Wiley Interdisciplinary Reviews: Computational Statistics*, 11(4). https://doi.org/10.1002/wics.1462
131. Molina-Solana, M., Ros, M., Ruiz, M. D., Gómez-Romero, J., & Martin-Bautista, M. J. (2017). Data science for building energy management: A review. *Renewable and Sustainable Energy Reviews*, 70, 598–609. https://doi.org/10.1016/j.rser.2016.11.132
132. Fleischmann, M., Feliciotti, A., & Kerr, W. (2021). Evolution of urban patterns: Urban morphology as an open reproducible data science. *Geographical Analysis* (Article in Press). https://doi.org/10.1111/gean.12302

133. Chambers, J. M. (2020). S, R, and data science. *Proceedings of the ACM on Programming Languages*, 4(HOPL). https://doi.org/10.1145/3386334
134. Lal, M. K., Kim, T. H., & Singleton, D. M. (2020). Data science use case for brownfield optimization—A case study. In *SPE Western regional meeting proceedings 2020 SPE western regional meeting 2020*, WRM 2020. https://doi.org/10.2118/200781-MS
135. Muller, M., Lange, I., Wang, D., Piorkowski, D., Tsay, J., Vera Liao, Q., Dugan, C., & Erickson, T. (2019). How data science workers work with data. In *Conference on human factors in computing systems—proceedings*. https://doi.org/10.1145/3290605.3300356
136. Jaimes, A. (2012). A human-centered perspective on multimedia data science: Tutorial overview. In *MM 2012—proceedings of the 20th ACM international conference on multimedia*, pp. 1537–1538. https://doi.org/10.1145/2393347.2396555
137. Nielsen, S. (2008). The effect of lexicographical information costs on dictionary making and use. *Lexikos*, 18, 170–189.
138. McQuillan, D. (2018). Data science as machinic neoplatonism. *Philosophy and Technology*, 31(2), 253–272. https://doi.org/10.1007/s13347-017-0273-3
139. Nasution, M. K. M. (2020). Industry 4.0. *IOP Conference Series: Materials Science and Engineering*, 1003(1). https://doi.org/10.1088/1757-899X/1003/1/012145
140. Nasution, M. K. M. (2021). Industry 4.0: Data science perspective. *IOP Conf. Series: Materials Science and Engineering*, 1122. https://doi.org/10.1088/1757-899X/1122/1/012037
141. Cannas, V. G., Ciano, M. P., Pozzi, R., & Rossi, T. (2020). Data science supporting lean production: A bibliometric study. *Proceedings of the Summer School Francesco Turco*, 2020.
142. Everest, G. C. (1986). *Database management: Objectives, system functions, and administration*. New York: McGraw-Hill Book Company.
143. Herold, K. (2003). An information continuum conjecture. *Minds and Machines*, 13(4), 553–566. https://doi.org/10.1023/A:1026204901999
144. Nasution, M. K. M., Sitompul, O. S., Elveny, M., & Syah, R. (2021). Data science: A review towards the big data problems. *Journal of Physics: Conference Series*, 1898(1). https://doi.org/10.1088/1742-6596/1898/1/012006
145. Concolato, C. E., & Chen, L. M. (2017). Data science: A new paradigm in the age of big-data science and analytics. *New Mathematics and Natural Computation*, 13(2), 119–143. https://doi.org/10.1142/S1793005717400038
146. Paul, P. K., & Dey, J. L. (2017, January). Data science Vis-À-Vis efficient healthcare and medical systems: A techno-managerial perspective. *2017 Innovations in Power and Advanced Computing Technologies, i-PACT*, 1–8. https://doi.org/10.1109/IPACT.2017.8245148
147. Chambers, J. M. (2020). S, R, and data science. *R Journal*, 12(1), 462–476. https://doi.org/10.32614/rj-2020-028
148. Song, I., & Zhu, Y. (2016). Big data and data science: What should we teach? *Expert Systems*, 33(4), 364–373. https://doi.org/10.1111/exsy.12130
149. Demchenko, Y., Belloum, A., Los, W., Wiktorski, T., Manieri, A., Brocks, H., Becker, J., Heutelbeck, D., Hemmje, M., & Brewer, S. (2016). EDISON data science framework: A foundation for building data science profession for research and industry. In *Proceedings of the international conference on cloud computing technology and science, cloudcom*, pp. 620–626. https://doi.org/10.1109/CloudCom.2016.0107
150. Wiktorski, T., Shirazi, A., Demchenko, Y., & Belloum, A. (2016). Quantitative and qualitative analysis of current data science programs from perspective of data science competence groups and framework. In *Proceedings of the international conference on cloud computing technology and science, cloudcom*, pp. 633–638. https://doi.org/10.1109/CloudCom.2016.0109

151. Silva, J., Portillo, R., Hernandez, A. E., Varela, N., Caraballo, H. M., Palma, H. H., Bilbao, O. R., & Castro, N. L. (2020). Data sciences and teaching methods—learning. *Lecture Notes in Electrical Engineering*, 637, 335–342. https://doi.org/10.1007/978-981-15-2612-1_32
152. Yeturu, K. (2020). Machine learning algorithms, applications, and practices in data science. *Handbook of Statistics*. https://doi.org/10.1016/bs.host.2020.01.002
153. Maheswari, P., & Narayana, C. H. V. (2020). Predictions of loan defaulter—A data science perspective. In *Proceedings of the 2020 international conference on computing, communication and security, ICCCS 2020*. https://doi.org/10.1109/ICCCS49678.2020.9277458
154. Alexander, R., Alexopoulos, M., Lyons, K., & Austin, L. (2020). Workshop on barriers to data science adoption: Why existing frameworks aren't working. In *CASCON 2019 proceedings—conference of the centre for advanced studies on collaborative research—proceedings of the 29th annual international conference on computer science and software engineering*. Center for Advanced Studies on Collaborative Research, pp. 384–385.
155. Patel, A., Debnath, N. C., & Bhusan, B. (Eds.). (2022). *Data science with semantic technologies: theory, practice and application*. John Wiley & Sons.
156. Cady, F. (2017). The data science handbook. *The Data Science Handbook*, 1–396. https://doi.org/10.1002/9781119092919
157. Zhong, W.-W., Ji, C., Guo, L.-W., Li, B., & Xu, X.-S. (2020). Thoughts and practices on studies of membrane processes for Chinese material medical based on data science. *Chinese Traditional and Herbal Drugs*, 51(1), 1–8. https://doi.org/10.7501/j.issn.0253-2670.2020.01.001
158. Nasution, M. K. M., Sitompul, O. S., Nababan, E. B., Nababan, E. S. M., & Sinulingga, E. P. (2020). Data science around the indexed literature perspective. *Advances in Intelligent Systems and Computing*, 1294, 1051–1065. https://doi.org/10.1007/978-3-030-63322-6_91

2 Big Data and its Future

Musa Milli and Mehmet Milli

CONTENTS

2.1 DEFINITION OF BIG DATA AND ITS CHARACTERISTICS

Over the last several decades, data production has increased tremendously for a variety of reasons. The current era is known as the data age because of (i) the increase in the potential of digital data storage (technological developments), (ii) the fact that people consume and create content like social media, (iii) the fact that physical quantities in our environment can be quickly and easily converted into digital values (the sensor driven Internet of Things), and (iv) the existence of data-intensive scientific studies.

When the concept of big data was first introduced, it was defined as a huge volume of data that can not handle conventional methods.[1] However, defining big data only as large volume data provides a very narrow perspective. Big data not only describes the data itself, but it also describes the systems that produce data, the methods that handle data, the platforms, the databases, and all the technologies, as well as the

DOI: 10.1201/9781003310785-2

value and opportunities derived from data. It was not possible to present, store, process, and analyze big data with traditional methods. Over time, researchers discovered features of big data other than large volume, such as veracity, variety, velocity, etc.[2, 3] While some researchers [4] talk about the 3V features of big data, the company Oracle has included the value feature of big data. The characteristics of big data known as 4V can be summarized as follows.

Volume refers to the large scale of the data that is a common property and also a major problem due to the lack of methods and technologies that can process and store this type of data. A few decades ago, data were generally collected manually, but today, data are usually collected automatically by sensors and data-generating systems. Other reasons for the increase in volume are data production, the transition from a centralized structure to a more distributed structure, and the ordinary people consuming and producing data.

There is no doubt that volume is the most important characteristic of big data. Size is a relative concept, and when it comes to large numbers, it can be difficult to understand the scale of size. In broader sense, Table 2.1 explains the volume feature of big data.

Digital data size climbed from a few zettabytes in 2013 to 33 zettabytes in 2020 and is predicted to reach 175 zettabytes in 2025.[5] The amount of data continues to increase yearly at greater speeds. The size of digital data is expected to reach the order of yottabytes in the coming years.

Variety describes the incoming data with different structures and types from different sources. This is another property and problem of big data because current machine learning methods and algorithms are adept at handling single type (quantitative, qualitative, continuous, binary, numeric, or categorical) of data at the same time. Today's advanced technology can collect and store data from many sources. In order to process each observation from various sources with different structures and reveal their relationships with each other, new methods, algorithms, and applications are needed.

The collection of heterogeneous data with different structures is a result of different data sources. Heterogeneity is one of the important features of big data, and it complicates the integration of data. In order to have better quality data sources and to obtain more reliable results from data, methods that perform data integration have been put forward. Semantic techniques have been frequently used to merge heterogeneous data with large structural differences and increase the value of the input data.[6, 7]

Velocity refers to the speed of incoming data. This property alone is sufficient to describe a process that will require reconsideration of data processing. The rapid generation of data and its high speed arrival to data storage centers has revealed the scalability problems of methods and algorithms in the storage and processing stages. Especially in systems that require real-time and nearly real-time response, the high speed of incoming data has been one of the important problems that algorithms had to overcome.

Big data processing tools are more scalable than conventional methods; however, they have natural limits, especially the central processing tools. When the velocity of data is high enough, the tools compromise the quality of outputs. Although it is

TABLE 2.1
Data Volume

Name	Symbol	Bytes (Approximate)	Bytes
Bit	0 or 1	2^0 bit	
Byte	Byte	1	2^0
Kilobyte	KB	1,000	2^{10}
Megabyte	MB	1,000,000	2^{20}
Gigabyte	GB	1,000,000,000	2^{30}
Terabyte	TB	1,000,000,000,000	2^{40}
Petabyte	PB	1,000,000,000,000,000	2^{50}
Exabyte	EB	1,000,000,000,000,000,000	2^{60}
Zettabyte	ZB	1,000,000,000,000,000,000,000	2^{70}
Yottabyte	YB	1,000,000,000,000,000,000,000,000	2^{80}
Brontobyte	BB	1,000,000,000,000,000,000,000,000,000	2^{90}
Geopbyte	GB	1,000,000,000,000,000,000,000,000,000,000	2^{100}

true that big data processing tools are more scalable, they do not have an infinitive processing capacity.

The model established may not be valid after a while due to the high speed of data arrival. Hyde et al. report that True-Positive observations can turn False-Positive when the incoming data changes statistically.[8] In fact, this is also true for the reverse; observations that are True-Negative may return False-Negative when the incoming data changes statistically. Therefore, the designed algorithm should keep the model up to date but keeping a model up to date may not be easy if the environment is highly dynamic.

Value refers to the data quality. It is the fundamental theorem in the field of data mining. The better the quality of the incoming data, the more accurate and high quality is the information obtained from this data. In order to interpret the data correctly, the factors that determine the quality of the data must be suitable, but the quality of the data can be adversely affected due to the unique characteristics of big data. From this point of view, low-value data is another reason that dealing with big data is difficult. As seen in Figure 2.1, accuracy, reliability, precision, completeness, format, and validity are features that determine the quality of the data.[9, 10]

2.2 BIG DATA VALUE CHAIN

2.2.1 Data Generation

Raw data goes through many stages (data acquisition, data analysis, data curation, data storage, data usage) until it turns into information; these stages are called the value chain of big data. Figure 2.2 shows the chart of big data known as the value chain. The increase in IoT devices that automatically generate data, rapid

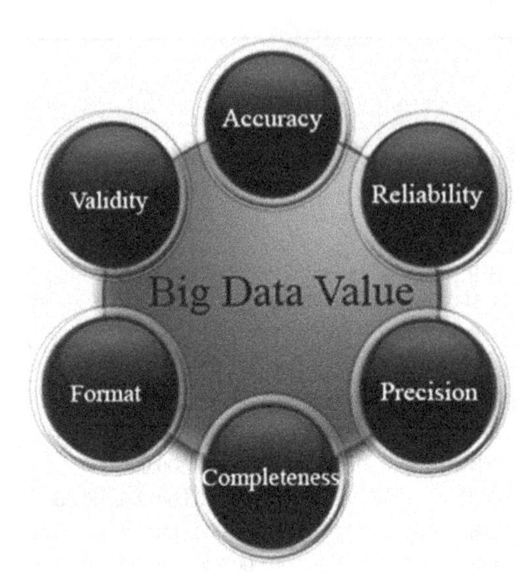

FIGURE 2.1 Parameters that determine the quality of big data (adapted from Milli et al., 2016)[10].

Data Acquisition	Data Analysis	Data Curation	Data Storage	Data Usage
• Structured data • Unstructured data • Event processing • Sensor networks • Protocols • Real-time • Data streams • Multimodality	• Stream mining • Semantic analysis • Machine learning • Information extraction • Linked Data • Data discovery • 'Whole world' semantics • Ecosystems • Community data analysis • Cross-sectorial data analysis	• Data Quality • Trust / Provenance • Annotation • Data validation • Human-Data Interaction • Top-down/Bottom-up • Community / Crowd • Human Computation • Curation at scale • Incentivisation • Automation • Interoperability	• In-Memory DBs • NoSQL DBs • NewSQL DBs • Cloud storage • Query Interfaces • Scalability and Performance • Data Models • Consistency, Availability, Partition-tolerance • Security and Privacy • Standardization	• Decision support • Prediction • In-use analytics • Simulation • Exploration • Visualisation • Modeling • Control • Domain-specific usage

FIGURE 2.2 The value chain of big data.[11]

digitalization in many fields (finance, education, e-commerce, etc.), individuals as digital content provider have resulted in the production of more numerical data than ever before.

Lately, the digitalization race has been rapidly continuing in many major sectors due to the ease and flexibility the Internet and computers brings to our lives. This has caused a daily exponential increase in the size of digital data, and researchers have begun to better understand the characteristics of big data known as the Vs.

Enterprises are able to provide valuable insight through good data management. Thanks to big data, they analyze transaction history and clickstream data of customers, and then they extract customers' profiles and behaviors. They use this knowledge to discover customers' hidden purchasing potentials in order to increase their profits. Recently, highly deployed recommendation systems in trade have provided significant

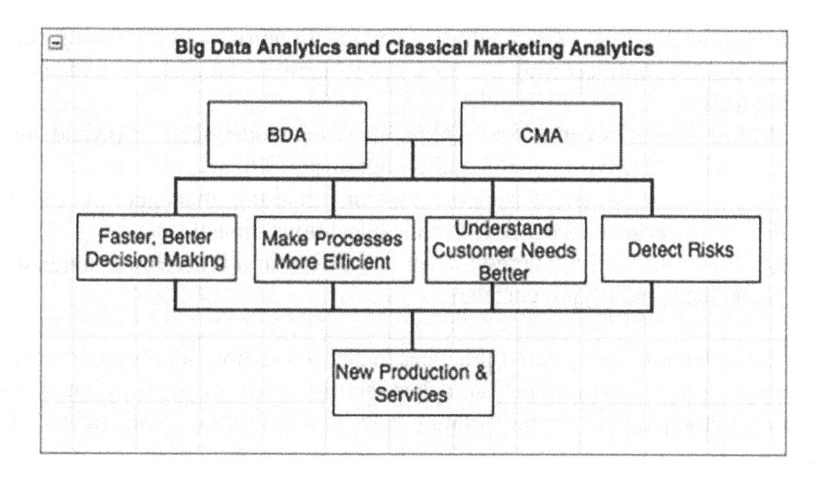

FIGURE 2.3 The impact of big data analytics and classical marketing analytics.[12]

strategic advantages to companies. Recommendation systems appear as applications that provide personalized services to market customers by extracting their profile information and improving their digital shopping experiences. Recommendation systems allow companies to manage the big data they collect from their customers in a creative manner.

Classical market analysis calculates risks and focuses on success with a small amount of data and in a narrower context. Processing mass volume data, performing customer analysis, and obtaining knowledge that will meet customers' expectations are among the capabilities of data-intensive systems developed for big data. Big data analytics methods and algorithms, shown in Figure 2.3, are used to collect all kinds of customer data, to better analyze customer needs, to calculate risks, and to make faster and more accurate decisions.

The two best examples in the field of commerce that generate value from big data are Walmart and Amazon. By examining the data they collect from their users, they offer customers better shopping experiences and, thus, increase customer loyalty and retention. In 2022, Walmart and Amazon are the top two companies on the National Retail Federation's "Top 100 Retailers 2022 List", and their retail sales are $459.51 and $217.79 billion dollars respectively.

Apart from manufacturing, the areas where big data is studied extensively are as follows:

Health sector: processing bioinformatic data, diagnosing disease, and monitoring patients; in decision support systems; and for the effective use of medical resources and the management of healthcare organizations.

Social media: link prediction, profile matching, opinion mining, and sentiment analysis.

Scientific research: future prediction and scientific improvements, genome project, big hadron collider, space research, telescope data, military experience, sonar images, unmanned aerial vehicle (UAV) images, etc.

Politics and government: strengthening national security and improving public services, online voting systems, crime analysis, blockchain, and smart contracts.

Agriculture sector: weather forecasting, improved modeling of crop yield, and increasing the environmental and economic performance of farming.

Business sector: decision support systems, making management easier, employee performance analysis, and determining annual salaries.

Finance sector: predicting credit score and determining investment methods, blockchain, and crypto currency.

Other fields that use big data best practices include education, supply chain management, engineering, food production, cyber security, real time monitoring systems, medicine sector, judgment, transportation, telecommunication, mobile devices, GPS data, etc.

2.2.2 Data Acquisition

After the data production process, the next step in the big data value chain is data acquisition. Data acquisition is the process of collecting data from various sources and delivering it to data warehouses or environments in which it will be processed, without changing it, by complying with the data integrity, cleaning, and filtering processes. In the big data context, data usually comes in a large size, high speed, high diversity, and low value,[13] and in order to make accurate inferences, it is necessary to undergo a preprocessing stage before it is processed and stored. Moreover, not all incoming data comes in the form of structured data that the computer and algorithms can easily perceive. For this reason, it is necessary to preprocess the data and represent it in a way that the algorithm can handle. For example, ontology-based systems that reveal conceptual aspects of the data are used to standardize the input data independent of the domain.[7, 14]

2.2.3 Data Storage

As the size of the data obtained and the variety of data increased, the expectations of storage systems increased and changed. Although the data produced in the first days of widespread use of digital systems was stored in sequential access files or random access files, databases and relational database management systems (RDBMS) were developed to handle the increase of data-intensive systems and computational performance expectations. Over the last years, when the RDBMS could no longer give the expected performance for big data, more scalable, non-relational databases (NoSQL) were developed; distributed database models (DFS) were emphasized, such as the Hadoop Distributed File System (HDFS).

2.2.4 Data Processing/Analysis

With the increase in the size of the data, processing methods were also forced to change. Traditional machine learning and reasoning algorithms were inadequate in processing such large data because many batch processing algorithms have $O(n^2)$ as computational complexity. This computational complexity was far from efficient for big data. In addition, due to the rapid production and flow of data, sometimes it wasn't possible to find x data in memory a second time; x data could have already been deleted.

In the age of big data, researchers face some constraints: (i) memory constraints (data is too large to fit in memory all at one time),[15] and (ii) knowledge to be obtained from data is required in real time (surveillance system, intrusion detection system, etc.). New and novel algorithms that can handle big data in real time have been developed by researchers, and traditional batch processing algorithms have been modified to handle big data.

2.2.5 Data Visualization

Data visualization is part of the big data value chain, and it refers to translating outputs obtained from machine learning algorithms to a human-understandable display such as a histogram, graph, tree (to represent hierarchy), etc. At this stage, the information obtained from the processing and analysis of the data is served as output. The output obtained from machine learning algorithms serves for another transaction or is represented and visualized in a way that the human mind can easily understand.

One of the most comprehensive definitions of data visualization is

> Visualization is the means by which humans understand complex analytics and is often the most crucial and overlooked step in the analytics process. As you increase the complexity of your data, the complexity of your final model increases as well, making effective communication and visualization of data even more difficult and critical to end users. Data visualization is the key to actionable insights.[16]

2.3 CHALLENGES BIG DATA RESEARCHERS FACE

2.3.1 Vs of Big Data

Size is not the only threating feature of big data. The data characteristics previously mentioned have brought new challenges for researchers, companies, and stakeholders to overcome while collecting, expressing, storing, processing, and visualizing data; these characteristics necessitate the development of new and different methods. There is a well-known rule in statistics: The more data we have, the closer we are to the real model. However, in practice, it is not that simple because of the features of big data. Gaining benefit from big data requires solving the big problems that come with big data.

If there is an instant need for information to be produced from the data (cyber security, surveillance systems, online recommendation systems, etc.), the generated data is processed as stream data in real time. In addition to the problematic natural characteristics of big data processing the data, the fact that big data is generally considered as stream data causes it to be more problematic when expressing and processing.

In this data world, a tremendous amount of data is generated continuously, and generally big data is defined as an endless sequence of data. So earlier methods and algorithms are unable to be directly applied to big data because algorithms that work with big data must have the ability to run forever.

Algorithms that can work with potentially infinite amounts of data need to have some capabilities:

- It must be scalable. Many algorithms that work on stationary data also work in $O(n^2)$. However, algorithms that work with big data usually do not have that much time.
- It should be able to obtain the result by passing through the data only once (single scan). Therefore, big data algorithms generally operate at O(n) or more efficient complexity times.
- It should be able to produce a result based on the data it has seen so far at any time *t*.
- Data coming with high velocity can change statistically in a short time. From this point of view, big data algorithms should be able to detect context drifts when working in a highly dynamic environment and update themselves according to the direction of the drift.
- It should have the potential to work with heterogeneous data containing multiple data types (binary, text, numeric, audio, image, video, etc.).
- It should be able to work with low quality data and extract information. The higher the value of the input data, the more accurate and consistent the results will be. However, data sources often produce input as raw data, and this raw state of data is unqualified.

2.3.2 Data Representation and Integration

A computer program consists of two main components: (i) algorithm and (ii) data structures.[17] Data structures are concerned with how to represent input data, while algorithms are concerned with how to manipulate data. Before generating information from the input data, preprocessing and presenting it appropriately (stack, list, queue, graph, tree, etc.) will allow the algorithm to produce more accurate results and/or run faster.

Data representation refers to the rearrangement of input data to improve the algorithms' processing abilities. The digital big data world include different types of data and information that is diverse in semantics, organization, granularity, density, and statistical properties.[18] "Presentation of data must be designed not to merely display singularity of data but rather reflect the structure, hierarchy, and diversity of the data, and an integration technique should be designed to enable efficient operations across different datasets".[19]

Raw data often requires preprocessing and is represented differently after performing a transformation. In order to increase the efficiency of the algorithm, the raw data is transformed into a structured data type (e.g., Extensible Markup Language (XML) data or RDF triples by establishing ontological relationships between the data). Large data is not easy to represent because it is dynamic and new incoming data may not fit into the old model. In addition, the sources that produce big data are diverse, so there are few semantic similarities between the data produced, and it is not easy to establish a semantic infrastructure with these data. Therefore, it is difficult to establish a representative model for big data with various features that will cover all the data.

Although it is difficult to establish a semantic relationship between data collected from different sources, semantic enrichment is one of the most widely used methods to integrate big data. A common infrastructure representation is required to integrate different types of data. One of these methods is ontology. However, like any method, ontology has some disadvantages. For instance, the main disadvantage of ontology-based systems is that they are strictly dependent on a specific language that it is based on. No matter what method is used, it is not very easy to integrate different varieties and types of data. Malik et al. (2016) mentioned the problems of data representation and integration at the level of data generation and collection.[20]

2.3.3 Data Quality and Preprocessing

The quality of the input data increases the output quality, and the quality of the output allows us to make more accurate inferences about the data. Generally, there are repetitive data, outliers, missing data in the raw data in the data sets, or the attributes of the data have different ranges in terms of the values they take, and normalization is needed. Therefore, data is usually preprocessed to increase the value of the inputs. While this preprocessing stage is easier and faster to overcome such problems when working with relatively small-sized and static data, things get a little more complicated when it comes to large, dynamic and stream data. This section has discussed what kind of problems the data preprocessing, which is needed to improve the data quality, will cause when big data is concerned.

2.3.3.1 Redundancy Data Cleaning

There may be many repetitive observations in the collected data. It is a well-known fact that many redundant records may be found in data collected by sensors, in particular. For example, if the sensors measure a stable physical value with very small intervals, there will be thousands of observations with the same value in the data produced. If there is a lot of repetitive observation in the incoming data, it is reduced; otherwise the effect of the repetitive observation on the results obtained will be high, that is, the repetitive data will dominate the result. The preprocess of deleting repetitive observations in the data set is called redundancy data cleaning.

If the amount of redundancy is too high, actual patterns and information that needs to be discovered in the data may be overlooked. If a high level of redundancy dominates the dataset, this redundancy renders the relations between different data patterns meaningless. High level redundancy dominates the dataset and makes the relations between different data patterns meaningless. It may cause real recordings to be perceived as outliers. For example, in the very common dataset KDDCUP 99 Intrusion Detection, most of the records describe innocent connections, but there are also a small number of connections that are attempted hacking. The proportion of innocent and hacking connections in the dataset is 95% and 5%, respectively. Deleting redundant records in such a scenario helps to make more meaningful attack-tagged records and reduces the likelihood of the applied method falling into overfitting.

Although for many real-life transactions, deleting redundant records and clearing data is a preferred method to increase the accuracy of the results, sometimes it is better not to use this step depending on the nature of the operation. Because deleting redundant records may cause a loss in data value, at least the frequency of occurrence of deleted events is lost. Therefore, it may not be appropriate to clean repetitive records for every transaction. For example, deleting redundant records in a process with statistical mean, mode, or median calculations of the data can be problematic. To give another example, it is not appropriate to delete redundant records when it is necessary to measure the popularity and trend of a phenomenon or item. Since the purpose is to extract the histogram of the phenomena or items, it is necessary to use the whole dataset in such tasks.

For big data, the number of redundant records can be much greater. Data redundancy elimination is an essential stage. When redundancy is applied in big data analysis, not only will the accuracy of the results increase, but also the speed of the algorithms to produce results will increase dramatically. Although redundancy is a problem for big data, offering solutions to the redundancy problem also provides opportunities for researchers to produce new algorithms. As a matter of fact, many methods have already been designed to prevent or remove redundancy in big data,[18] and it still stands as a hot research topic at the point of providing researchers with opportunities.

2.3.3.2 Missing Data and Imputation

For a variety of reasons, missing parts may remain during data collection. For example, in a sensor network that is placed in the field, one of the sensors in the network may malfunction, its energy may run out, it may encounter an obstacle that cuts off the Wi-Fi signal, or the sensor may not be able to send partial or complete data to the center due to human-induced errors. In such cases, there is missing data in the dataset. Although a certain amount of missing data is tolerable for many algorithms, as the amount of missing data increases, the accuracy of the results produced by the algorithms decreases. Even some algorithms are not capable of working with missing data. Therefore, the missing data must be filled in using a suitable method to ensure that the algorithms achieve quality results. Systematically filling in missing data using specific methods is called data imputation. Data imputation is a kind of data recovery operation.

As mentioned earlier, high-value results can be produced if the value of the input is high. Another parameter that determines the value of the input is the completeness of the input. Therefore, the value of the input must be increased to produce accurate results. Running algorithms with missing data causes results to be biased depending on the proportion of missing data.[21] In order to properly fill in missing data, it is necessary to review the reasons for missing data and their relationship with other data. The relationship of missing data with other data on both a row and column basis can be summarized as follows:[22, 23]

1. Missing completely at random (MCAR): In this type of data deficiency, the missing cell of the missing record is neither related to other records nor related to other attributes of that record. So the missing observations

represent a random subset of all observations. The missing values have similar statistical values as the existing values. Completely random errors do not cause bias.
2. Missing at random (MAR): It describes the existence of a systematic relationship between the other observed values and the missing value, and the missing data can be explained using other observed values.
3. Missing not at random (MNAR): The loss of data is completely systematically related to the data itself. For example, in a survey of individuals, people with extreme weight may avoid specifying their weight. Such a lack of data falls under this category.

There are different imputation techniques to improve the quality of the data, depending on the type of missing data: mean imputation, substitution, hot deck imputation, cold deck imputation, regression imputation, maximum likelihood imputation, stochastic regression imputation, interpolation and extrapolation, k-nearest neighbor based imputation, hybrid imputation techniques, etc.

The aspect of data imputation that concerns the big data concept is that performing data imputation will add additional computational complexity no matter which method you use. If this complexity is large, it may push us to make a choice in terms of accuracy and time. After all, time complexity and optimality are trade-off parameters in many digital processes. Since the critical parameter of each transaction is different, there will always be questions about which one to choose or to what extent. For this reason, it would not be appropriate to say that this is the best method, platform, or algorithm for every situation.

Moreover, most of the systems that produce big data expect real-time responses, and the selected data imputation method should not harm the real-time responsiveness of the algorithm.

Although many methods and algorithms have been proposed to fill in missing data, data imputation is problematic when working with big data, which makes this a potential field of study for researchers.

2.3.3.3 Outlier Detection

Outlier is a term used to describe observations that are significantly different from the overall of other observations in the data. In other words, outliers represent data that is not likely to be in the dataset but is somehow inside that dataset. Outliers may be caused by an error or by large but natural deviations in the population. Furthermore, statistical variation of data, fraudulent behavior, partial deterioration in the devices that collect the data, human-induced errors, and bit corruptions caused by transmission can be the causes of outliers.

Outliers are one of the main factors that directly affect the quality of the data, whether they occur naturally or as a result of an error or corruption. The presence of a large number of outliers in the dataset constitutes an obstacle to the achievement of precise results by the algorithm. There are many studies on detecting, managing, and deleting outliers.[24, 25]

The process of detecting outliers in big data can be more problematic. Some outlier detection mechanisms run by taking an integer as a parameter. If there is enough data, it means there are enough outliers to bypass the outlier determination parameter in question. Therefore, data sizes should also be considered when setting such outlier determination parameters.

In addition, the definition of big data is generally a broad framework that includes stream data in online systems that continue to flow. For the stream data in question, outlier detection mechanisms used in stationary data may not work consistently. According to the statistical values of data coming over time, an observation determined as an outlier at the time of t_1 may be included in a group at the time of t_2 or vice versa.

2.3.3.4 Normalization

In order to increase the quality of the data and produce accurate results, one of the preprocessing operations is to normalize the data. The normalization process can be described as shifting the attributes of the data that take values at different intervals to the same interval values. Normalizing the data is one of the necessary preprocessing steps to obtain accurate results.

At first glance, it may not seem problematic that the data has attributes with values at different intervals. However, if Euclidean and similar distance measures are used as similarity measures, the attribute with a large variance value will have more power to affect the result. In other words, the attribute with a high interval will dominate the effect of other attributes on the result. Moreover, if the attribute with a high variance value is not sufficient to distinguish the data, it may cause the algorithm to obtain incorrect results.

If the power of the data's attributes to distinguish between observations is unknown, it is obviously necessary to shift the attribute values to the same intervals in order for the algorithm to give equal weight to these attributes. Typically, all the attributes intervals are shifted to the 0–1 range so that the attributes have equal impact.

Parallel systems and parallelized algorithms are often used to process big data fast. It may be necessary to reconsider the normalization algorithms developed for sequential systems to be able to use them in parallel-running distributed systems.

The normalization process for stationary data does not describe a very complex process. The data is passed over once, max and min values are found, and then all other values are shifted to the 0–1 range according to these max and min values. But for the stream data, normalization is not that simple. The term stream data describes incoming data that is ongoing to flow and incomplete. In other words, the arrival process of the data is still ongoing. Stream data is even described as a potentially endless sequence of data. Therefore, it does not seem possible to find the max and min values among data whose arrival is still incomplete. It is possible to find partial min and max values only from the data that has come so far. Researchers [26] already working on stream data declare that they are working on normalized data. But in real life, data is not produced as normalized.

It is unknown if there is a way to completely normalize the data that is incomplete and that continues to arrive. However, designing partial normalization algorithms in this area remains open research.

2.3.3.5 Other Challenges/Issues

The growth of the volume of data comes with additional challenges for big data stakeholders. There are also difficulties other than those mentioned above.

Abstraction of data: One of the ways to deal with big data, especially the volume and velocity characteristics of big data, is to summarize the data. However, the versatility and diversity of the data also reduces the potential for summarizing the data to be representative. Especially when dealing with stream data, data summarization techniques are frequently used to design scalable algorithms. In these studies, time-based (time-weighted) data summarization techniques are used to create smaller-sized data to represent big data.

Data integration: A common infrastructure representation is required to integrate two different types of data. One of the methods used at the point of data representation and integration is ontologies. Ontology-based systems are techniques used to bring data to a common representation standard and to integrate different types of data.

Scalability and high dimension: One of the first problems that comes to mind when talking about big data will be the complexity of the algorithm to be designed. Algorithms with complexity below $O(n^2)$ are considered successful when dealing with stationary data, while algorithms with complexity of $O(nlogn)$ can be considered inefficient when it comes to big data. Algorithms dealing with big data, especially stream data, have already adopted the single scan working principle, so their complexity targets are $O(n)$ linear. One of the factors affecting the runtime of an algorithm is the complexity of the algorithm; the other is the amount of data and the number of dimensions that observations contain. Some attributes in the data do not help to solve the problem; they are irrelevant. In order to reduce the data size, researchers have conducted research to design more scalable systems using techniques such as dimension reduction and feature selection. In addition, if the algorithm has a high parallelization potential, parallel processing can also be done to respond quickly.

Interoperability and platform independent: Data in different structures can be encountered in many areas. It is one of researchers' highest goals to be able to operate these data together and for a designed system to work with all kinds of data. However, it is not an easy goal when working with big data.

2.4 BIG DATA PLATFORMS AND TECHNOLOGIES

New platforms and technologies have been produced to cope with big data at all stages from the creation of data to the acquisition of knowledge. In this section, the technologies developed to produce, store, and process big data will be discussed.

2.4.1 Big Data Generation

In the previous sections of this chapter, the factors that create big data have been discussed. There are dozens of factors that create big data, but the most important are developing technology in every field, rapid digitalization, social media, sensor networks, and the success of data-intensive decision-making systems.

Currently, a rapid digitalization process continues in every field. For example, in the financial sector, institutions can do most of their business online. They have chosen to provide online service to their customers in order to provide more effective corporate management. In addition, the emergence of new technologies, such as blockchain, are factors that cause an increase in the amount of data in the digital world. Blockchain is not only encountered in the distributed finance (DeFi) sector as a cryptocurrency but also in the field of public administration, protection of intellectual property rights in digital environments, and business management and organization. We are faced with the design of systems using blockchain technology infrastructure.

Non-fungible token (NFT) is a system that uses the blockchain infrastructure, grants the originality of digital products (sounds, video, drawings, games, or anything represented in digital platforms) that can be exchanged and stored in a distributed ledger just like cryptocurrencies, and ensures the protection of copyright. Thanks to NFT technology, a digital product with regulated usage rights can be traded.

Decentralized autonomous organization (DAO) and smart contract are important areas where blockchain infrastructure is used. Smart contracts and these smart contracts-based DAO infrastructure are decentralized autonomous entities and defined by the use of blockchain technology to provide a secure digital ledger for tracking digital interactions over the Internet. It is secured against forgery by reliable timestamping and propagation of a distributed database.[27–29] One of the most important features that DAO brings to the public domain is that it eliminates trusted third parties with smart contracts. Thus, DAO facilitated the management and monitoring of business processes by reducing bureaucracy in the business world and the public sector. To summarize, DAO has paved the way for the complete digitalization of business processes.

For example, thanks to smart contracts, two companies can sign an agreement without being tied to any intermediary institution; the parties can sell cars without the need for a notary public; the parties can also perform the title deed transactions without the need for a title deed institution. In fact, in an extreme example, with blockchain infrastructure, people can create and store their marriage contracts without the need for anyone else through digital marriage certificates.

2.4.2 Big Data Storage

Storage mechanisms not only store data, but also control the data flow, run administrative processes on the data, and have the capacity to effectively retrieve the data when requested. Considering these features, databases are major players that will determine the efficiency of the system when processing big data. For this reason, more efficient storage technologies have been developed that can store big data that has a high volume, variety, and velocity.

In the past decades, different methods for big data in storage have been studied. NoSQL database models have become the emerging storage techniques of recent times due to their flexibility, scalability, and high performance. Extensive information about NoSQL database models has been reported by Sicari et al.[30] In this context there are four main NoSQL databases for big data: (i) key-value, (ii) column-oriented, (iii) document-oriented, and (iv) graph-oriented.

As in every field, a single method is not always superior to other methods in big data storage. While method A is more effective in some situations and conditions, method B may be more effective when conditions change. Therefore, the four storage methods previously mentioned have advantages and disadvantages compared to each other. For example, the key-value model employs hash tables in order to store, sort, lookup, and retrieve the data efficiently. But it lacks customized queries, and it does not support complex queries. If you do not need complex queries, the key-value databases are an appropriate solution with highly scalable abilities. However, if complex queries are required, document-oriented databases are becoming better options. Because compared to key-value based model, they offer more flexibility for building nested queries.

In the past decades, many technologies and platforms have been proposed for big data storage. Some of these are MongoDB, Redis, Bigtable, Cassandra, Neo4j, HBase, etc.

2.5 CONCLUSION

The chapter discusses the features of big data and the problems that these features bring to people who are faced with big data. In addition, algorithms and methods developed to solve these problems are emphasized. Although the advantages of big data have been studied extensively in recent years, during the transformation of big data into information, there are advantages and challenges. Researchers and companies' R&D units are struggling to eliminate the handicaps that come with big data.

Today, with the rapid development of technology and the success of data-intensive systems in decision-making processes, the need for data has increased. Not only has the need for data has increased, but the need for methods, algorithms, and platforms to produce, transport, store, process, and display this data has also increased. In line with this need, researchers, corporate managers, and stakeholders of big data have spent their resources in line with these studies for decades. However, the data is not as few, singular, and stable as it was 20 years ago. With the developing technology and achievements, data collection systems are faster and more diverse. Therefore, new algorithms and platforms that can handle this new type of data have been studied.

Knowledge is power, and data are the first stop on the path to realizing this power. The higher the quality of the data produced, the easier it will be to process and store, and the information obtained will be more accurate and useful as a result of this process. Therefore, it is essential to develop not only systems that process information but also system designs that produce, store, and retrieve it correctly when desired.

REFERENCES

1. Goldston, D. (2008). Big data: Data wrangling. *Nature*, 15, 455. https://doi.org/10.1038/455015a
2. Sharma, S., Kumar, N., & Kaswan, K. S. (2021). Big data reliability: A critical review. *Journal of Intelligent & Fuzzy Systems*, 40(3), 5501–5516.

3. Au-Yong-Oliveira, M., Pesqueira, A., Sousa, M. J., Dal Mas, F., & Soliman, M. (2021). The potential of big data research in healthcare for medical doctors' learning. *Journal of Medical Systems*, 45(1), 1–14.
4. Woldt, J., Prasad, S., & Ozgur, C. (2020). Big data and supply chain analytics: Implications for teaching. *Decision Sciences Journal of Innovative Education*, 14, 155–176.
5. Rydning, D. R. J. G. J., Reinsel, J., & Gantz, J. (2018). The digitization of the world from edge to core. *Framingham: International Data Corporation*, 16.
6. Jirkovský, V., & Obitko, M. (2014). Semantic heterogeneity reduction for big data in industrial automation. *ITAT*, 1214.
7. Castro, A., Villagrá, V. A., García, P., Rivera, D., & Toledo, D. (2021). An ontological-based model to data governance for big data. *IEEE Access*, 9, 109943–109959.
8. Hyde, R., Angelov, P., & MacKenzie, A. R. (2017). Fully online clustering of evolving data streams into arbitrarily shaped clusters. *Information Sciences*, 382, 96–114.
9. Wang, R. Y., & Strong, D. M. (1996). Beyond accuracy: What data quality means to data consumers. *Journal of Management Information Systems*, 12(4), 5–33.
10. Milli, M., Şentürk, F., Çınaroğlu, S., & Çınaroğlu, İ. (2016). Büyük Veri Kavramı ve Karakteristik Özellikleri. *Akademik Bilişim*, 1, 183–188.
11. Curry, E., Becker, T., Munné, R., Lama, N. D., & Zillner, S. (2016). The BIG project. In *New horizons for a data-driven economy*. Cham: Springer, pp. 13–26. https://doi.org/10.1007/978-3-319-21569-3_2
12. Saidali, J., Rahich, H., Tabaa, Y., & Medouri, A. (2019). The combination between big data and marketing strategies to gain valuable business insights for better production success. *Procedia Manufacturing*, 32, 1017–1023.
13. Lyko, K., Nitzschke, M., & Ngonga Ngomo, A. C. (2016). Big data acquisition. In *New horizons for a data-driven economy*. Cham: Springer, pp. 39–61.
14. Aktaş, Ö., Milli, M., Lakestani, S., & Milli, M. (2020). Modelling sensor ontology with the SOSA/SSN frameworks: A case study for laboratory parameters. *Turkish Journal of Electrical Engineering and Computer Sciences*, 28(5), 2566–2585.
15. Milli, M., & Bulut, H. (2022). SubtStream: Online subtractive stream clustering algorithm. *Concurrency and Computation: Practice and Experience*, 34(15), e6968.
16. DATACONOMY. Retrieved May 15, 2022, from http://dataconomy.com/2017/05/big-data-data-visualization/
17. Wirth, N. (1985). *Algorithms & data structures*. Englewood Cliffs, NJ: Prentice-Hall, Inc.
18. Liew, C. S., Abbas, A., Jayaraman, P. P., Wah, T. Y., & Khan, S. U. (2016). Big data reduction methods: A survey. *Data Science and Engineering*, 1(4), 265–284.
19. Hu, H., Wen, Y., Chua, T. S., & Li, X. (2014). Toward scalable systems for big data analytics: A technology tutorial. *IEEE Access*, 2, 652–687.
20. Malik, K. R., Ahmad, T., Farhan, M., Aslam, M., Jabbar, S., Khalid, S., & Kim, M. (2016). Big-data: Transformation from heterogeneous data to semantically-enriched simplified data. *Multimedia Tools and Applications*, 75(20), 12727–12747.
21. Vergouw, D., Heymans, M. W., van der Windt, D. A., Foster, N. E., Dunn, K. M., van der Horst, H. E., & de Vet, H. C. (2012). Missing data and imputation: A practical illustration in a prognostic study on low back pain. *Journal of Manipulative and Physiological Therapeutics*, 35(6), 464–471.
22. Bhaskaran, K., & Smeeth, L. (2014). What is the difference between missing completely at random and missing at random? *International Journal of Epidemiology*, 43(4), 1336–1339.
23. Wang, C., Shakhovska, N., Sachenko, A., & Komar, M. (2020). A new approach for missing data imputation in big data interface. *Information Technology and Control*, 49(4), 541–555.

24. Cao, J., & Hu, R. (2021, December). Scalable outlier detection using distance projections. In *2021 IEEE international conference on big data (big data)*. New York: IEEE, pp. 4431–4440.
25. Alghushairy, O., Alsini, R., Soule, T., & Ma, X. (2020). A review of local outlier factor algorithms for outlier detection in big data streams. *Big Data and Cognitive Computing*, 5(1), 1.
26. Mansalis, S., Ntoutsi, E., Pelekis, N., & Theodoridis, Y. (2018). An evaluation of data stream clustering algorithms. *Statistical Analysis and Data Mining: The ASA Data Science Journal*, 11(4), 167–187.
27. Vigna, P., & Casey, M. J. (2015, January 27). *The age of cryptocurrency: How bitcoin and the blockchain are challenging the global economic order*. New York: St. Martin's Press. ISBN 9781250065636.
28. Hodson, H. (2013, November 20). Bitcoin moves beyond mere money. *New Scientist*, 220(2945).
29. Wright, A., & De Filippi, P. (2015). Decentralized blockchain technology and the rise of lex cryptographia. SSRN 2580664.
30. Sicari, S., Rizzardi, A., & Coen-Porisini, A. (2022). Security&privacy issues and challenges in NoSQL databases. *Computer Networks*, 108828.

24. Cao, [illegible] R. (2021, December). [illegible] ... tions. In 2021 IEEE International ... pp. [illegible].
25. Alghushairy, O., Alsini, R., [illegible], T., & Ma, X. (2020). A review of local outlier factor algorithms for outlier detection in big data streams. Big Data and Cognitive Computing, [illegible].
26. Munadhil, [illegible] ... (2020). [illegible] ... stream clustering algorithms ... [illegible]
27. [illegible]
28. [illegible]
29. [illegible]
30. [illegible]

3 Smart Warehouse Testbed

From Conceptual Framework to a Real Project

Ngoc-Huan Le, Ngoc-Bich Le, Manh-Kha Kieu, Xuan-Hung Nguyen, Vu-Anh-Tram Nguyen, Tran-Thuy-Duong Ninh, Duc-Canh Nguyen and Narayan C. Debnath

CONTENTS

DOI: 10.1201/9781003310785-3

3.1 INTRODUCTION

Warehouses are utilized to store products for commercial purposes. For the Vietnamese market, warehouses play a significant role in the supply chain of food and agricultural products. In the world market, warehouse service management determines the success or failure of e-commerce companies. Through a survey of the Vietnamese market, most warehouses are built and managed traditionally. Thus, a traditional warehouse has many limitations, such as inefficient space management, perishable materials, inefficient operations, over-handling material, and inefficient material handling equipment.[1]

Intelligent autonomous cyber-physical systems [2] and the rise of e-commerce are two characteristics of advanced technology development that are influenced by Industry 4.0 (I4.0). There is evidence of technological innovations adaption in warehouses and the global supply chain.[3] A smart warehouse should incorporate best practices and contemporary technologies to operate as efficiently as possible.[4–6] The critical research difficulties for smart warehouse applications are human activity detection, time-efficient communication scheduling, robust location, and multi-robot collaboration.[7] Papcun et al. (2019) investigated a new trend of dynamic slotting as a best practice to reduce delivery time in place of assigning fixed places per product.[8] Several algorithms have been used in the intelligent warehouse management system to implement functions such as area capacity selection, optimal picking, and optimal product positioning.[9]

Deep learning and machine learning are two examples of artificial intelligence (AI) algorithms that have been successfully used across the board in recent years.[10, 11] In the field of manufacturing, AI is seen as a game-changer. AI has the potential to change performance in all aspects of manufacturing operations. More than 50% of large European companies used at least one AI in their industrial processes.[12] This study used a SWOT analysis to evaluate AI applications' strengths, weaknesses, opportunities, and threats in warehouse management and operation, especially in Vietnam. Using SWOT analysis, the impact of the application of AI for smart warehouse development was investigated. Economic, technological, market demands, resources, investment strategies, the environment, and other factors were all addressed to suggest solutions. Three solutions are groups include (1) solutions to support investment decisions and limit the risks created through the testing of innovative approaches and algorithms, which confirmed the feasibility and estimated post-investment engineering economics problems; (2) solutions for AI applications in the field of intelligent warehouses development; and (3) solutions for developing AI resources for smart warehouses. In addition, from the previous study,[10–12] it is clear that using AI to address optimization problems (location, mobility, building space, energy, inventory, etc.) has considerable potential for improving operational efficiency and reducing warehouse management and running expenses. Then, to grow the logistics sector in the 4.0 era, unique AI-based solutions to address the existing problems were provided.

Logistics in Vietnam typically includes four areas: transportation, freight transportation, warehousing, and other value-added services. In some reports, authors argue that warehouses, including warehouse service in Vietnam, are crucial in enhancing logistics competitiveness.[13] The limitation in warehouse service has been simple storage, processing, value-adding, and piece-picking.[14] Store-related services are one of the leading outsourcing services, with 20% of businesses outsourcing from 51% to 75%.[15] Different target customers can offer various warehousing services such as dry storage, cold storage, bonded warehouses, and distribution centers.[16] This is considered a warehouse trend.

The warehousing demand in Vietnam received significant attention due to the dramatic development of the manufacturing and e-commerce sectors. For more detail, across Vietnam, 327 industrial parks have been developed, and in Dong Nai and Binh Duong, the growth in warehousing is 21% and 54%, respectively.[17] In 2018 and 2019, the industries of shipping, warehousing, and allied services grew favorably. The Southern zone is leading in the number of crucial warehouse distributions in Vietnam, such as Saigon Newport with 616,650 sqm; Mapletree with 529,632 sqm, or Sotrans with 178,700 sqm.[18]

Nevertheless, the information technology (IT) application level is the challenge and barrier to developing warehousing services. The foundation of the system is information technology. There has been remarkable attention to the role of the application of information technology to facilitate the logistics industry, especially the circulation and goods distribution.[19] The role of IT is to provide information to relevant stakeholders—suppliers, customers, partners, and other critical participants—for efficient, effective, accurate, and timely management of logistics centers.[20] The trend in consolidated warehouses, including order management, managing and tracking suppliers, managing data, and providing end-to-end customer data via the online platform, has been booming in recent years, and integrated logistics are typical.[21]

In modern warehouses, moving from traditional fixed automatic guided vehicle (AGV) systems to wireless control and monitoring presents several new advantages and opportunities. In Culler et al. (2016), the authors developed a new algorithm to use a camera system (Microsoft Kinect, IP camera) to guide the robot, process video, and monitor real-time.[22] In Yang et al. (2020), the authors proposed an solution to effectively coordinate the AGV group in the warehouse based on a global vision when the number of AGVs increases significantly.[23] To improve the safety and operational efficiency of the warehouse, a real-time location system (RTLS) that can locate and track forklifts and other mobile entities in the warehouse was investigated.[24]

Additionally, the use of AI improves management, logistics, and coordination capacities in warehouse operations.[25] A traditional warehouse might be converted into a smart environment for automation in 2019 by using AI, the Internet of Things (IoT), and cloud computing.[26] A sensor system and Wi-Fi network were utilized to monitor and analyze the environmental factors affecting the fruits warehousing quality.[27]

Most researchers pay attention to quantitative research methods and mathematical modeling without understanding the complexities of the real environment.[28] The smart warehouse project at Eastern International University (EIU) aims to teach, learn, research, and connect university and industry. In terms of education, this project will improve the learning and teaching quality at three EIU schools, namely the Becamex Business School, School of Engineering, and School of Computing and Information Technology. The project gives students in three schools the opportunity to apply the theory into practice to make real-life decisions. Regarding the university-industry connection, this project is a testing ground for firms to evaluate solutions or cutting-edge technology before actually applying them in real life. This is considered a way to optimize the manufacturing process while saving costs and avoiding risks simultaneously. In this project, we design and control a smart warehousing system that can store up to 1372 packages. So far we have completed the prototype phase with the design and control of the central rack system that can store 196 packages. This chapter will be encapsulated in the prototype version of the system.

3.2 SWOT ANALYSIS

3.2.1 Strengths

3.2.1.1 Keeping Up with Emergent Needs

eCommerce. The Covid epidemic impact and the growth of the middle income have created a shift in the community's buying habits and accelerated the development of Vietnam's eCommerce system. For the producers to expand and ensure a secure future amid the continuing epidemic, there will therefore be a continued need for more warehouses, particularly in Binh Duong, Dong Nai, and Long An.[29] The growth of eCommerce has made it easier for manufacturers, industry, transportation, real estate, and warehousing to work together across the nation. This tendency will boost the market for factories and warehousing to keep up with the new economic wave.[30]

Industrial centers rearrangement. Covid's epidemic condition remains complex around the world and is leading to rapid growth in eCommerce and an influx of relocations from China. Low capital requirements and the need to diversify production across multiple bases to start a new production in Vietnam are revitalizing well-situated warehousing, transportation, and enterprise zones for capitalizing in Vietnam.[30] During the Covid epidemic, it is anticipated that good return products including electronics, semiconductors, medical gadgets, and items from international corporations like Pegatron, Foxconn, Sharp, Nintendo, and Lenovo will continue to proliferate in Vietnam.[31]

The first indicator is the expanding industrial park (IP) in Vietnam. Nationwide, 374 industrial plants were established on more than 114,000 hectares of land, with an average share increase of nearly 73% to 99% in the South and North. In the first half of 2020, the IPs and processing zones attracted approximately $6 billion of foreign direct investment (FDI) with nearly 335 FDI projects. In September 2020, manufacturing gained registered FDI of $2.88 billion in the North, $1.64 billion in the South, and $227 million in the Central. In addition, nine additional industrial zones in Dong Nai, two projects with more than 500 ha in Hai Phong, the new 238 ha industrial park in Bac Ninh province, 177 ha Song Lo 1 IP in Vinh Phuc province, and 1800 ha Viet Phat IP in Long An province are expected to open in 2020.[31]

According to the Vietnam Logistics Business Association, the Vietnamese economy is worth $4.42 trillion and is growing at a rate of 1416% every year. To improve the consolidation of the master plan, several provinces and localities will draw investment in new manufacturing sectors of infrastructure, businesses, and warehouses in 2021.[30]

3.2.1.2 Governmental Engagement in Digitalization is Strong

Under Decision No. 703, which was approved by the Government on June 7, 2019, the Vietnamese government has highlighted the role of technological applications.[19] The foundation of distribution and how various elements of the value chain are connected is IT. An emerging concept in warehouse consolidation is the integration of all operations, including order management, supplier management, data management, and end-to-end data management. Although the level of IT implementation is indeed a problem, it is crucial that the Vietnamese government concentrate on enhancing IT.

Weak technological integration continues to be a major obstacle to the logistics sector's competitiveness in Vietnam.[15] Technical consequences are starting to be applied in the logistics sector. Numerous businesses are making use of technological solutions to help specialized industries. For nearly 70% of the businesses polled, warehousing software solutions are a crucial tool. The following are the softwares applied in logistics industry: logistics management, ERP, SRM, and CRM, with respective percentages of 50.7%, 35.3%, 30.4%, and 27.3%.[32]

3.2.1.3 AI is Receiving Significant Investment, Particularly from the Commercial Sector

The AI industry saw growth of over 70% in 2018 compared to 2017 and reached $200 billion.[33] The amount of investment for 2020 was $67,854 million as opposed to $48,851 million in 2019, according to the AI Index Report 2021.[34] Particularly, majority of investment activities were in the private sector with $38,659 in 2019 and $42,238 million in 2020. The total amount invested grew 5.3 times over 2015 levels. In the next 10 years, AI is likely to become the most revolutionary technology. One of the major technological advances recognized by Vietnam is AI. The Ministry of Planning and Investment has been tasked by the government with fostering AI development and creating a national strategy for the Fourth Revolution. AI is included as one of the plan's top focuses for investment and growth.[35] Artificial intelligence is no longer a scientific theory but a socioeconomic challenge for Vietnam's development, according to Deputy Prime Minister Vu Duc Dam.[36]

3.2.1.4 High Warehouse Demand

Rising warehouse costs driven by the surge in demand is anticipated to upsurge from 1.5% to 4% annually. Also, after the Covid pandemic is over, factory and warehouse demand is expected to increase from 4% to 11%.[37] As a result, investors are seeking out Vietnam's industry and logistic rental properties, including its warehouses and refrigerated warehouses.

Due to the pandemic impact, the recent wave of relocations, and the growth of eCommerce, there is a high demand for rental developments for ready-built factories (RBF), warehouses, and built-to-suit solutions.[31] In particular, GLP, the biggest warehouse development company in Asia, announced a $1.5 billion investment. Mirae Asset Daewoo Co., Ltd. in South Korea and Naver Corporation also announced warehousing investment plans totaling $37 million. GLP, LOGOS, and JD.com, the three global giants in the industry, have also made investments in the North and the South since 2020. Vietnamese real estate developers CenLand Joint Stock Company and Vingroup also expanded their ecosystem, adding logistics, real estate, and warehouses with the Cen Cuckoo brand, as well as two new industrial parks in 2021 to safeguard the future of high demand.[30, 31]

3.2.1.5 The Demand for AI Workers is Growing as a Result of Competitive Pay

The "wave" of artificial intelligence is something that all key scientific and engineering universities are attempting to ride. The AI Index Report 2018 [33] indicates that there are more undergraduates studying this subject. From 2012 to 2018, this figure nearly quadrupled in the United States. In 7 years, the number of students enrolling in AI courses at Tsinghua University in China has increased by 16 times. In Vietnam, there are 400,000 IT human resources, and 50,000 IT students graduating from more than 153 training institutions each year. Particularly in the top two IT industries with the highest salaries, where data scientists make $3,531 and Python programming engineers make $2,900, the average AI hiring income is $1,958.[38] The AI ecosystem is becoming more diverse in communities such as clubs, faculties, schools, laboratories, VietAI, VinAI, and Quy Nhon AI.

3.2.2 Weaknesses

3.2.2.1 There is a Lack of Infrastructure for Digital Transformation and AI Technology

The data and computational infrastructure form the basis of AI. Today, Vietnam's large universities, as well as several significant AI research institutions and corporations, are home to the majority of the country's huge computing and large data acquisition machines. Additionally, systems for data exchange and computer infrastructure that may foster community have not been developed systematically or simultaneously.

3.2.2.2 Technology is a Weak Motivator to Alter Thinking

There is psychological resistance to making the transition to the new paradigm from the conventional model. It's a difficult process, and the changeover to a new model necessitates consistency in the governing structure, transformational strategy, and implementation plan. Calculating the resources and risks involved in changing the company model is also crucial. Companies may experience organizational change anxiety when integrating AI and digital transformation. In a workforce with multiple generations, for instance, it might be challenging to change ingrained work practices. The fear of change is influenced by several factors, including large investments and the need for return on investment (ROI). Thinking is a hurdle that needs to be removed before using AI and digital transformation. Leaders who have undergone a true mental transformation will pave the way.

3.2.2.3 Absence of Coordinated and Organized Investment

Investment in smart warehousing system development and the use of AI in smart warehousing management and operation are still insufficient, disjointed, and unstructured. However, because AI uses huge data to its advantage, it is systematic and collaborative. Currently, huge corporations like Amazon, Alibaba, Walmart, Vinamilk, and DHL are the only ones targeted by smart warehouse apps both globally and in Vietnam.

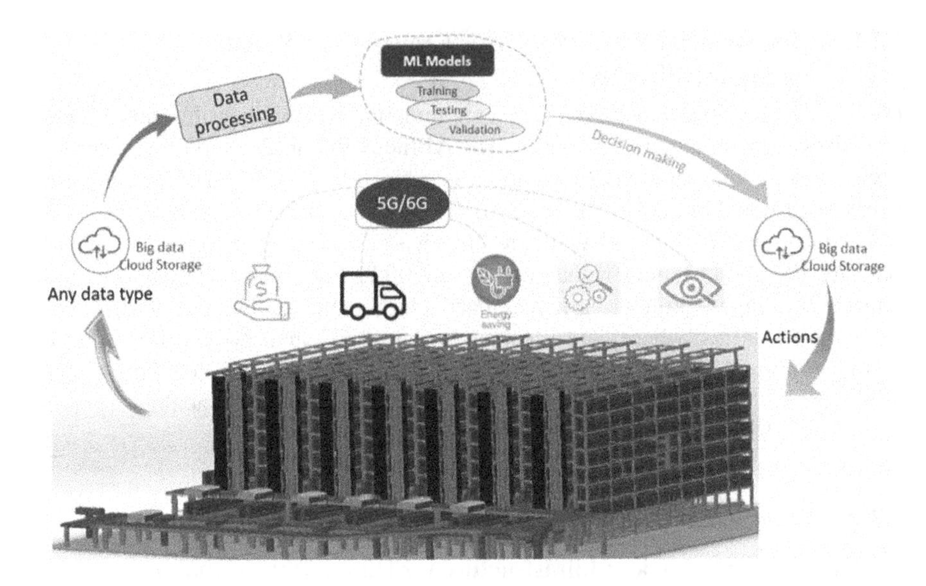

FIGURE 3.1 Possibilities of using AI in smart warehouses.

3.2.3 Opportunities

The possibilities of using AI in intelligent warehouses is depicted in Figure 3.1. They consist of (1) reducing operational power; (2) lowering logistics expenses; and (3) benefits from the perspective of automated systems.

3.2.3.1 Reducing Operational Power

The energy of warehousing operations has become the main topic of green warehousing and attracts numerous research on reducing the supply chain's carbon footprint. Energy consumption in the warehouse is a complex issue and involves many stages. Reducing energy consumption in warehousing is based on cost savings, profit improvement or marketing strategies, and regulatory compliance issues.[15] Furthermore, the optimization of energy consumption in warehouse operation is to reduce logistical costs and increase efficiency.

Consequently, the application of AI to the operation of the warehouse system will help reduce operational power thanks to optimum administration and processing algorithms. Based on the previous operating parameter results, these algorithms will find the best parameters to save energy on their own. According to Amazon's data, warehouse productivity has increased by 40% as a result of the company's intelligent goods handling and operation system, which efficiently incorporates people and machines. At the same time, operating costs have decreased to less than 10% of revenue. This demonstrates very clearly how effective smart warehouses are for businesses.

3.2.3.2 Lowering Logistics Expenses

Outsourcing transportation services by businesses in Vietnam ranges from 0% to 100%, per the Vietnam Logistics Report 2019. For companies with outsourcing proportions between 51% and 75% and 755 to 100%, respectively, the maximum option of warehouse leasing is up to 20% and the lowest is 3.6%. But between 10% and 41% of a company's overall logistics costs are spent on warehousing and warehouse management.[39] It is clear that smart warehouses have a lot of potentials, especially those powered by AI to reduce logistical costs. In particular, manual storage management still prevails in conventional warehouses. Some other warehouses employ semi-manual management and operation techniques. In both circumstances, the placement of commodities is based on some basic guidelines to keep the operation and management as simple as possible. The use of AI models to determine the best retrieval solution, together with the ability to evaluate more variables and operating conditions, will considerably improve the optimization of the commodities arrangement process.

3.2.3.3 Benefits from the Perspective of Automated Systems

It is evident that automation is the essential foundation to implement AI. Therefore, the application of AI will inherit the benefits that automation creates. Investors can get various advantages from smart warehouse automation, including decreased labor expenses, faster pick-up and shipping times, fewer product delivery failures, and improved management effectiveness. Many technologies and methods are currently being used in smart warehouse management and operation to create a smarter warehouse.

3.2.4 Threats

3.2.4.1 Cyberattack Risk and Security

Hackers can effectively carry out AI system assaults across the network by employing AI and taking into account the peculiarities of information transfer and administration through the network. Because every communication relies on an Internet connection, the likelihood of security issues caused by a lack of security is very significant. Additionally, the strength of the present cryptographic algorithms is insufficient to provide a certain level of security.

3.2.4.2 Limited Workforce Resources

The AI index 2021 annual report [34] has demonstrated the disparity in AI skill levels across various nations. The majority of them are concentrated in industrialized nations like the United States, the European Union, or China. There is a severe lack of high-level AI human resources in Vietnam, with present availability only meeting a tenth of market demand. Despite the very attractive remuneration, it might be challenging for firms to find qualified AI candidates because they are so hard to come by. This is because the supply of AI experts is mainly foreign graduates from prestigious universities. Some large enterprises in Vietnam have also launched policies to recruit AI talent, but the number still does not meet the general growth rate.

3.3 CURRENT METHODS AND SUGGESTED REMEDIES

SWOT's advantage is that its components can be used to develop valuable strategies. Combining the four elements of the SWOT analysis yields four strategies: (1)

attach strategy (SO); (2) improve strategy (WO); (3) defend strategy (ST); and (4) exit strategy (ST). This chapter concentrates on two proactive methods for minimizing weaknesses and maximizing strengths to seize chances.

3.3.1 Testbed as a Pilot toward Investment Strategy: WO Approach (Enhance)

Among the challenges of implementing AI in the warehouse system, researchers highlight investment and management costs. These costs are additional to the cost of many implementation failures. Proven models for innovation co-development among supply chain partners, which emphasize risk mitigation, cost, and benefits sharing, have been developed. By cutting down costs of implementing new technologies, decreasing the cost of technology transfer and coaching, and enhancing perceivability, integrating with supply chain partners including suppliers, consumers, and service providers will stimulate innovation.[40] However, because AI is such a difficult field, special requirements need to be met. The testbed facility must be supplied with cutting-edge tools and innovations that are adaptable enough to mimic a range of AI investing circumstances and provide businesses with the assurance and information they need to make informed choices.

3.3.1.1 System/Algorithm Assessment

The testbed serves as a foundation for assessing the accuracy and repeatability of theory, simulation findings, computational methods, technological innovations, and implementations. The testbed offers an interactive workspace without the risks and repercussions related to testing in a real-world setting. The testbed may incorporate software, hardware, networking devices, and innovative modules or actual solutions.[41]

3.3.1.2 Put a Cap on Investing Risks

Prospective warehousing will be built from a modular that has integrated elements that can communicate and operate in an energy-conscious manner when handling and storing traditional materials. Nevertheless, the extensive use of embedded systems, wireless technology, and logistics software creates a new level of complication in developing and evaluating such systems. Although idea and simulation solutions for every field are offered, the installation of realistic devices frequently reveals several issues.[42]

3.3.1.3 Analyzing the Effectiveness of Investments

Because of limited working space and minimal management costs, warehouse systems are always moving toward innovation toward digitalization. Testing and validation are necessary before integrating complicated alternatives (such as detection methods, virtual-reality technology, self-driving cars, etc.) into company operations. In a sizable, persistent, and tough working environment, this task is challenging. The authors designed a warehouse automation experiment with a focus on the evaluation of mobile robot units on a modest scale in a warehouse environment.[43] For testing, evaluating, and validating integration with the actual platform, the author employed the manufacturing department's Cyber-Physical Production System testbeds.[44]

3.3.2 AI-driven Strategy: WO Approach (Enhance)

Data, AI algorithms, and robots, according to our findings, are essential AI resources for manufacturing companies as they develop AI capabilities. Therefore, the following topics will be discussed: systematic data collection, effective AI algorithms, and robotics that may employ intelligent information from AI algorithms.

3.3.2.1 IoT and AI Blending

We can gather a sizable amount of data from numerous sources thanks to IoT. As a result, Industry 4.0 can revolutionize industries and business operations through the use of AI and IoT. IoT is the ideal platform for AI to develop intelligent objects, intelligent industries, and intelligent decisions with or without human involvement. Throughout this combination, IoT gathers data from devices communicating with one another online while AI develops the tool's intelligence using its own data sets and perspectives.

3.3.2.2 Optimization of Operational Power

In order to model functional CO2 refrigerant-based commercial chillers for direct integration into a worldwide electricity control system, the authors used an AI-based technique.[45] When using artificial intelligence to improve operational procedures, the output should be set up so that the amount of electricity consumed is within acceptable bounds. For instance, to reduce energy consumption, set up should include optimizing the positioning of goods, the movement of AGVs, the equipment uptime, and so forth. For the real-time routing of vehicles, Opalic et al. (2020) utilized a deep reinforcement learning (DRL) approach.[46] The authors have expanded their multi-goal optimization studies of real-time planning of AGVs based on that outcome (power consumption, device usage, maintenance costs, etc.).

3.3.2.3 Optimizing the Routing of AGVs or Forklifts

AGVs or forklift scheduling can ultimately benefit from AI techniques. In Li et al. (2018), the authors developed a deep reinforcement learning (DRL) technique for warehouse management that enables a vehicle to select from a variety of activities and travel to the closest task by combining the knowledge of navigation and positioning, sensors, computer vision, and communication.[47] The AGV route planning for the AGV sorting system has also employed the DRL technique.[48] For automated driving, a new layout technique for deep reinforcement learning systems was developed and proved.[49, 50] Additionally, Salunkhe et al. (2018) developed a novel method leveraging DRL for the real-time scheduling of AGVs.[45]

3.3.2.4 Warehousing Management Optimization

The authors were aware of the effective usage of AI in Alibaba's Smart Warehouse methods for product storage, order processing, and order loading while researching machine learning and artificial intelligence in warehousing administration. Alibaba's

Smart Warehouse has put out a variety of AI approaches to swiftly assess and apply data for tactical decision-making and real-time optimization.[51]

3.3.2.4.1 The Smart Warehouse of Alibaba

In the goods storing process. The storing of commodities from automated tridimensional storehouses (ATSs) involves artificial intelligence. At ATSs, stock data is detected and updated concurrently for information on the total weight, 3D dimensions, and package identification. Then, using past data, the AI program determines the best place to store packages.

During the selecting of the order. The three AI applications used in the handling process are lifting AGVs, Order-to-Man (O2M) AGVs, and Goods-to-Man (G2M) AGVs. When an order is received, the warehouse management system (WMS) uses the warehouse storage data and the loading algorithm to determine the proper packing box. The essential items will be delivered by the AGVs to the stations, where human employees will pick them up and put them in the proper order boxes.

During the packing of the order. Following confirmation of the order, the delivery box is packed by the 3D packing process's guidelines. The AI system ensures that the item will be packed securely.

3.3.2.5 Basic to Advanced Steps to Develop an AI Project that Supports the Operation of a Physical Smart Warehousing System [15]

Step 1: Introduce simple-to-implement AI applications into a real-world warehouse setting. This is the fundamental stage in processing actual information from the warehousing system.

Step 2: Lay a strong data governance foundation. To lay the groundwork for future scalability and deployment, businesses should create data governance mechanisms that detail the critical steps in data production, governance, and analysis.

Step 3: Scale AI technologies throughout the intelligent warehouse. Once the AI platform is ready, AI apps can be installed and spread across the warehouse's many levels and in a variety of industries.

3.3.3 Resource Development for AI: SO Approach (Attack)

Take advantage of existing strengths, including a strong investment wave in AI, a solid assurance to the government digitalization, and huge market demand with attractive salaries to develop AI resources for Vietnam.

3.3.3.1 Shape an AI Community and Network for Smart Warehousing in a Methodical and Coordinated Manner

With AI, the winner is the person who keeps up with the data. This demonstrates the enormous significance of data in the applications of AI. It has been noted that industries such as health care, medicine, education, finance, and natural language processing all have great data infrastructure. Vietnam has been creating AI networking for many different industries to strengthen the power of the society in data generation. The Vietnam-Australia Artificial Intelligence Cooperation Network

(Vietnam-Australia AI) is the most current and was started in August 2021. These are a good start. However, to be effective, it is necessary to build networks and communities systematically and synchronously in specialized fields. One of them is the AI network and ecosystem for smart warehousing systems.

3.3.3.2 Preparation for AI Resources

AI education for engineers in automation and mechatronics. The aforementioned foundation of data, automation, and digitalization is also the cornerstone of computer infrastructure. This is a fantastic chance for students majoring in automated systems and mechatronics to learn AI and develop AI resources. Specifically, the integration of AI competencies for these two disciplines will be very practical given the knowledge base of automation and mechatronics. Additionally, there are enough schools with a focus on automation and mechatronics that produce a large number of graduates each year.

AI training for logistics bachelors. Beside mechatronics and automation engineers, logistics graduates are also potential candidates. Recently, multi-capacity combined and developed to support digital transformation and the digital economy are receiving significant attention and application. Since smart warehousing is a growing industry, it is obvious that incorporating AI capabilities for logistics students is both beneficial and extremely relevant. In addition, institutions pay close attention to new majors like logistics because of the high market demand. Particularly throughout and after the effects of the Covid epidemic, the logistics sector is and will continue to express its strengths.

3.4 EIU SMART WAREHOUSE TESTBED

The EIU Smart Warehouse Project is an application testing facility that uses a 1:10 scale replica of a physical model that performs like an actual system. The following are hardware and software options:

(1) Storage racks, AGVs, conveyor systems, RFID sensors, and controllers (PLCs) are just a few examples of the hardware solutions for smart warehouses
(2) Smart warehouse management software: Solution for connecting the physical controller system and management software
(3) Smart warehouse management and operating model: The key performance indicator (KPI) method is used to assess the effectiveness of the warehouse and the approach to increasing operational effectiveness.

The study's scope will be constrained to the following areas because of the study's financial, physical, and temporal constraints.

(1) Package load of three models with corresponding colors is green: 0.5 kg; yellow: 1 kg; and red: 1.5 kg
(2) Model size is 1/1000 compared to real size
(3) Model capacity: 7 x 7 x 7 x 2 x 2 = 1,372 packages

(4) Package size: 12 x 12 x 20 (cm^3)
(5) Pallet size: 12 x 12 (cm^2)
(6) Surrounding dimensions of the system: 2.5 x 4.0 x 1.6 (m^3)
(7) Upon request, a WMS with fundamental features for smart warehouse management is available.

3.4.1 AGV

3.4.1.1 X and Y Axis

Rack and pinion are used to transfer the X-axis motion. A timing belt, seen in Figure 3.2, transmits the Y-axis.

3.4.1.2 Pick_Place Module

The project requires the AGV to transport parcels on two side-by-side racks. The pick_place module must be created by the aforementioned specifications. The module is controlled by a stepper motor. When the motor rotates forward (from left to right), active pulley 1 rotates clockwise. Timing belts are positive drives used to drive a double pulley (pulley 2, Figure 3.3a; non-reversing open belt). The remaining groove of pulley 2 combines with pulley 3 to create a reversing crossed belt with the two ends of the belt fixed on the 2nd floor (middle assembly, Figure 3.3b). The velocity of the 2nd floor is equal to the cross-belt velocity. The movement of the 2nd floor leads to the movement of the two pulleys 4 and 5 which move in a left-to-right direction. The 3rd floor (top assembly) is pulled by the pulley 4 tension created by one end of the rope fixed at the 1st floor (bottom assembly) and one end fixed in the same direction relative to the other end at the 3rd floor. Pulley 5 helps keep the rope tight and prevents the 3rd floor from moving in the opposite direction. The speed of stage 3 is two times faster than the speed of stage 2. When the motor changes direction, the motor rotates counterclockwise, and the functions of pulleys 4 and 5 are swapped, Figure 3.3c.

The following equations are applied to get the distance traveled per rotation. Equation 1 gives the pitch circle diameter of a timing pulley:

$$D_w = \frac{z * p}{\pi} \tag{3.1}$$

FIGURE 3.2 Motor, gearbox, and servo driver in 3D view; AGV in 3D perspective.

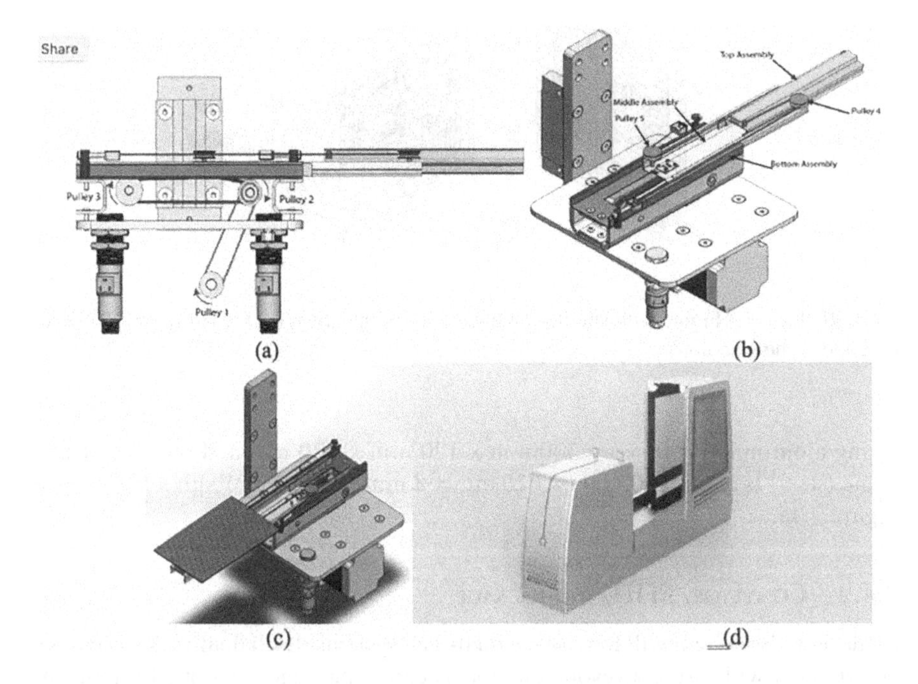

FIGURE 3.3 (a), (b), and (c) pick_place module in 3D perspective; (d) the AGV cover in 3D perspective.

Where z = 20, P = 2

Equation 2 provides the length (c) of a single rotation:

$$c = D_w * \pi \tag{3.2}$$

The travel distance of the 2nd floor is: $s_2 = 75mm$

Number of revolutions is given in Equation 3:

$$n_M = \frac{S_2}{c} = \frac{S_2}{z * P} = \frac{75}{2 * 20} = 1.875(Revolutions) \tag{3.3}$$

3.4.1.3 AGV Cover

To ensure beauty, as well as fire safety during the operation, the equipment (programmable logic controller (PLC), motor, driver, etc.) will be encapsulated in one cover. When designing the cover, the following factors were considered: technical requirements, aesthetics, operating environment, and heat dissipation. Figure 3.3d is the final accepted version.

3.4.2 Rack System

As described earlier, each AGV serves to pick/place packages for two shelves placed in two parallel directions of the AGV's movement. Each shelf's dimensions

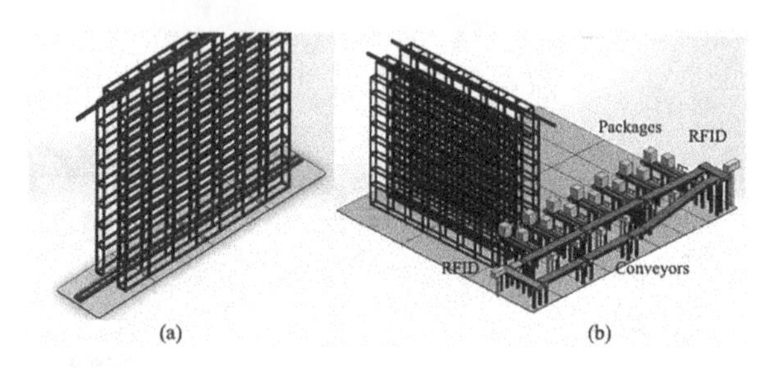

FIGURE 3.4 (a) Frame rack and floor plane in 3D perspective; (b) the conveyor, RFID, and package in the warehouse.

(using aluminum profile) are 300 mm x 120 mm x 200 mm and the size of floor plane (steel) is about 3600 mm x 620 mm x 8 mm (Length x Width x Height); see Figure 3.4a.

3.4.3 Conveyor, RFID, and Package

By design, the system will have seven dual-rack systems. Each dual-rack system will be arranged with two conveyors, one for receiving packages and one for returning packages. Each conveyor may accommodate up to five packages, which helps the primary conveyor system run more efficiently.

Additionally, a conveyor system with a 65-package capacity is put directly next to the receive/return conveyors (Figure 3.4b) to enhance the system's flexibility and save staff by transporting packages in and out of the system without the need for employees. This conveyor system acts as a buffer used to store packages from the warehouse. The goods will then be kept at new locations using the conveyor system after the system uses the RFID writer to write the new code on them.

There are up to 1,372 color-coded packages with three different weights (0.5 kg [green], 1 kg [yellow], and 1.5 kg [red]) that can be stored in the seven dual-rack systems. One RFID tag is used to identify one package. The variable package weights make it easier to adjust trip times and measure how much power the system uses under various loads.

3.4.4 WMS and Control System

The EIU warehouse management system (E_WMS) facilitates inventory tracking, sorting process optimization, and delivery/receiving tracking. From the time a delivery request is made until the products are discharged from the warehouse, the system oversees and supervises warehouse activities. E_WMS's main goal is to create a paperless warehouse that would optimize the movement and placement of products.

Figure 3.5 is a flowchart of the system's operations from the user's request to the E_WMS until the package is stored or brought to a return location.

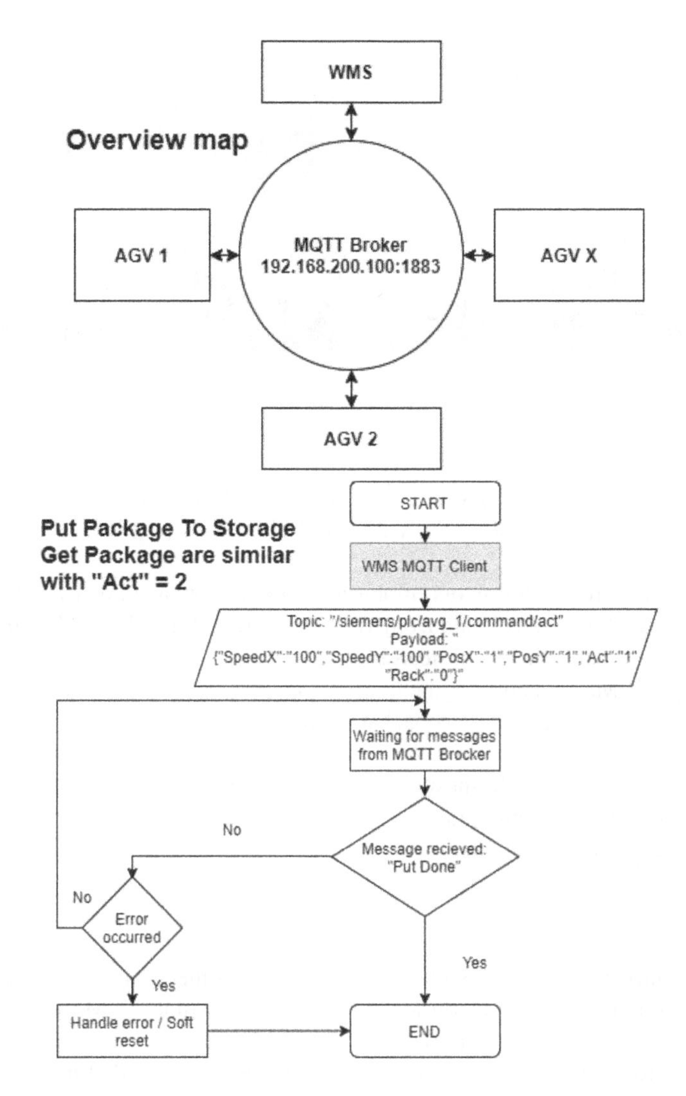

FIGURE 3.5 The flowchart of the system's operations: overview map and putting package to storage.

3.5 RESULTS AND DISCUSSIONS

3.5.1 The Full System

The finished dual racking system meets all specifications.

(1) The maximum speed of an AGV is 0.3 m/s.
(2) The pick_place module runs stably; the programmable logic controller (PLC) is a product of Siemens, and the motor and driver are products of Mitsubishi; the control module works smoothly and without any errors.

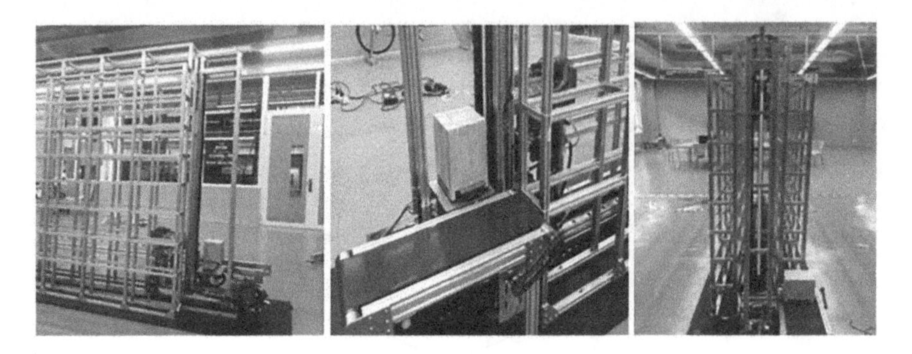

FIGURE 3.6 The whole system from a 3D perspective.

(3) To improve the system's aesthetics, the central dual-rack system (as seen in Figure 3.6) is built of 20 x 20 mm2 aluminum profiles with recessed connections.
(4) The receive/return conveyors employ a 12v DC motor and have a small design.
(5) When sending control data from WMS (Figure 3.7) to PLC and vice versa, the central controller runs steadily.
(6) The E_WMS system originally succeeded in completing its core task's objectives: assign codes to packages in accordance with the fundamental principles (e.g., goods from the same firm should be placed on the same rack; closer to first, farthest from last); indicate which cells are occupied, vacant, or waiting; and practical statistics.
(7) AGV case design is finished.

3.5.2 Contributions

The EIU Smart Warehouse provides a dynamic environment for research, education, and training activities for undergraduate and graduate students. Moreover, the EIU Smart Warehouse can help the university increase the community's attention through this project. Regarding research and education purposes, the EIU Smart Warehouse offers benefits to multiple stakeholders.

3.5.2.1 Students

Based on the model types of testbeds,[52] suggested activities, and the benefits students could generate from testbed-based learning, are shown in Table 3.1.

The EIU Smart warehouse provides the best opportunity to learn in a real environment on actual industrial devices and to face real-world problems similar to those they will face in the workplace. Students encounter topics such as motion control; PLC systems; industrial communication networks; supervisory control and data acquisition (SCADA) systems; optimizing the sorting process; warehouse design, warehouse planning, and warehouse operation; warehouse inventory management; warehouse cost optimization; and warehouse performance measurement.

FIGURE 3.7 The E_WMS interface.

TABLE 3.1
Benefits for Students

	Co-design	Test and Learn	Evidence Hub	Edtech Network
Activities	Test products/processes and provide feedback and design ideas	Participate in testing	Participate in testing	Participate in testing
Benefit	Gain insights into how to develop products, processes, and solutions; understand the real business demand	Apply the specific knowledge to improve or provide solutions	Advantages of learning technology have been proven	The network of lecturers, institutes, and entrepreneurs has joined the testbed

3.5.2.2 Professors

The EIU Smart Warehouse operates as an interdisciplinary environment that will create new courses and new academic activities. This will benefit traditional subjects that are only treated theoretically, which is usually what happens in engineering

TABLE 3.2
Benefits for Professors

	Co-design	Test and Learn	Evidence Hub	Edtech Network
Activities	Test products/processes and provide feedback and design ideas	Participate in testing	Participate in testing	Experience, expertise, and insights
Benefits	Build new skills and a market-driven mindset	Learn what the best way to apply the product/technology is and whether it works theoretically and practically	Benefit from the advice; benefit from the educational value of the testbed	Insights and best practices based on others' practical experience; inspiration and confidence

schools. From the perspective of types of testbed models, Table 3.2 lists activities and benefits for lecturers.

3.5.2.3 Researchers

The EIU Smart Warehouse allows for new solutions to be tested and developed in real and live production environments and integrates various solutions from applied research projects. Topics that researchers can address on the system include applied computer vision, AI for the sorting process, motion optimal control, Internet of Things, energy optimization, transaction time optimization, optimization of warehouse technical structure, optimization of warehouse operational structure, optimization of warehouse management, and typical warehousing operations.

3.5.2.4 Practical

The primary purpose of conducting testbeds is to integrate the objectives from various stakeholders to see how an innovative technology works in reality and how that new technology can enhance competitive advantage for companies.[53, 54] Therefore, the smart warehouse at EIU opens up numerous opportunities for firms to test new solutions and better understand the academic environment. University members, including faculty and students, can gain opportunities to work on real-life projects. Consequently, it can promote more efficient technology transfer between universities and enterprises.

In addition, the EIU Smart Warehouse is expected to attract resources for cutting-edge technology or particular fields that local authorities have emphasized. This testbed also lowers the risk level of research and development for firms, especially when developing new products. Instead of investing a large amount of money in new technology or products, companies can now find a safer way to "repeat, fail and influence the necessary regulatory and policy changes".[55] This creates more opportunities for adjustment if required.

The final stakeholder to take advantage of the EIU Smart Warehouse is the local economy as the project is expected to maximize the potential and value of local research. Thus, it attracts more investment and makes public services more efficient.

3.6 CONCLUSIONS

In the context of Vietnam, this chapter offers suggestions for implementing AI in the creation of smart warehouses. SWOT analysis has been used to evaluate strengths, weaknesses, opportunities, and threats from economic, technical, market needs, resource, investment trends, and environmental viewpoints. From the results of the SWOT analysis, including using strengths and opportunities to overcome weaknesses and limiting risks and taking advantage of strengths to promote opportunities, the following solution groups were proposed: (1) solutions to support investment decisions and limit the risks due to the test innovative approaches and algorithms, thereby confirming the feasibility and estimating of post-investment engineering economics problems; (2) solutions for AI applications in the field of intelligent warehouses development; (3) solutions for developing AI resources for smart warehouses.

The system—consisting of the single rack system, AGV, E_WMS, and the controller—is finished and functioning normally as the prototype phase draws to a close. The AGV can move at a speed of 0.3 m/s. The initial objective of directing students and faculty to investigate the real system was accomplished by the system. Due to its ability to automatically organize and manage inventories, the system is receiving significant attention from local businesses and governments. As the pandemic situation improves, the entire system, which can accommodate up to 1372 packages and contains the full E_WMS, is quickly speeding up.

Following the government's development orientation, the EIU Smart Warehouse project is a useful tool for the academy and industry. In terms of educational purposes, students will benefit from a highly applicable learning environment, particularly on topics about I4.0 and digital transformation. Additionally, lecturers will be offered a great environment to investigate the in-depth study of both supply chain management and warehousing management.

3.7 ACKNOWLEDGMENT

This research is financially supported by Eastern International University, Binh Duong Province, Vietnam.

REFERENCES

1. Kamali, A. (2019). Smart warehouse vs. Traditional warehouse. *CiiT International Journal of Automation and Autonomous System*, 11(1), 9–16.
2. Tekinerdogan, B. (2017). *Engineering connected intelligence: A socio-technical perspective*. Wageningen: Wageningen University & Research. doi:10.18174/401115
3. Le, N., Nguyen, D., Nguyen, X., Kieu, M., Nguyen, V., & Ninh, T. et al. (2022). An intelligent algorithmic approach for data collection in a smart warehouse testbed. In *The 8Th international conference on advanced machine learning and technologies and applications (AMLTA2022)*, pp. 557–566. doi:10.1007/978-3-031-03918-8_46
4. Jabbar, S., Khan, M., Silva, B. N., & Han, K. (2018). A REST-based industrial web of things' framework for smart warehousing. *The Journal of Supercomputing*, 74(9), 4419–4433.

5. Le, N. H., Phan, M. D. K., Nguyen, D. C., Nguyen, X. H., Kieu, M. K., Nguyen, V. A. T., Ninh, T. T. D., Phan, M. D. K., Debnath, N. C., & Le, N. B. (2022). Multilayer communication based controller design for smart warehouse testbed. In *3rd international conference on emerging technologies in data mining and information security (IEMIS 2022)*, Springer Nature Singapore Pte Ltd. 152 Beach Road, #21-01/04 Gateway East, Singapore 189721, Singapore. pp. 345–353. doi:10.1007/978-981-19-4676-9_29
6. Le, N. H., Kieu, M. K., Nguyen, V. A. T., Ninh, T. T. D., Nguyen, X. H., Nguyen, D. C., Debnath, N. C., & Le, N. B. (2022). Implementation of a multi-disciplinary smart warehouse project with applications. In *3rd international conference on emerging technologies in data mining and information security (IEMIS 2022)*, Springer Nature Singapore Pte Ltd. 152 Beach Road, #21-01/04 Gateway East, Singapore 189721, Singapore. pp. 403–411. doi:10.1007/978-981-19-4676-9_34
7. Liu, X., Cao, J., Yang, Y., & Jiang, S. (2018). CPS-based smart warehouse for industry 4.0: A survey of the underlying technologies. *Computers*, 7(1), 13.
8. Papcun, P., Cabadaj, J., Kajati, E., Romero, D., Landryova, L., Vascak, J., & Zolotova, I. (2019, September). Augmented reality for humans-robots interaction in dynamic slotting "chaotic storage" smart warehouses. In *IFIP international conference on advances in production management systems*. Cham: Springer, pp. 633–641.
9. Cogo, E., Žunić, E., Beširević, A., Delalić, S., & Hodžić, K. (2020, March). Position based visualization of real world warehouse data in a smart warehouse management system. In *2020 19th international symposium INFOTEH-JAHORINA (INFOTEH)*. New York City, United States: IEEE, pp. 1–6. doi:10.1109/INFOTEH48170.2020.9066323
10. Copeland, B. J., & Proudfoot, D. (2007). Artificial intelligence: History, foundations, and philosophical issues. In *Philosophy of psychology and cognitive science. Handbook of the philosophy of science, philosophy of psychology and cognitive science*. North-Holland, Elsevier. pp. 429–482.
11. Tran, V., Kieu, M., Nguyen, X., Nguyen, V., Ninh, T., & Nguyen, D., et al. (2022). Reinforcement learning for developing an intelligent warehouse environment. In *The 8th international conference on advanced machine learning and technologies and applications (AMLTA2022)*, 11–20. doi:10.1007/978-3-031-03918-8_2
13. Blancas, L. (2014). Rapid growth, challenges and opportunities in Vietnam's logistics limited connectivity. [PDF]. https://drive.google.com/file/d/1qdvAfnwfGcMYosUlmWIRg3Grr64wKHC0/view
14. Arnold, J., Arvis, J. F., Mustra, M. A., Horton, B., Carruthers, R., & Ojala, L. (2010). *Trade and transport facilitation assessment: A practical toolkit for country implementation*. Washington, D.C., United States: The World Bank. ISBN 978-0-8213-8412-1.
15. Ministry of Industry and Trade. (2019). *Vietnam logistics report 2019: Logistics enhances the value of agricultural products* [PDF]. https://gosmartlog.com/wp-content/uploads/2019/12/Bao-cao-logistics-viet-nam-2019.pdf
16. StoxPlus. (2017). Preview on Vietnam warehousing market 2017. Retrieved July 31, 2022, from http://biinform.com/Reports/EE5-preview-on-vietnam-warehousing-market-2017–2770.html
17. Whelan, S. (2019). Vietnam warehousing boosted by the booming of manufacturing and eCommerce. Retrieved July 31, 2022, from https://vietnaminsider.vn/vietnam-warehousing-boosted-by-the-booming-of-manufacturing-and-eCommerce/
18. FiinGroup. (2017). *Vietnam logistics market 2017* [PDF]. www.amchamvietnam.com/wp-content/uploads/2017/09/StoxPlus_LOGISTICS_Amcham_Deck_final.pdf
19. Prime Minister. (2020). *The decision number 703/QD-Ttg*. Hanoi: Decision of Prime Minister.
20. Hoa, H. T. T., Van Khoang, N., Lien, B. T. B., Dao, T. Q., & Hang, T. T. (2020). A study on the selection of the logistics Centre location—Vietnam-based logistics sector. *Management*, 8(2), 121–127.

21. Linh, P. N. M., & Huong, N. T. T. (2020). The supply chain and logistics of Vietnam in the context of international economic integration. *International Business Research*, 13(7), 27–44.
22. Culler, D., & Long, J. (2016). A prototype smart materials warehouse application implemented using custom mobile robots and open source vision technology developed using emgucv. *Procedia Manufacturing*, 5, 1092–1106.
23. Yang, Q., Lian, Y., & Xie, W. (2020). Hierarchical planning for multiple AGVs in warehouse based on global vision. *Simulation Modelling Practice and Theory*, 104, 102124.
24. Halawa, F., Dauod, H., Lee, I. G., Li, Y., Yoon, S. W., & Chung, S. H. (2020). Introduction of a real time location system to enhance the warehouse safety and operational efficiency. *International Journal of Production Economics*, 224, 107541.
25. Yerpude, S., & Singhal, T. K. (2018). Smart warehouse with internet of things supported inventory management system. *International Journal of Pure and Applied Mathematics*, 118(24), 1–15.
26. Pandian, A. P. (2019). Artificial intelligence application in smart warehousing environment for automated logistics. *Journal of Artificial Intelligence*, 1(02), 63–72.
27. Sowmya, T. K., Agadi, S. V., Saraswathi, K. G., Nirvani, P. B., & Prajwal, S. (2018). 0 Implementation of IoT based smart warehouse monitoring system. *International Journal Of Engineering Research Technology (IJERT) ICRTT*, 6(15).
28. Kersten, W., Blecker, T., & Ringle, C. M. (2018). The road to a digitalized supply chain management. In *Proceedings of the Hamburg international conference of logistics (HICL)*, epubli GmbH: Berlin. no. 25, p. 339, ISBN 978-3-746765-35-8.
29. Ngoc, B. (2021). Warehousing continues to rise unabated in local market warehousing continues to rise unabated in local market. Vietnam Investment Review.
30. Savills. (2022). Vietnam industrial real estate market forecast. Retrieved July 31, 2022, from https://industrial.savills.com.vn/2021/07/industrial-and-logistics-real-estate
31. Savills. (2022). Warsaw research forum. Retrieved July 31, 2022, from www.savills.com/research_articles/255800/187576-1
32. Ministry of Industry and Trade. (2020). *Vietnam logistics report 2020: Reduced logistics costs* [PDF].
33. Shoham, Y., Perrault, R., Brynjolfsson, E., Clark, J., Manyika, J., Niebles, J. C., Lyons, T., Etchemendy, J., Grosz, B., & Bauer, Z. (2018). AI index 2018. Retrieved July 31, 2022, from https://hai.stanford.edu/ai-index-2018
34. Zhang, D., Mishra, S., Brynjolfsson, E., Etchemendy, J., Ganguli, D., Grosz, B., . . . Perrault, R. (2021). *The AI index 2021 annual report*. Retrieved March 2021, from https://arxiv.org/ftp/arxiv/papers/2103/2103.06312.pdf
35. Prime Minister. (2021). National strategy for 4th Industrial Revolution. Retrieved July 31, 2022, from https://en.baochinhphu.vn/national-strategy-for-4th-industrial-revolution-11140283.htm
36. Media Center. (2021). Trends in Vietnam's logistics and warehousing market: Refrigerated warehouse. Retrieved July 31, 2022, from www.phumy3sip.com/media-center/general-new/trends-in-vietnams-logistics-and-warehousing-market-refrigerated-warehouse
37. VietnamWorks. (2022). VietnamWorks công bố Báo cáo thị trường nhân lực ngành Công Nghệ Thông Tin 2020 và Ra mắt thương hiệu VietnamWorks InTECH. Retrieved July 31, 2022, from https://intech.vietnamworks.com/article/vietnamworks-cong-bo-bao-cao-thi-truong-nhan-luc-nganh-cong-nghe-thong-tin-thap-nien-2010-va-nam-2020-ra-mat-thuong-hieu-vietnamworks-intech
38. The European Parliament and the Council of the European Union. (2022). Directive 2009/125/EC of the European parliament and of the council. Retrieved July 31, 2022, from https://eur-lex.europa.eu/legal-content/EN/TXT/HTML/?uri=CELEX:32009L0125&from=EN

39. Siagian, H., Tarigan, Z. J. H., & Jie, F. (2021). Supply chain integration enables resilience, flexibility, and innovation to improve business performance in COVID-19 era. *Sustainability*, 13(9), 4669.
40. National Institute of Standards and Technology. (2015). Measurement challenges and opportunities for developing smart grid testbeds. In *10th Carnegie Mellon conference on the electricity industry*. Retrieved April 1, 2015, from https://research.ece.cmu.edu/electriconf/slides_2015/Boynton%20talk%20CMU%20Conference%2020150401%20 0100pm.pdf
41. Falkenberg, R., Masoudinejad, M., Buschhoff, M., Venkatapathy, A. K. R., Friesel, D., ten Hompel, M., . . . Wietfeld, C. (2017, September). PhyNetLab: An IoT-based warehouse testbed. In *2017 federated conference on computer science and information systems (FedCSIS)*. New York City, United States: IEEE, pp. 1051–1055. doi: 10.15439/2017F267
42. Ridolfi, M., Macoir, N., Vanhie-Van Gerwen, J., Rossey, J., Hoebeke, J., & De Poorter, E. (2019, April). Testbed for warehouse automation experiments using mobile AGVs and drones. In *IEEE INFOCOM 2019-IEEE conference on computer communications workshops (INFOCOM WKSHPS)*. Paris, France: IEEE, pp. 919–920. doi:10.1109/INFCOMW.2019.8845218
43. Monostori, L. (2014). Cyber-physical production systems: Roots, expectations and R&D challenges. *Procedia Cirp*, 17, 9–13.
44. Hu, H., Jia, X., He, Q., Fu, S., & Liu, K. (2020). Deep reinforcement learning based AGVs real-time scheduling with mixed rule for flexible shop floor in industry 4.0. *Computers & Industrial Engineering*, 149, 106749.
45. Salunkhe, O., Gopalakrishnan, M., Skoogh, A., & Fasth-Berglund, Å. (2018). Cyber-physical production testbed: Literature review and concept development. *Procedia Manufacturing*, 25, 2–9.
46. Opalic, S. M., Goodwin, M., Jiao, L., Nielsen, H. K., Pardiñas, Á. Á., Hafner, A., & Kolhe, M. L. (2020). ANN modelling of CO2 refrigerant cooling system COP in a smart warehouse. *Journal of Cleaner Production*, 260, 120887.
47. Li, M. P., Sankaran, P., Kuhl, M. E., Ganguly, A., Kwasinski, A., & Ptucha, R. (2018, December). Simulation analysis of a deep reinforcement learning approach for task selection by autonomous material handling vehicles. In *2018 Winter simulation conference (WSC)*. New York City, United States: Gothenburg, Sweden, pp. 1073–1083. doi:10.1109/WSC.2018.8632448
48. Kamoshida, R., & Kazama, Y. (2017, July). Acquisition of automated guided vehicle route planning policy using deep reinforcement learning. In *2017 6th IEEE international conference on advanced logistics and transport (ICALT)*. New York City, United States: Bali, Indonesia, pp. 1–6. doi:10.1109/ICAdLT.2017.8547000
49. Hillebrand, M., Lakhani, M., & Dumitrescu, R. (2020). A design methodology for deep reinforcement learning in autonomous systems. *Procedia Manufacturing*, 52, 266–271.
50. Andersen, P. A., Goodwin, M., & Granmo, O. C. (2020). Towards safe reinforcement-learning in industrial grid-warehousing. *Information Sciences*, 537, 467–484.
51. Zhang, D., Pee, L. G., & Cui, L. (2021). Artificial intelligence in E-commerce fulfillment: A case study of resource orchestration at Alibaba's smart warehouse. *International Journal of Information Management*, 57, 102304.
52. Batty, R., Wong, A., Florescu, A., & Sharples, M. (2019). *Driving EdTech futures: Testbed models for better evidence*. London: Nesta.
53. Eliasson Lilja, C. (2018). Smart manufacturing systems-a testbed for decision support tool requirements. Master's thesis.
54. Nesta. (2022). Why use a real-world testbed? Retrieved July 31, 2022, from www.nesta.org.uk/feature/why-use-real-world-testbed/
55. Arntzen, S., Wilcox, Z., Lee, N., Hadfield, C., & Rae, J. (2019). *Testing innovation in the real world*. London: NESTA.

4 Empirical Study on Sentiment Analysis

S. Divya Meena, Nouluri Vamsi Krishna, Meghana Nagaraj Cilagani, P. Anushri Sowmya, Thoom Purna Chander Rao, Pedaballi Rajeswari and J. Sheela

CONTENTS

4.1 INTRODUCTION

Using natural language processing, text analysis, and statistics, sentiment analysis gauges customer sentiment. The finest companies comprehend the feelings of their clients and take into account their words, body language, and intentions.[1] Customer sentiment may be shown by tweets, comments, reviews, and other websites where your brand is discussed. Today's engineers and executives must be proficient in sentiment analysis, which is the field of using software to analyse these emotions.[2] Similar to how they have in many other areas, advances in deep learning have brought sentiment analysis to the forefront of cutting-edge algorithms. Words are extracted and categorised into positive, negative, and neutral categories using natural language processing, statistics, and text analysis.[3] We proposed a research based on

DOI: 10.1201/9781003310785-4

sentiment analysis in comparison with previous research, chose suitable challenges from each exploration, and clarified their consequences for feeling precision.[4] The relevance and consequences of sentiment analysis issues in sentiment evaluation are discussed in this overview, which is based on two correlations among 47 studies.[5–42] Sentiment audit frameworks and sentiment analysis issues are the focus of this investigation. As a result of this work, area reliance [43] is now a significant consideration in the interpretation of sentiment concerns. Also, the experimental results show that the proposed model could provide advertisers with suitable targets for diffusing advertisements continuously and thus efficiently enhance advertising effectiveness. The succeeding correlation is therefore determined by the feeling examination problems that apply to the precision rate. The findings emphasize the significance of sentiment concerns in gauging feelings and how to choose the optimal test to increase precision.[44] We discovered a link between the extent to which sentiment approaches are used in hypothetical and specialized situations to deal with emotional challenges.

This chapter centres around the main difficulties in the sentiment assessment stage that have a huge impact in sentiment scores and extremity locations. This chapter sums up the keys of sentiment difficulties concerning the kind of evaluation structure. It further categorises issues into two groups and makes it easier to manage them and focus on the level of critical importance. This investigation focuses on these emotional issues, as well as the factors that influence them and their relevance.

4.2 EMPIRICAL STUDY

In this sentiment analysis we utilized two sorts of examinations. In our first examination, we gathered difficulties from 37 exploration papers.[45–82] The goal of this correlation is to determine the relationship between sentiment challenges and audit construction, as well as the impact on sentiment results. To deal with this, we needed to consider three formulated sentiments: structured sentiments,[37–41] semi-structured sentiments,[43–52] and unstructured sentiments.[54–73]

(i) **Structured sentiment**. In structured sentiment analysis, entire opinion tuples are extracted from texts, but over time, this process has been broken down into smaller and smaller sub-tasks, such as targeted polarity classification or target extraction.[39] Structured sentiment analysis endeavours to extricate full assessment tuples from a text. Yet over the long run, this undertaking has been partitioned into more modest sub-assignments.

(ii) **Semi-structured sentiment**. Semi-structured sentiments are written in a free arrangement by the essayist with few limitations. There could be no legitimate separation of positives and negatives, and the substance may involve a couple of sentences in which every sentence presents more features or conceivable speculations.[47]

(iii) **Unstructured sentiments.** Unstructured sentiments are a casual and free message design that doesn't follow a formal format; the author doesn't have any limitations. There could be no appropriate separation of positives and negatives, and the substance may involve a couple of sentences in which every sentence contains features or possible notions.[59] The model under those unstructured reviews might perhaps give more abundant and ordered evaluation information than its accomplice.

(iv) **Explicit feature**. In the off chance that an element shows up in a fragment of audit sentences, this element is known as an explicit feature in an item. In this instance, the fragment image is brilliant, and the image is an unequivocal component.[69]

(v) **Implicit feature.** In the off chance an element doesn't show up in review, but is inferred, this is known as an implicit feature in an item. The implicit features, which are implied by some words or phrases, are so significant that they can express the users' opinion and help us to better understand the users' comments. As for the significance of the opinion examination, this study talks about the connection between the survey design and feeling investigation challenges.[72] We analyse the feeling challenge that shows up in an additional kind of sentiment framework.

Table 4.1 represents an examination of 41 reviewed reports on sentiment investigation.[5–37] The correlation results articulate that there is a fundamental component that is significant and pertinent to audit structure. This component is space arranged and needs to hold a direction of theme area and provisions to decide the test for the examination. The other outcome is invalidation, the main test that has the best effect in many feeling investigations and assessments in organized, semi-organized, unstructured reviews.[83] Be that as it may, the examination inadequacy requires updatable exploration to arrive continually to address the reasonable difficulties effectively and rapidly. The second comparison clarifies the outline of opinion difficulties and how to work on the exactness of them, dependent on the past works. Its objective is distinguishing the main difficulties in feeling and how to further develop its outcomes to be applicable to the pre-owned methods.[84] Table 4.1 recognizes the use of every strategy. The hypothetical moves utilise numerous methods to work on the outcomes with addressing the specific feeling difficulties. The most noteworthy procedure utilisation in the hypothetical sort is parts of speech. The bag-of-words procedure is the subsequent method[85, 86] and another factor is the Maximum entropy technique. The outcomes are diverse in technical sentimental challenge types; the most noteworthy utilisation method is the n-gram procedure because it depends on expressions and articulations. The vocabulary-based strategy is the one that is used the least.

Table 4.2 analyses a few boundaries applicable to sentiment analysis challenges. These boundaries are dictionary type, space situated, and dataset. This strategy utilised the exactness consequences.[95–98] This examination sums up impact of sentimental challenges arrangements in breaking down and assessing feeling investigation precisely. The words 'opinion' and 'extreme' can be found in pre-owned vocabularies. The extreme differences in opinion characterise the extreme levelling. This grouping of extremity is isolated to a few classes or levels that go progressively from two levels to five levels.[99–101] The examination's qualities are:

TABLE 4.1
Challenges in Sentiment Analysis (SA)

Study ID	Domain Aligned	Challenge Type	SA Challenge	Review Structure
[87]	-	T	Dependence on a domain	S and objectives expression
[88]	-	T	Spam and fake detection + negation	SS
[89]	-	T	Negation + huge lexicon	SS
[90]	N, online -collaborative media	Technical	Huge lexicon	S and US
[91]	Y, social media	Technical	Bipolar words	US
[92]	N, online client testimonial	T	Finding spam and fake emails	US
[93]	-	T	NLP overhead (emotions)	US
[94]	N, broader sense domain	T	Negation	SS NN/ADJ/VB and ADV-clauses and phrases
[46]	Y, online news review	T	World knowledge	SS nouns, US
[47]	Y, social media	T	NLP overhead (emotions)	US
[48]	-	T	Negation	SS adjectives only
[49]	Y, research journals	Technical + T	Negation + world knowledge + feature extraction + lexicon	S
[50]	Y, products	Technical	Keyword or feature extraction	SS
[51]	Y, Facebook and Twitter	T	Overheads in NLP (short abbreviation)	US
[52]	Y, tweets	Technical	Huge lexicon	US
[53]	Y, with aspect level	Technical	Extracting features or keyword	SS
[54]	Y, CNET, IMDB movie review	Technical	Huge lexicon	SS
[55]	Y	Technical + T	Negation + bipolar	SS, sentences or topic documents
[56]	-	T	NLP overheads (Ambiguity)	Structured adjective only
[57]	Y, tweets	Technical	Bipolar words	US
[58]	Yes, movie review	T	Negation	US
[59]	Yes	T	Dependence on a domain	US in combination with a taxonomy of emotional concepts that is predefined

[60]	Y, movie and product domains	Technical	Bipolar words	US
[61]	Y, TripAdvisor	Technical	Extracting features or keyword	SS
[62]	Y	T	Dependence on a domain	US, online customers reviews
[63]	Y, tweets NLP	T	overheads (sarcasm) + negation	US
[64]	Y, ecommerce and online security	T	Spam and fake detection	SS
[65]	N mutli-domain	T	Dependence on a domain	SS
[66]	-	T and technical	Negation plus entity traits or keywords	S or SS
[67]	Y, tweets	T	Negation with overheads (sarcasm) from NLP	US
[68]	Y, tweets	T	NLP overheads (sarcasm)	US
[69]	Y	T	Dependence on a domain	S, news articles
[70]	Y, tweets	T	NLP overheads (short abbreviations)	US
[71]	Y, product reviews	T	Detection of spam and fakes	US
[72]	N, online customer reviews	T	Detection of spam and fakes	US
[73]	Yes health/medical domain	T	Negation	SS
[74]	Yes movies	T	Negation + dependence on a domain	SS ADV, ADJ
[75]	-	T	Dependence on a domain	US, emotion reviews
[76]	Yes movies	T	Negation + dependence on a domain	SS ADV, ADJ
[77, 78]	Y, social media	T	Spam and fake detection	US
[79]	Y	T	Dependence on a domain	US, Twitter

T: theoretical; S: structured; SS: semi-structured; US: unstructured; NN: noun; ADJ: adjective; VB: verb; ADV: adverb

TABLE 4.2
Challenges in Sentiment Analysis (SA) Based on Dataset

Study ID	SA Challenge	Technique Used	Domain Aligned	Lexicon Type	Data Set	Accuracy
[102]	Bilingual terms	Combining features (n-grams) and preprocessing methods (unsupervised stemming and phonetic transcription)	Aligned	English Facebook	10,000 Facebook postings are included in the sample	69%
[103]	Negation + domain dependence	BOW term frequencies	Aligned	Two wordlists	2000 reviews of movies, 1000 of them positive and 1000 of them negative	With a greater recall of 65% 83 percentage points
[95]	Spam/fake reviews	POS tagging and the n-gram method share similarities	Internet user testimonials	LIWC	800 opinions	Nearly 90%
[96]	Field dependence	SemEval-2013	Y	English MPQA and tweets	Additional to the well-known MPQA, there were 2000 positive words and 4700 negative terms	Improve accuracy and F-measurement by roughly 13% from the baseline to reach 69%.
[104]	Overheads in NLP (emotions)	(Maximum entropy, naive Bayes, and SVM)	Multi-domain N	Microblogging lexicon	1,600,000 training tweets, 800,000 emoticon-containing tweets, 800,000 emoticon-containing tweets that are both good and negative	Naive Bayes accuracy increased from 81.3% to 82.7%, while MaxEnt accuracy increased from 81.3% to 82.7% (from 80.5% to 82.7%); SVM, on the other hand, has declined (from 82.2% to 81.6%)
[96]	Field dependence	Lexicon built on WordNet	Y	News reviews	Articles from newspapers (the set of 1292 quotes)	82% improve the baseline 21%

[88]	Huge lexicon	Distinguishing between previous and contextual polarity	N	Multiperspective Question Answering (MPQA) Opinion Corpus1	In 425 documents (8,984 sentences), there are 15,991 subjective terms)	75.9%
[78,77]	NLP overheads (multilingual)	A technique that integrates information retrieval, natural language processing, and machine learning	Y	English, Dutch, and French texts	Texts located on blogs, reviews, and forums on the Internet	83%, 70%, and 68%
[80]	Domain dependence	Dependency, Sentiment-LDA, Markov chain	Y	Hownet, Senti-wordnetMPQA	Words having a positive or negative score; MPQA 4152 2304; HowNet 2700 2009; English translation of the Chinese SentiWordNet 4800 2290; positive/negative words lexicon of MPQA subjectivity	70.7
[47]	NLP overheads (emotions)	Fine-grained emotions	Y	Chinese lexicon	Tweets about the Sichuan earthquake totaled 35,000	80%,
[97]	Domain dependence	WEKA5 N naive Bayes and support vector machines	Multi-domain	46 English Story	Movie review (MR) data and multi domain data are the two datasets	90%
[87]	Domain dependence	Deep sentiment analysis is a technique similar to machine translation	N	Japanese	Polar sentences that reflect virtue and badness in a particular area	94% (25% to 33%)
[48] [75, 76]	Negation domain dependence	Part of speech (POS); the space vector model uses the Emotion Dependency Tuple (EDT-enhanced (BOW) TF-IDF and cross entropy	40 different topics N	Open NLP Chinese	Dutch language COAE2014 dataset	Negation: 71.23% (accuracy increases by 1.17%); 60%

(Continued)

TABLE 4.2 (Continued)
Challenges in Sentiment Analysis (SA) Based on Dataset

Study ID	SA Challenge	Technique Used	Domain Aligned	Lexicon Type	Data Set	Accuracy
[91]	Bipolar words	n-gram (uni and bi-grams)	Y	HL and MPQA lexicon	1,600 Facebook messages in the data set	70%
[92]	Spam and fake reviews	Lexicon and a shallow dependency parsed together	N, online customers reviews	SentiWordNet and MPQA	Store#364	85.7% for the mood technique, but 76.7% for the word counting method
[82]	Feature and grained	Applications for fine POS tagging	- N, 7	SentiStrength	7 applications from the Google Play Store and Apple App Store	91%
[98]	General knowledge	Adding the results of the sentiment lexicons' word polarity scores	Y	Context dependent lexicon	6500 responses to video game reviews	Upgrading precision by 60% to 80%
[99]	Negation	Dependency and parse tree	Y	English in the health/medical field	1000-sentence data set	Using four separate methodologies, from 79.2% and 82%
[81]	NLP overheads plus domain dependence (multilanguage)	POS tagging is necessary for the lexicon-based method	N	Tool for Arabic opinion mining that is 16 domain lexicon-based	Handle emoticons, chat language, Arabizi	93.9%
[105]	Massive lexicon	Lexicon based technique	Y	6,74,412 tweets	The dictionary's word polarities are established in accordance with a certain field	73.5%
[94]	Domain dependence	n-gram	N, 7 domains	Chinese reviews b	560 Chinese review	65%
[46]	Negation	POS (part of speech), word sense disambiguation, sentiment analysis	Y	WORDNET	There are 1000 positive and 1000 negative reviews for the English film 1000	98:7%

[49]	World knowledge plus feature extraction from a lexicon plus negation	Improved BOW model	Y, scientific papers	New lexicon	1000, 5000, and 10,000 data points from three datasets (training set, test set, and verified set)	83.5%
[89]	Domain dependence + extracting features or keywords	Instead of terms, use character n-grams.	Y	Hotel reviews in Germany.	A total of 1559 hotel reviews were gathered from the Internet	83%
[90]	SVM with a large lexicon bag-of-words	SVM bag-of-Words	Senti Movie reviews from Y, CNET, and IMDB	Senti Movie	The first dataset is software reviews, while the second is movie reviews	82.30%

- The ability to comprehend hot area research
- Demonstrating the most significant difficulties to accuracy results
- Evaluating the prevalence of each sentiment analysis approach
- Investigating the link between domain reliance, lexicon type, and accuracy outcomes. The consequences of the examination were vital in picking appropriate methods to tackle feeling difficulties and arrive at the most elevated precision.

The examination's decisions in Table 4.2 incorporates the connections in middle of the sentiment analysis and the significance of essence. Although the negative is the most influenced in many attitude types, the results in association in Table 4.1 indicate that there is a significant number of examinations.[106–108] That means that the research area with the bipolar words has the lowest average accuracy. At that point, area reliance and NLP overheads have the subsequent positions. What's more, the negation challenge has third position. Nullification has the most elevated exactness rate that can uphold the consequence of the main examination on the grounds that investigates in sentiment don't have to comprehend the negative surveys whether expressed or certain.[100–109] Our recommendation is to build an investigation into the words 'minimal record' and 'accuracy' because they are bipolar terms in research.

4.3 LEVELS OF SENTIMENT ANALYSIS

Sentiment analysis has been studied on a variety of levels. The easiest way to identify sentiments and opinions is to look at the text, phrase, or aspect level.[32–34] The classification of sentiment analysis is displayed in Figure 4.1.

4.3.1 Sentiment Analysis at the Aspect Level

This level conducts fine-grained analysis to uncover sentiments about specific entity characteristics. Consider this: "The iPhone 11's photography is incredible." The

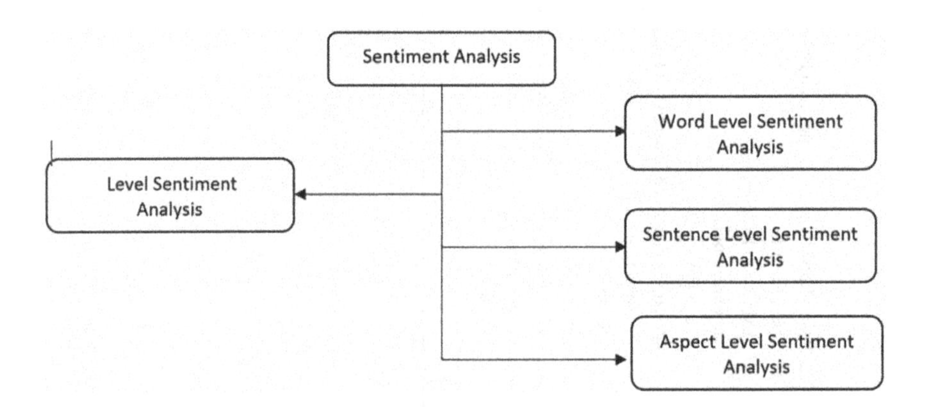

FIGURE 4.1 Classification of sentiment analysis.

review focuses on the 'camera,' a feature of the entity 'iPhone 11,' and the result is positive. Work at this level assists in determining what individuals like and dislike.[35] Rather than the sentiment of paragraphs or words, it is concerned with the attributes of entities (e.g., product qualities). An implicit or explicit characteristic can be extracted from a text using sentiment analysis, according to Cho et al. (2014).[36] A review of implicit aspect extraction approaches was proposed in this area by the authors.

Numerous real-world applications necessitate this level of detail investigation. For example, in order to develop a product, firms assess which components or features appeal to customers. Aspect-based Adaptive Chen et al.[37] presented lexicons as a method for sentiment analysis based on aspects. The authors developed two strategies for creating two dynamic lexicons; the A statistical technique and a genetic algorithm are used to provide aspect-based sentiment analysis. Automatic updates are possible with a dynamic lexicon, which also allows for more accurate ratings of terms that are relevant to the current context and to categorise the aspects in reviews. The suggested lexicons were used in conjunction with a selection of well-known static lexicons drawn from works of literature.

To boost performance, two or more levels of sentiment analysis could be combined instead of performing a single level of sentiment analysis. Using a technique developed by Hu and Liu,[38] it was suggested that product comments on YouTube be analysed using sentiment analysis at the sentence and aspect levels. It has been hypothesised by the authors that sentiment polarity on the sentence level is dependent on and influences sentiment polarity on the aspect level, and that the combined method is capable of dealing with the issues associated with both levels of sentiment analysis. After preprocessing the comments, the author's emotional response was extracted at the sentence and aspect levels using a Bidirectional Encoder Representations from Transformers (BERT)-based model.[39] The outcomes of the analysis were combined to provide statistics reports for the target product.

4.3.2 Sentence-Level Sentiment Analysis

In sentence-level sentiment analysis, the overall meaning of the text is emphasised. Analysis of the sentence is the primary goal in identifying whether a statement indicates a favourable, unfavourable, or neutral attitude.[40] To do this, it is necessary to distinguish between objective phrases that convey factual information and subjective statements that convey thoughts and views. This level of study has been addressed in a number of ways. According to Li and Tsai,[41] sentence-level sentiment analysis can be improved by grouping sentences into several categories. Depending on the quantity of targets in a phrase's first occurrence, they were able to classify it into one of three categories: low, medium, or high (sentence with nontarget, one-target, or multi-target). Each form of text was fed into the model separately for its categorisation, which was done using a one-dimensional convolutional neural network. Analysis of sentiment is important at both the sentence and document levels. It does not create opinions on all elements of an object, however, since it does not specify precisely what individuals like or hate about an object and because it does not describe precisely what individuals like or dislike about an object.[21]

4.3.3 Document-Level Sentiment Analysis

The technique seeks to evaluate whether a work as a whole demonstrates a negative or positive attitude or outlook.[42] All of the documents are classified according to the general feeling the opinion bearer has against a certain entity, as expressed in their words and actions directed towards that object (e.g., single product). This type of classification is most successful when the file is generated by a single individual, and it is not suited for publications that estimate or compare many things at the same time, as is the case with most research articles. There have been a number of different approaches to sentiment analysis at the document level presented. Li and Shiu [43] developed a domain-independent framework for document-level sentiment analysis (DFDS), which was published in the *Journal of Documentary Studies* and included Rhetorical Structure Theory (RST)-based weighting criteria. To compute the emotion ratings of the sentences in each of the texts, the authors divided the articles into rhetorical structure trees, which they then compared to two well-known lexicons. The ratings of the sentences were calculated by the researchers using weighting procedures; they were able to establish the polarity of the document's overall mood.[86, 104, 105] It is likely that the text may contain some conflicting sentiments that will have an impact on the final decision. In a wide variety of fields, the use of sentiment analysis is quite advantageous.

4.4 RELATED FIELDS

There are a few subjects that fall within the subject of SA and have recently piqued the interest of scholars. Three of these subjects are discussed in greater depth in the next subsection.

4.4.1 Emotion Detection

According to neuroscientists, an opinion is a provisional phrase that conveys a boldness to an entity—sentiment reveals feelings or emotions, whereas emotion reflects an attitude toward an entity. In natural language processing, sentiment analysis has been described as a job that identifies different points of view on an entity. Because the distinction between opinion, sentiment, and emotion is fuzzy, they described it as a combination of the three.[110–113] Emotion detection (ED) may be classified as a SA task. SA is more interested in describing positive or negative concepts, while ED is more concerned with distinguishing distinct emotions from text. ED can be implemented as a sentiment analysis job that uses either a machine learning technique or a lexicon-based approach, depending on the situation. Happiness, grief, anger, fear, trust, disgust, surprise, and anticipation, according to Jain et al.,[110] are the eight essential emotions that serve as archetypal examples of human behaviour.

It is a probability distribution of mutual actions if the subject and object of an event frequently engage in mutual activities. Priyanka et al.[114] who used text-mining technologies to analyse emotion in human language, were the first to suggest ED at the sentence level. A system for detecting event emotions was developed by Lu and Lin using web-based text mining, semantic role labelling, a number of reference entity pairings, and custom algorithms.[115] Neither large-scale lexical sources nor knowledge bases were used in their investigation. In their demonstration, they established that their method gave acceptable results for recognising joyful, negative, and neutral emotions. They revealed that the difficulty in detecting emotions is based on the setting.

As an example, Dang et al.[45] presented a strategy that integrated machine learning with a lexicon-based approach. They discussed an approach in their presentation that was created on facts that make sense contained in the emotion corpus knowledge repository (EmotiNet). According to them, emotions are not necessarily portrayed through the use of emotive words such as cheerful, but rather through the description of real-world situations that readers associate with a certain sensation. To achieve their aim, they employed the Support Vector Machine (SVM) and SVM-SO algorithms. They demonstrated that the EmotiNet-based technique is the best for detecting emotions in settings with no affect-related words. Techniques based on common sense can be used to identify emotions in texts such as those in the ISEAR emotion corpus, as demonstrated by the researchers. They found that using EmotiNet outperformed other methods, such as supervised learning with a smaller training set or lexical knowledge.[116]

Jain et al.[111] have developed an Affect Analysis Model (AAM). Affect analysis (AA) is the process of recognising the emotions elicited by a specific semiotic modality and categorising them accordingly. The AAM is composed of five steps: the symbolic cue, the syntactical structure, the word-level, phrase-level, and sentence-level analyses. AAMs were used in a number of applications, which were detailed in Jain's research.[112–114] Another effort reported by Prabu et al.[112] is sentence classification using fine-grained attitude categories. They developed a compositionality-based technique for attitude analysis and a fresh approach to verb semantics. They looked through 1000 sentences (www.experienceproject.com). This was a website where individuals could share their personal narratives. Their study revealed that their method produced trustworthy findings in the textual attitude analysis task. Using a corpus-based method, affect emotion terms such as those proposed by Angel et al.[117] could be employed. In their study, they presented a bootstrapping strategy for recognising paraphrases and extracting them from nonparallel corpora based on contextual and lexical criteria. They started with just a few seeds. It was through the use of their method that they uncovered patterns for six different sorts of emotions. They employed text to extract excerpts from annotated blogs and other data sources, which they then used to create their final product. For their effort, they drew on information from real-time journals and blogs, text effects, fairy tales, and annotations from blogs. Using their data set, they demonstrated that their technique worked effectively.[118] Aozora Bunko Japanese stories were used in the study by Kaushik and Mishra,[50] who conducted text-based affect analysis. Their research focused on how person/character-related factors affect story recognition. They employed anaphoric expression analysis to extract the emotion subject from a phrase, and then they used the affect analysis technique to analyse each character's emotional condition during each section of the story.

Tang et al.[49] pioneered the study of AA through mail and books. He examined the Enron email corpus and discovered significant disparities in how men and women utilise emotion terms in work-related emails. Annotations on the words' correlations with positive/negative and the eight major emotions were manually annotated utilising crowdsourcing to build an annotated lexicon. He employed this tool to analyse and monitor emotional terms in books and communications. Using books and fairy tales, he introduced the concept of emotional word density. According to Mohammad, fairy tale stories featured a considerably broader range of emotional word density than books.

4.4.2 Building Resources

When constructing resources, the goal is to employ polarity to annotate opinion statements in dictionaries, corpora, and lexica. Although it may contribute to the improvement of SA and ED, creating resources is not a SA activity. The key obstacles faced in this category's study were word ambiguity, multilingually, granularity, and disparities in viewpoint expression across textual media,[119] which was introduced by McDonald et al.[20] While doing their research, they used a random walk technique to construct a sentiment lexicon for different domains at the same time. They used three different types of sentiment data in their study. Their findings revealed that using their proposed technique improved the efficiency of automated sentiment lexicon development for domains. The concept of creating corpus was proposed by Wilson et al.[34] They introduced Opinion Mining-ML, a new XML-based approach for labelling textual expressions conveying viewpoints on current events deemed important. Along with Emotion-ML and WordNet, it is a new standard. There were two elements to their effort. First, they developed a standardised approach for annotating emotional claims in manuscript that was completely self-governing of any application domain.[120] Additionally, they investigated domain-specific variation, which was based on the usage of a domain-specific ontology of support. They began with a data set of food blogs and extracted them using a query-oriented technique. They assessed their idea using a fine-grained examination of conflict among different authors. Their findings suggested that their idea constituted an efficient technique capable of covering considerable complexity while maintaining good accord among different individuals.

This fine-grained annotation technique was developed by Li and Tsai [41] for categorising subjectivity in unconventional literary genres. This research focused on document, phrase, and element-level annotations. EmotiBlog, a collection of 270,000 blog posts published in three languages—Spanish, English, and Italian—was also displayed. Natural language processing challenges were used to verify the model's sturdiness and applicability. They used ISEAR to test their model on a variety of corpora. Their research yielded positive outcomes. EmotiBlog was used to classify sentiment polarity and identify emotions. These researchers proved that the resources they used improved the efficiency of systems created specifically for this purpose.

Li and Shiu [43] presented a study about the building dictionary. They developed a semi-automated method for constructing emotion dictionaries in a variety of languages in their research. After constructing emotion dictionaries of the highest calibre in two languages, they mechanically translated each dictionary into a third language. In order to get the most out of your target language vocabulary, you should look for words that appear on both lists. Morphological inflection and the subjective nature of human annotation and judgement were the main topics of their investigation. They gathered information from the news and compared triangulated lists to non-triangulated machine translation word lists to verify their process.

4.4.3 Transfer Learning

Transfer learning is the technique of using knowledge from an auxiliary domain to enhance learning in the target domain. Moving information from a Wikipedia page to a tweet or an English search to an Arabic search are two examples. There are many aspects of cross-domain learning that can be covered by transfer learning. Text classification,[115] sentiment analysis,[116] named entity identification,[117] part-of-speech tagging,[118] and other text mining tasks can all benefit from it. This can be done by transferring sentiment categorizations from one area to another or by establishing a bridge between two different fields of study. To discover high-frequency domain-specific (HFDS) characteristics, Tsai et al.[21] developed an entropy-based technique and a weighting model that considered both the aspects and occurrences. Lower weights were assigned to HFDS characteristics and events that had the same label with the appropriate pivot feature. A Chinese data collection was utilised to compile appraisals of education, equities, and computers. In the study, they found that their proposed model could alleviate some of the harmful impacts of having HFDS. Their model might be a better choice for SA applications that need a lot more classification than what is currently available in training data.

An approach proposed by Bosco et al.[22] for sentiment categorization in various domains is a two-stage process. An initial linkage between source and target domains was set up so that they could access the most reliable tagged content in their target domain. The target-domain data was tagged in the second stage by applying the inherent structure of the labelled documents to it. They worked on reviews of books, hotels, and notebooks from a certain domain based on data from China. They demonstrated that their proposed scheme has the potential to increase sentiment categorization accuracy across a variety of different areas. The stochastic agreement regularization method is used to classify cross-domain polarity.[119] It's a probabilistic agreement approach that works by reducing the Bhattacharyya distance between trained models with two distinct perspectives. It generalises the models out of each point of view by limiting the amount of disagreement they may have on unlabelled cases from a scientific approach. Maks and Vossen [24] reported work on the challenge of cross-domain text subjectivity categorization, which employed the stochastic agreement regularization technique as a foundation. Using the agreement-constrained co-training method and multiple views of a problem, they came up with three new algorithms. Three well-known data sets were utilised to analyse movie reviews and questions. They demonstrated that their suggested approach outperforms the stochastic agreement regularization technique.

For combined modelling of several data sources, diversity among distinct data sources is an issue. Turney [32] attempted to overcome this challenge because joint modelling is vital for transfer learning. Their paper proposed a regularised shared subspace learning framework that may take advantage of the mutual strengths of linked data sources while avoiding the consequences of each source's changeability. For their effort, they drew on data from well-known social media and news sites including Blogspot, Flickr, and YouTube. They were able to show that their strategy was superior to others.

4.5 APPLICATIONS OF SENTIMENT ANALYSIS

Sentiment analysis may help with everything from detecting client opinions to assessing individual mental wellbeing [44, 45] through social platforms As a result of technological breakthroughs such as big data,[47, 48] cloud computing, and blockchain,[50] sentiment analysis can be used in virtually any business. The following subsections, for example, cover the most typical sentiment analysis applications.

4.5.1 Intelligence for Business

There are numerous advantages of using sentiment analysis in the world of business intelligence. Organizations may apply sentiment analysis data to improve goods, study customer comments, or create a new marketing strategy.[51] For example, in the realm of business intelligence, the most common use of sentiment analysis is to examine consumer perceptions of products or services. Buyers, not just product manufacturers, may utilise this research to evaluate products and make well educated selections. Dang et al.[45] studied Amazon evaluations of local cuisine over 8 years. Using the NRC Emotion Lexicon, customers' reviews were divided into eight emotional responses (rage, fright, faith, eagerness, grief, astonishment, disgust, and joy) and two sentiments (positive and negative). Their research discovered that sentiment analysis may be useful in detecting consumer attitudes and mitigating risks in order to keep customers delighted.

Market and forex fluctuations were also predicted using sentiment research. Mudinas and colleagues [52] investigated the impact of news emotion on Bitcoin and traditional currency returns, volume, and fluctuation. Ravenpack News Analytics 4.01 was used to look at high-frequency intra-day data (15 minutes) across a 7-year period (2012–2018) to see how unscheduled news about Bitcoin and six other currencies affected mood. According to the authors, outdated currencies respond swiftly and dramatically to business news wire transfers. The results of the Bitcoin experiment were different from those of the forex experiment, showing that Bitcoin responds to information differently than other currencies do. Bitcoin and digital currencies (also known as cryptocurrencies) are terms for a novel technology known as blockchain, which is a fully decentralised ledger that allows peer-to-peer assets (such as digital money) to be transferred effectively and firmly deprived of the use of third-party intermediaries such as banks and legal teams.[50] Participants in the blockchain network use peer-to-peer general agreement procedures to validate transactions in real time. A small number of studies have used sentiment analysis to estimate the value of digital currencies, but Gezici et al.'s [53] research is an exception. Cryptocurrency experts employed a specific cryptocurrency terminology to do sentiment analysis on

Twitter to estimate the price returns of several familiar digital currencies. In Miao et al. [2009] [54] proposed a hypothetical outline for automatically recognising forged news on social media by means of blockchain technology principles and methodologies.

4.5.2 System for Recommending

A procedure that attempts to present users with appropriate material (movies, music, or product recommendations) is known as a recommender system.[55] A good recommender system can bring in a lot of money in some industries. As a result, using sentiment analysis for these systems[56–58] can aid in the generation of more accurate recommendations. Based on the sentiment analysis of microblogs, an intelligent movie recommendation system, KBridge, was created by Liebrecht and colleagues.[59]

The sentiment based matrix factorization with reliability (SBMF+R) approach suggested by Balahur and Steinberger [60] is a novel approach for utilising reviews for trustworthy suggestions. They built a sentiment lexicon and utilised it to translate appraisals into sentiment assessments. In the additional step, handler dependability metrics were produced, which included user consistency and comments on reviews. A probabilistic matrix factorization of the rating, reviews, and feedback followed.[61] Using social networks as a model of responsive e-learning, the authors showed that the use of large data sets and sentiment analysis has the potential to completely transform the e-learning landscape. Using the proposed sentiment analysis, we can identify social aspects impacting learners that are important in the establishment of an effective learning rhythm.

4.5.3 Government-Provided Intelligence

People comment on a variety of things in addition to goods and services, including politics, religious beliefs, social issues, and even products and services themselves, as researchers Yang et al.[62] found out by analysing Twitter posts for sentiment on Brexit outcomes and other similar topics. Monitoring public reaction to the effective execution of specific measures is immensely valuable. Ott et al.[63] examined local government tweets in the United States to see whether emotion (tone) affects citizen engagement with government via social media platforms.

Sentiment Political Compass (SPC) was used by Ott et al.[63] to measure newspapers' attitudes on political parties. The goal of this study was to determine how political preferences expressed in newspapers affect the evolution of voter opinion. More than 740,000 political entities were extracted from 180,000 newspaper articles over an 18-month period during the German Federal Elections by crawling the data from 25 newspapers. These numbers are employed to investigate the relationship between media and political groups. In some instances, sentiment analysis should be conducted in real time to monitor the public mood. The use of additional technological solutions, such as big data, in real-time sentiment analysis,[66] on the other hand, is required. There is no need to employ sentiment analysis and big data in tandem for real-time analysis.[48, 67] The term "big data" refers to an enormous amount of complex data that is currently being handled ineptly. Dadvar et al.[65] utilised big data techniques. An adaptive system that analyses social media posts in real time to extract user opinions was developed. This method is divided into three stages: creating word embedding dictionaries for each item, categorising postings, and adjusting sentiment weights before running a prediction algorithm. They compared their method to various sentiment analysis tools after evaluating the post-classification performance of the analysed tweets in terms of whether or not they were viewed as positive or negative.

4.5.4 Medical and Healthcare Domain

The use of sentiment analysis in the field of medicine has received a lot of interest recently. In order to provide better treatment, this programme assists healthcare providers in collecting and analysing data concerning diseases, poor medication responses, epidemics, and patient emotions.[68] However, as Cambria [69] shown, sentiment analysis is challenging to utilise in this sector due to a number of issues, including terminology. Li et al.[70] used tweets about patient experiences as a unique public health surveillance method and found and studied them. Twitter's public streaming application programming interface (API) was used to collect over 5.3 million breast cancer-related tweets. Researchers have used sentiment analysis to analyse breast cancer patients' tweets. They found that good experiences with patient care, mobilising support, and increasing awareness were shared. The study shows that social media might be a useful tool for patients to communicate their wants and concerns.

As previously stated, additional technologies may be used with sentiment analysis to assist its adoption in a range of industries by addressing some of the challenges that sentiment analysis encounters, such as data imbalance and the requirement for processing resources. For example, cloud computing technology may be utilised to do sentiment analysis using computationally expensive approaches, as well as for a variety of other tasks. By utilizing resources that are mutually and dynamically scalable and that can be delivered and released with minimum participation from the service provider, on-demand network access to computing services (hardware and software) can be offered.[49] It was developed by Al-Kabi et al.[71] to be compatible with a health monitoring system that recognises emotions. On a wearable computer, physiological data is collected and tracked. The gathered data is analysed and assessed using machine learning algorithms in order to foresee participants' physiological or psychological circumstances.

4.6 CONCLUSION AND FUTURE WORK

According to this chapter, sentiment analysis issues have a significant impact on sentiment appraisal. Using data from 47 studies, two comparisons were made. The key contrast is between the structure of sentiment reviews and the associated difficulties. The result demonstrates domain dependence, which is a key factor to consider while attempting to comprehend sentiment issues. The main distinction is the implied or clear relevance; the negation challenge has become more and more common in all kinds of evaluations. This comparative result makes it simpler to see how each sentiment issue affects the various review structure kinds. We discovered that the issue kind and review format define the suitable difficulties for the evaluation sentiment reviews. The other comparison is based on sentiment analysis difficulties related to the accuracy rate. These findings provide light on the importance of sentiment challenges in analysing feelings, as well as how to select the ideal test to maximise the level of accuracy. We examined the relationship between the proportion of sentiment management techniques applied to sentiment problems in conceptual and practical categories. Another discovery explains why research on conceptual type sentiment difficulties is a prominent area of focus. The average sentiment challenge accuracy rate declines as more research are done. The average of exactness rates decreases as the number of tests in an emotion challenge increases. The long-term effort will consist of continually expanding the comparison circle with fresh findings.

REFERENCES

1. Pang, B., & Lee, L. (2008). Opinion mining and sentiment analysis. *Foundations and Trends® in Information Retrieval*, 2(1–2), 1–135.
2. Dobrescu, A. B. (2011). Methods and resources for sentiment analysis in multilingual documents of different text types. Doctoral dissertation, Universitat d'Alacant-Universidad de Alicante.
3. Niles, I., & Pease, A. (2003, June). Linking lixicons and ontologies: Mapping wordnet to the suggested upper merged ontology. In *Proceedings of the 2003 International Conference on Information and Knowledge Engineering (IKE 03)*, Las Vegas, 2003, pp. 23–26.
4. Strapparava, C., & Valitutti, A. (2004, May). Wordnet affect: An affective extension of wordnet. *Lrec*, 4(1083–1086), 40.
5. Esuli, A., & Sebastiani, F. (2006, May). Sentiwordnet: A publicly available lexical resource for opinion mining. In *Proceedings of the fifth international conference on language resources and evaluation (LREC'06)*, (pp. 417–422). European Language Resources Association (ELRA) Genoa, Italy.
6. Baccianella, S., Esuli, A., & Sebastiani, F. (2010). Sentiwordnet 3.0: an enhanced lexical resource for sentiment analysis and opinion mining. In *Lrec* Vol. 10, No. 2010, pp. 2200–2204.
7. Yi, J., Nasukawa, T., Bunescu, R., & Niblack, W. (2003, November). Sentiment analyzer: Extracting sentiments about a given topic using natural language processing techniques. In *Third IEEE international conference on data mining*. pp. 427–434. IEEE.
8. Li, S. K., Guan, Z., Tang, L. Y., & Chen, Z. (2012). Exploiting consumer reviews for product feature ranking. *Journal of Computer Science and Technology*, 27(3), 635–649.
9. Kamps, J., & Marx, M. (2002, June). Notions of indistinguishability for semantic web languages. In *International semantic web conference*. Berlin, Heidelberg: Springer, pp. 30–38.
10. Crammer, K., & Singer, Y. (2003, January). Ultraconservative online algorithms for multiclass problems. *Journal of Machine Learning Research*, 3, 951–991.
11. Rabelo, J. C., Prudêncio, R. B., & Barros, F. A. (2012, October). Using link structure to infer opinions in social networks. In *2012 IEEE international conference on systems, man, and cybernetics (SMC)*. pp. 681–685. IEEE.
12. Wilson T., Wiebe J., Hoffmann P. Recognizing contextual polarity in phrase-level sentiment analysis *Proceedings of human language technology conference and conference on empirical methods in natural language processing*. - HLT '05, Association for Computational Linguistics, Morristown, NJ, USA (2005), pp. 347–354.

13. Moreo, A., Romero, M., Castro, J. L., & Zurita, J. M. (2012). Lexicon-based comments-oriented news sentiment analyzer system. *Expert Systems with Applications*, 39(10), 9166–9180.
14. Feldman, R. (2013). Techniques and applications for sentiment analysis. *Communications of the ACM*, 56(4), 82–89.
15. Cambria, E., Schuller, B., Xia, Y., & Havasi, C. (2013). New avenues in opinion mining and sentiment analysis. *IEEE Intelligent Systems*, 28(2), 15–21.
16. Tang, H., Tan, S., & Cheng, X. (2009). A survey on sentiment detection of reviews. *Expert Systems with Applications*, 36(7), 10760–10773.
17. Chen, H., & Zimbra, D. (2010). AI and opinion mining. *IEEE Intelligent Systems*, 25(3), 74–80.
18. Bollen, J., Mao, H., & Zeng, X. (2011). Twitter mood predicts the stock market. *Journal of Computational Science*, 2(1), 1–8.
19. Kumar, S., Morstatter, F., & Liu, H. (2014). *Twitter data analytics*. New York: Springer, pp. 1041–4347.
20. McDonald, R., Crammer, K., & Pereira, F. (2005, June). Online large-margin training of dependency parsers. In *Proceedings of the 43rd annual meeting of the association for computational linguistics (ACL'05)*, Ann Arbor, Michigan, pp. 91–98.
21. Tsai, A. C. R., Wu, C. E., Tsai, R. T. H., & Hsu, J. Y. J. (2013). Building a concept-level sentiment dictionary based on commonsense knowledge. *IEEE Intelligent Systems*, 28(2), 22–30.
22. Bosco, C., Patti, V., & Bolioli, A. (2013). Developing corpora for sentiment analysis: The case of irony and senti-tut. *IEEE Intelligent Systems*, 28(2), 55–63.
23. Poria, S., Gelbukh, A., Hussain, A., Howard, N., Das, D., & Bandyopadhyay, S. (2013). Enhanced SenticNet with affective labels for concept-based opinion mining. *IEEE Intelligent Systems*, 28(2), 31–38.
24. Maks, I., & Vossen, P. (2012). A lexicon model for deep sentiment analysis and opinion mining applications. *Decision Support Systems*, 53(4), 680–688.
25. Miao, Q., Li, Q., & Dai, R. (2009). AMAZING: A sentiment mining and retrieval system. *Expert Systems with Applications*, 36(3), 7192–7198.
26. Saleh, M. R., Martín-Valdivia, M. T., Montejo-Ráez, A., & Ureña-López, L. A. (2011). Experiments with SVM to classify opinions in different domains. *Expert Systems with Applications*, 38(12), 14799–14804.
27. Sobkowicz, P., Kaschesky, M., & Bouchard, G. (2012). Opinion mining in social media: Modeling, simulating, and forecasting political opinions in the web. *Government Information Quarterly*, 29(4), 470–479.
28. Pennebaker, J. W., Francis, M. E., & Booth, R. J. (2001). Linguistic inquiry and word count: LIWC 2001. *Mahway: Lawrence Erlbaum Associates*, 71(2001), 2001.
29. Narayanan, R., Liu, B., & Choudhary, A. (2009, August). Sentiment analysis of conditional sentences. In *Proceedings of the 2009 conference on empirical methods in natural language processing*, Singapore, pp. 180–189.
30. Balahur, A., Hermida, J. M., & Montoyo, A. (2011). Building and exploiting emotinet, a knowledge base for emotion detection based on the appraisal theory model. *IEEE Transactions on Affective Computing*, 3(1), 88–101.
31. Hatzivassiloglou, V., & Wiebe, J. (2000). Effects of adjective orientation and gradability on sentence subjectivity. In *COLING 2000 volume 1: The 18th international conference on computational linguistics,* Germany.
32. Turney, P. D. (2002). Thumbs up or thumbs down? Semantic orientation applied to unsupervised classification of reviews. arXiv preprint cs/0212032.
33. Pang, B., Lee, L., & Vaithyanathan, S. (2002). Thumbs up? Sentiment classification using machine learning techniques. arXiv preprint cs/0205070.

34. Wilson, T., Hoffmann, P., Somasundaran, S., Kessler, J., Wiebe, J., Choi, Y., ... Patwardhan, S. (2005, October). OpinionFinder: A system for subjectivity analysis. In *Proceedings of HLT/EMNLP 2005 interactive demonstrations*, Canada, pp. 34–35.
35. Penalver-Martinez, I., Garcia-Sanchez, F., Valencia-Garcia, R., Rodriguez-Garcia, M. A., Moreno, V., Fraga, A., & Sanchez-Cervantes, J. L. (2014). Feature-based opinion mining through ontologies. *Expert Systems with Applications*, 41(13), 5995–6008.
36. Cho, H., Kim, S., Lee, J., & Lee, J. S. (2014). Data-driven integration of multiple sentiment dictionaries for lexicon-based sentiment classification of product reviews. *Knowledge-Based Systems*, 71, 61–71.
37. Chen, L., Qi, L., & Wang, F. (2012). Comparison of feature-level learning methods for mining online consumer reviews. *Expert Systems with Applications*, 39(10), 9588–9601.
38. Hu, M., & Liu, B. (2004, August). Mining and summarizing customer reviews. In *Proceedings of the tenth ACM SIGKDD international conference on Knowledge discovery and data mining*, Washington, U.S.A, pp. 168–177.
39. Maas, A., Daly, R. E., Pham, P. T., Huang, D., Ng, A. Y., & Potts, C. (2011, June). Learning word vectors for sentiment analysis. In *Proceedings of the 49th annual meeting of the association for computational linguistics: Human language technologies*, United States, pp. 142–150.
40. Taddy, M. (2013). Measuring political sentiment on Twitter: Factor optimal design for multinomial inverse regression. *Technometrics*, 55(4), 415–425.
41. Li, S. T., & Tsai, F. C. (2013). A fuzzy conceptualization model for text mining with application in opinion polarity classification. *Knowledge-Based Systems*, 39, 23–33.
42. Figueiredo, F., Rocha, L., Couto, T., Salles, T., Gonçalves, M. A., & Meira Jr, W. (2011). Word co-occurrence features for text classification. *Information Systems*, 36(5), 843–858.
43. Li, Y. M., & Shiu, Y. L. (2012). A diffusion mechanism for social advertising over microblogs. *Decision Support Systems*, 54(1), 9–22.
44. Wang, S., Li, D., Song, X., Wei, Y., & Li, H. (2011). A feature selection method based on improved fisher's discriminant ratio for text sentiment classification. *Expert Systems with Applications*, 38(7), 8696–8702.
45. Dang, Y., Zhang, Y., & Chen, H. (2009). A lexicon-enhanced method for sentiment classification: An experiment on online product reviews. *IEEE Intelligent Systems*, 25(4), 46–53.
46. Balahur, A., & Steinberger, R. (2009). Rethinking sentiment analysis in the news: From theory to practice and back. *Proceeding of WOMSA*, 9, 1–12.
47. Tang, D., Qin, B., Liu, T., & Shi, Q. (2014). Emotion analysis platform on chinese microblog. arXiv preprint arXiv:1403.7335.
48. Mukherjee, S., & Bhattacharyya, P. (2012, March). Feature specific sentiment analysis for product reviews. In *International conference on intelligent text processing and computational linguistics*. Berlin, Heidelberg: Springer, pp. 475–487.
49. Tang, J., Wang, X., Gao, H., Hu, X., & Liu, H. (2012). Enriching short text representation in microblog for clustering. *Frontiers of Computer Science*, 6(1), 88–101.
50. Kaushik, C., & Mishra, A. (2014). A scalable, lexicon based technique for sentiment analysis. arXiv preprint arXiv:1410.2265.
51. Husaini, M. A., Koçyiğit, A., Tapucu, D., Yanikoglu, B., & Saygın, Y. (2012, September). An aspect-lexicon creation and evaluation tool for sentiment analysis researchers. In *Joint European conference on machine learning and knowledge discovery in databases*. Berlin, Heidelberg: Springer, pp. 804–807.
52. Mudinas, A., Zhang, D., & Levene, M. (2012, August). Combining lexicon and learning based approaches for concept-level sentiment analysis. In *Proceedings of the first international workshop on issues of sentiment discovery and opinion mining*, United States, pp. 1–8.

53. Gezici, G., Yanıkoğlu, B., Tapucu, D., & Saygın, Y. (2012). New features for sentiment analysis: Do sentences matter? In *1st international workshop on sentiment discovery from affective data, SDAD 2012*. CEUR Workshop Proceedings, UK.
54. Li, F., Huang, M., & Zhu, X. (2010, July). Sentiment analysis with global topics and local dependency. In *Twenty-Fourth AAAI conference on artificial intelligence,* USA, Vol. 24, No. 1, pp. 1371–1376.
55. Liebrecht, C. C., Kunneman, F. A., & van Den Bosch, A. P. J. (2013). *The perfect solution for detecting sarcasm in tweets# not,* Atlanta, Georgia. https://repository.ubn.ru.nl/bitstream/handle/2066/112949/112949.pdf.
56. Abbasi, A., Zhang, Z., Zimbra, D., Chen, H., & Nunamaker Jr, J. F. (2010). Detecting fake websites: The contribution of statistical learning theory. *Mis Quarterly*, 435–461.
57. He, Y., Lin, C., & Alani, H. (2011, June). Automatically extracting polarity-bearing topics for cross-domain sentiment classification. In *49th annual meeting of the association for computational linguistics: Human language technologies*. Portland, Association for Computational Linguistics OR, USA.
58. Wiegand, M., Balahur, A., Roth, B., Klakow, D., & Montoyo, A. (2010, July). A survey on the role of negation in sentiment analysis. In *Proceedings of the workshop on negation and speculation in natural language processing*, Uppsala, Sweden, pp. 60–68.
59. Liebrecht, C. C., Kunneman, F. A., & van Den Bosch, A. P. J. (2013). *The perfect solution for detecting sarcasm in tweets# not*. https://repository.ubn.ru.nl/bitstream/handle/2066/112949/112949.pdf
60. Balahur, A., & Steinberger, R. (2009). Rethinking sentiment analysis in the news: From theory to practice and back. *Proceeding of WOMSA*, 9, 1–12.
61. Kiritchenko, S., Zhu, X., & Mohammad, S. M. (2014). Sentiment analysis of short informal texts. *Journal of Artificial Intelligence Research*, 50, 723–762.
62. Yang, Y., Shen, H. T., Ma, Z., Huang, Z., & Zhou, X. (2011, June). L2, 1-norm regularized discriminative feature selection for unsupervised. In *Twenty-second international joint conference on artificial intelligence,* AAAI Press, Barcelona Catalonia Spain.
63. Ott, M., Choi, Y., Cardie, C., & Hancock, J. T. (2011). Finding deceptive opinion spam by any stretch of the imagination. arXiv preprint arXiv:1107.4557.
64. Jia, L., Yu, C., & Meng, W. (2009, November). The effect of negation on sentiment analysis and retrieval effectiveness. In *Proceedings of the 18th ACM conference on information and knowledge management*, Hong Kong, China, pp. 1827–1830.
65. Dadvar, M., Hauff, C., & de Jong, F. (2011, February). Scope of negation detection in sentiment analysis. In *Proceedings of the Dutch-Belgian information retrieval workshop (DIR 2011)*. pp. 16–20. University of Amsterdam.
66. Chunping, O., Wen, Z., Ying, Y., Zhiming, L., & Xiaohua, Y. (2014). Topic sentiment analysis in Chinese news. *International Journal of Multimedia and Ubiquitous Engineering*, 9(11), 385–396.
67. Hu, X., Tang, J., Gao, H., & Liu, H. (2014, December). Social spammer detection with sentiment information. In *2014 IEEE international conference on data mining*. Shenzhen, China IEEE, pp. 180–189.
68. Xiang, B., & Zhou, L. (2014, June). Improving twitter sentiment analysis with topic-based mixture modeling and semi-supervised training. In *Proceedings of the 52nd annual meeting of the association for computational linguistics (volume 2: Short papers)*, Baltimore, Maryland, pp. 434–439.
69. Cambria, E. (2013, November). An introduction to concept-level sentiment analysis. In *Mexican international conference on artificial intelligence*. Berlin, Heidelberg: Springer, pp. 478–483.
70. Li, F. H., Huang, M., Yang, Y., & Zhu, X. (2011, June). Learning to identify review spam. In *Twenty-second international joint conference on artificial intelligence*, Association for Computational Linguistics, Barcelona Catalonia Spain.

71. Al-Kabi, M. N., Gigieh, A. H., Alsmadi, I. M., Wahsheh, H. A., & Haidar, M. M. (2014). Opinion mining and analysis for Arabic language. *International Journal of Advanced Computer Science and Applications*, 5(5), 181–195.
72. Guzman, E., & Maalej, W. (2014, August). How do users like this feature? A fine grained sentiment analysis of app reviews. In *2014 IEEE 22nd international requirements engineering conference (RE)*. Karlskrona, Sweden, IEEE, pp. 153–162.
73. Thelen, M., & Riloff, E. (2002, July). A bootstrapping method for learning semantic lexicons using extraction pattern contexts. In *Proceedings of the 2002 conference on empirical methods in natural language processing (EMNLP 2002)*, EMNLP, pp. 214–221.
74. Montoyo, A., Martínez-Barco, P., & Balahur, A. (2012). Subjectivity and sentiment analysis: An overview of the current state of the area and envisaged developments. *Decision Support Systems*, 53(4), 675–679.
75. Khan, F. H., Bashir, S., & Qamar, U. (2014). TOM: Twitter opinion mining framework using hybrid classification scheme. *Decision Support Systems*, 57, 245–257.
76. Wiebe, J. (2000). Learning subjective adjectives from corpora. *Aaai/iaai*, 20.
77. Seki, Y., Kando, N., & Aono, M. (2009). Multilingual opinion holder identification using author and authority viewpoints. *Information Processing & Management*, 45(2), 189–199.
78. O'Leary, D. E. (2011). Blog mining-review and extensions: "From each according to his opinion". *Decision Support Systems*, 51(4), 821–830.
79. Li, Y. M., & Li, T. Y. (2013). Deriving market intelligence from microblogs. *Decision Support Systems*, 55(1), 206–217.
80. Li, N., & Wu, D. D. (2010). Using text mining and sentiment analysis for online forums hotspot detection and forecast. *Decision Support Systems*, 48(2), 354–368.
81. Duric, A., & Song, F. (2012). Feature selection for sentiment analysis based on content and syntax models. *Decision Support Systems*, 53(4), 704–711.
82. Xu, K., Liao, S. S., Li, J., & Song, Y. (2011). Mining comparative opinions from customer reviews for competitive intelligence. *Decision Support Systems*, 50(4), 743–754.
83. Xu, K., Liao, S. S., Li, J., & Song, Y. (2011). Mining comparative opinions from customer reviews for competitive intelligence. *Decision Support Systems*, 50(4), 743–754.
84. Balahur, A., Hermida, J. M., & Montoyo, A. (2012). Detecting implicit expressions of emotion in text: A comparative analysis. *Decision Support Systems*, 53(4), 742–753.
85. Kontopoulos, E., Berberidis, C., Dergiades, T., & Bassiliades, N. (2013). Ontology-based sentiment analysis of twitter posts. *Expert Systems with Applications*, 40(10), 4065–4074.
86. Song, L., Xin, C., Lai, S., Wang, A., Su, J., & Xu, K. (2022). CASA: Conversational aspect sentiment analysis for dialogue understanding. *Journal of Artificial Intelligence Research*, 73, 511–533.
87. Abbasi, A. (2003). Intelligent feature selection for opinion classification. *Technology*, 54(14), 1269–1277.
88. Neviarouskaya, A., Prendinger, H., & Ishizuka, M. (2011). SentiFul: A lexicon for sentiment analysis. *IEEE Transactions on Affective Computing*, 2(1), 22–36.
89. Zhang, L., Ghosh, R., Dekhil, M., Hsu, M., & Liu, B. (2011). Combining lexicon-based and learning-based methods for Twitter sentiment analysis. *HP Laboratories, Technical Report HPL-2011*, 89, 1–8.
90. Lapponi, E., Read, J., & Øvrelid, L. (2012, December). Representing and resolving negation for sentiment analysis. In *2012 IEEE 12th international conference on data mining workshops*. Lisboa, Portugal, IEEE, pp. 687–692.
91. Cambria, E., & Hussain, A. (2012). Sentic computing. *Marketing*, 59(2), 557–577.
92. Flekova, L., Preoţiuc-Pietro, D., & Ruppert, E. (2015, September). Analysing domain suitability of a sentiment lexicon by identifying distributionally bipolar words. In *Proceedings of the 6th workshop on computational approaches to subjectivity, sentiment and social media analysis*, Lisboa, Portugal, pp. 77–84.

93. Peng, Q., & Zhong, M. (2014). Detecting spam review through sentiment analysis. *Journal of Software*, 9(8), 2065–2072.
94. Mohammad, S., & Turney, P. (2010, June). Emotions evoked by common words and phrases: Using mechanical Turk to create an emotion lexicon. In *Proceedings of the NAACL HLT 2010 workshop on computational approaches to analysis and generation of emotion in text*, Los Angeles, CA, pp. 26–34.
95. Weichselbraun, A., Gindl, S., & Scharl, A. (2013). Extracting and grounding contextualized sentiment lexicons. *IEEE Intelligent Systems*, 28(2), 39–46.
96. Bollegala, D., Mu, T., & Goulermas, J. Y. (2015). Cross-domain sentiment classification using sentiment sensitive embeddings. *IEEE Transactions on Knowledge and Data Engineering*, 28(2), 398–410.
97. Liu, H., He, J., Wang, T., Song, W., & Du, X. (2013). Combining user preferences and user opinions for accurate recommendation. *Electronic Commerce Research and Applications*, 12(1), 14–23.
98. Spina, D., Gonzalo, J., & Amigó, E. (2013). Discovering filter keywords for company name disambiguation in twitter. *Expert Systems with Applications*, 40(12), 4986–5003.
99. Verma, S. (2022). Sentiment analysis of public services for smart society: Literature review and future research directions. *Government Information Quarterly*, 101708.
100. D'Aniello, G., Gaeta, M., & La Rocca, I. (2022). KnowMIS-ABSA: An overview and a reference model for applications of sentiment analysis and aspect-based sentiment analysis. *Artificial Intelligence Review*, 1–32.
101. Wankhade, M., Rao, A. C. S., & Kulkarni, C. (2022). A survey on sentiment analysis methods, applications, and challenges. *Artificial Intelligence Review*, 1–50.
102. Martín-Valdivia, M. T., Martínez-Cámara, E., Perea-Ortega, J. M., & Ureña-López, L. A. (2013). Sentiment polarity detection in Spanish reviews combining supervised and unsupervised approaches. *Expert Systems with Applications*, 40(10), 3934–3942.
103. Hung, C., & Lin, H. K. (2013). Using objective words in SentiWordNet to improve word-of-mouth sentiment classification. *IEEE Intelligent Systems*, 28(02), 47–54.
104. Feng, Z., Zhou, H., Zhu, Z., & Mao, K. (2022). Tailored text augmentation for sentiment analysis. *Expert Systems with Applications*, 117605.
105. Zhu, T., Li, L., Yang, J., Zhao, S., Liu, H., & Qian, J. (2022). Multimodal sentiment analysis with image-text interaction network. *IEEE Transactions on Multimedia*, 1–11.
106. Liang, B., Su, H., Gui, L., Cambria, E., & Xu, R. (2022). Aspect-based sentiment analysis via affective knowledge enhanced graph convolutional networks. *Knowledge-Based Systems*, 235, 107643.
107. Li, H., Chen, Q., Zhong, Z., Gong, R., & Han, G. (2022). E-word of mouth sentiment analysis for user behavior studies. *Information Processing & Management*, 59(1), 102784.
108. Wang, W., Guo, L., & Wu, Y. J. (2022). The merits of a sentiment analysis of antecedent comments for the prediction of online fundraising outcomes. *Technological Forecasting and Social Change*, 174, 121070.
109. Wu, S., Liu, Y., Zou, Z., & Weng, T. H. (2022). S_I_LSTM: stock price prediction based on multiple data sources and sentiment analysis. *Connection Science*, 34(1), 44–62.
110. Jain, P. K., Quamer, W., Saravanan, V., & Pamula, R. (2022). Employing BERT-DCNN with Sentic knowledge base for social media sentiment analysis. *Journal of Ambient Intelligence and Humanized Computing*, 1–13.
111. Jing, N., Wu, Z., & Wang, H. (2021). A hybrid model integrating deep learning with investor sentiment analysis for stock price prediction. *Expert Systems with Applications*, 178, 115019.
112. Prabu, P., Sivakumar, R., & Ramamurthy, B. (2021). Corpus based sentimenal movie review analysis using auto encoder convolutional neural network. *Journal of Discrete Mathematical Sciences and Cryptography*, 24(8), 2323–2339.

113. Khanam, R., & Sharma, A. (2021, December). Sentiment analysis using different machine learning techniques for product review. In *2021 international conference on computational performance evaluation (ComPE)*. IEEE Explore, USA, pp. 646–650. doi:10.1109/ComPE53109.2021.9752004
114. Priyanka, K., Janakiraman, S., & Priya, M. D. (2021). Aspect level sentimental analysis of opinion mining–a review. *Materials Today: Proceedings*, 56–71.
115. Ali, M. M. (2021). Arabic sentiment analysis about online learning to mitigate covid-19. *Journal of Intelligent Systems*, 30(1), 524–540.
116. Liu, S., & Lee, I. (2021). Sequence encoding incorporated CNN model for email document sentiment classification. *Applied Soft Computing*, 102, 107104.
117. Angel, S. O., Negrón, A. P. P., & Espinoza-Valdez, A. (2021). Systematic literature review of sentiment analysis in the Spanish language. *Data Technologies and Applications*, 55(4), 461–479.
118. Sazzed, S. (2021, September). A hybrid approach of opinion mining and comparative linguistic analysis of restaurant reviews. In *Proceedings of the International Conference on Recent Advances in Natural Language Processing* (RANLP 2021). Ruslan Mitkov, Galia Angelova (Eds.), INCOMA Ltd., pp. 1281–1288.
119. Akkaya, C., & Zhang, X. (2001). Sentiment classification in a nutshell. In *Proceedings of ACL-05, 43rd meeting of the association for computational linguistics*. Ann Arbor, US: Association for Computational Linguistics.
120. Boshoff, C., & Van Eeden, S. M. (2001). South African consumer sentiment towards marketing: A longitudinal analysis. *South African Journal of Business Management*, 32(2), 23–33.

5 Forecasting on Covid-19 Data Using ARIMAX Model

Noman Islam, Enayat Raza, Sheraz Mohsin, Ahsar Ansari, Razeen Shuja and Darakhshan Syed

CONTENTS

5.1 INTRODUCTION

The recent global outbreak of the coronavirus, also known as Covid-19, is probably the highlight of 2020. Covid-19 had a huge impact on humans, and it was declared a global pandemic by the World Health Organization (WHO) due to its fast outbreak across the world.[1, 2] Its first case appeared in December 2019 in Wuhan, China;[3] Covid-19 has now spread to all countries and has cost millions of lives across the globe. According to the WHO, close engagement with an infected individual by handshake, access to droplets released during sneezing or coughing, and travelling to an afflicted area and acquiring the virus in one way or another are the main person-to-person transmission mechanisms.[4] Covid-19 symptoms are extremely variable and range from being severely debilitated to being asymptomatic; those who are infected can develop minor to extremely severe respiratory infections. In the majority of symptomatic patients, high fever, cough, sore throat, and muscle aches

DOI: 10.1201/9781003310785-5

are the main symptoms.[5–15] Pneumonia, micro-coagulopathies, and septic shock appear in severe instances. Death may result from an abrupt medical deterioration.[16] Infectious Covid-19 symptoms and effects are more common in older adults as well as those with pre-existing medical disorders, such as insulin resistance, chronic pulmonary problems, or malignancies.[17]

SARS and MERS, two prior coronavirus outbreaks, were not as infectious or tenacious as Covid-19. The earliest stages of the outbreak were filled with uncertainty and a lack of openness, which made things worse. Today, 185 countries are affected by the epidemic with no cure in reach. The virus that is currently causing death due to respiratory illness is very contagious. The propagation of the virus has been unevenly distributed among nations as a result of variations in epidemiological circumstances and diagnostic infrastructure. Basically the countries are globally impacted by covid-19. But advanced countries like Spain, Italy, France, Germany, and the United States (US) are the severely impacted countries. The US is at the top of thc list, as compared to the other underdeveloped countries like Brazil, Russia, India, and Spain, in that order[18] Several vaccines include Moderna, Pfizer, Sputnik, CanSino, Sinovac, Sinopharm, AstraZeneca, etc. are introduced for Covid-19. But the reason that makes Covid-19 so lethal is that none of them is proven, i.e., not tested on humans prior.[19]

People with weak immunity are affected the most. Some of the symptoms of Covid-19 are fever, cough, tiredness, sore throat, headache, skin rashes, and bowel symptoms. There are neurological and cardiac impacts also found in patients affected by Covid-19. The symptoms usually develop after a few days.[20, 21] However, in a large number of patients, the symptoms don't appear at all. Another threatening factor is that people of an older age usually suffer more severely from Covid-19. People with a travel history from affected countries are generally associated with greater risk. The transmission of Covid-19 is usually from droplets from breath. Face masks can be used to avoid the spread of the virus. In addition, proper hand washing should be used to prevent the disease.

Several testing methods have been developed to test the disease. This includes polymerase chain reaction (PCR) tests. In addition, machine learning techniques have been discussed in literature to detect the Covid-19 through chest X-rays. The prognosis of Covid-19 is great and most people generally recover. A small number of people (3–4%) require hospitalization. Another challenge associated with Covid-19 is the variants that emerge due to slight genetic changes.[22]

Considering the governments' assurances that there are more testing and healthcare centers, the percentage of impacted cases is neither flattening nor declining.[23, 24] Twenty thousand new cases are being admitted daily, and the Covid-19 outbreak is causing a lot of worry. Who will contract the illness tomorrow? How many people will pass away tomorrow? When will the infection curve infuse or become flat? How many individuals will be impacted during the outbreak's height? Are there any mathematical models that can provide the answers to these queries? Given the situation, it is crucial to assess Covid-19's geographic distribution so that decision-makers in government, medicine, and the general public can be better equipped to handle the epidemic.[25] Considering the challenging nature of the disease, this chapter gathered publicly available Covid-19 data set of a developing country from different useful resources [26, 27] that have the latest records for Covid-19 for every country in the world. According to the information, Covid-19 affected more than 200 countries with 114,087,847 confirmed cases out of which 89,633,588 recovered; 2,531,297 deaths were confirmed by 27 February 2021. The United States, India, and Brazil were most affected by the Covid-19 with more than 50 million cases reported.

The objective of this chapter is to analyze the death rate due to the Covid-19 virus in a developing country. Section 5.2 presents the literature review. Section 5.3 will present the potential issues faced by the forecasting models. Section 5.4 presents the methodology. This is followed in Section 5.5 with implementation details and results. Section 5.6 concludes with a discussion on future work.

5.2 RELATED WORKS

In the last two decades, investigation has been concentrated on statistical challenges related to a potential diagnosis of outbreaks of contagious diseases. The difficulties come with early identification and potential epidemic progression for implementing the proper preventive actions. Biosurveillance is the term for this field's explosive expansion.[28, 29]

The main purpose of artificial intelligence and machine learning based frameworks is to increase the surveillance and diagnostic accuracy of non-infectious disorders. Additionally, machine learning techniques are frequently employed to analyze and forecast the Covid-19 chance of survival and the patient's release time based on medical/clinical data. Lai et al. (2020) took into account the daily aggregate lists, rate of death, and collaborative condition of the nations' economies and medical systems when analyzing the Covid-19 scourge hypothesis.[17] Using information from the Johns Hopkins dashboard, Punn et al. (2020) suggested using machine learning and deep learning frameworks to comprehend the Covid-19 epidemic.[30] To determine when the Covid-19 virus would stop spreading, Dandekar and Barbastathis (2020) suggested using a combination framework that combines first-standard statistical circumstances of the epidemic and an information-driven brain organization.[31] For four locations—Wuhan City, China; Italy; South Korea; and the United States—they employed a neural network architecture to make predictions. Finally, they forecasted the growth pattern of the existing illness in the United States and anticipated that infection would stop by 20 April 2020. According to the WHO guidelines for preparing for Covid-19 outbreaks, it can penetrate the body through the mouth, nose, or eyes. For these reasons, the guidlines included a few recommended precautions for all situations to prevent contracting the infection, such as avoiding direct contact with the face while washing hands and washing hands thoroughly with gels, wipes, or soaps for at least 20 seconds. Additionally, it was recommended to keep a physical distance of at least 1.5 meters or work remotely to lessen the risk of infection.[32] Previous research created techniques for producing timely and precise forecasts of Covid-19 infections. These investigations, however, lacked several promising characteristics, and that was mainly responsible for their poor predictive performance and for their inability to forecast reported cases of Covid-19 at the maximum potential efficiency.

Machine learning in medical sciences emerged dramatically in the last 50 years. It is defined as the process of training a model based on the data previously available. Two popular approaches to machine learning are supervised learning and unsupervised learning. The supervised learning is based on having the features of the data along with the ground truth available. Unsupervised learning is based on using only the features to train the model. Various approaches to supervised learning are support vector machines, logistic regression, decision trees, etc.[33] Unsupervised learning techniques are clustering, principal component analysis and association rules mining, etc.[34, 35] Many statistical models can forecast fundamental understanding and insights by observing data trends. The two domains that benefited the most in the field of medicine from machine learning applications are diagnosing of disease and predictions about the outcome. Predictive models are mainly used in predicting to understand correlations between patient's disease and drug dosage.[24, 36, 37] This work lies in the second domain which uses time-series forecasting, one of the many techniques of prediction.

Predictive methods, especially time-series forecasting, have been commonly used in the recent research related to Covid-19.[13] Tominaga (2020) used a logarithmic quadratic regression model for daily data count of Covid-19 cases in Tokyo and for other countries.[38] Yang et al. in March 2020 unified China's most updated Covid-19 epidemiological data and used the susceptible-exposed-infectious-removed (SEIR) model to predict the trend and derive the curves.[39] The researchers proposed an efficient time-series mutation prediction model (Tempel) for the mutation prediction of influenza A viruses.[39] Furthermore, Pandey et al. (2020) analyzed the coronavirus outbreak in India on the data accumulated through John Hopkins University from 30 January 2020 to 30 March 2020 and predict the number of cases for the next 2 weeks using the SEIR model and the regression model.[40]

Similarly, Al-Jameel et al. (2021) published research that used three classification algorithms—logistic regression (LR), random forest (RF), and extreme gradient boosting (XGB)—on 287 Covid-19 patients samples and predicted the survival rate of Covid-19 patients where the virus was identified in early stages and patients quarantined at home and took precautions.[41] Ahmad and Asad (2020) collected data of 137 days from 25 February to 10 July 2020 of coronavirus patients in Pakistan and implemented an artificial neural network with rectifying linear unit-based technique to predict about the number of cases for next 7 days after training their model on whole collected data.[42]

In addition, Khan et al. (2020) published their study in which they operated vector autoregressive time series models to Pakistan Covid-19 data and forecasted new daily confirmed cases, deaths, and recovery cases.[43] Satu et al. (2021) designed a cloud-based machine learning model by using several regression-based models for short term forecasting and predicted the daily number of coronavirus patients in Bangladesh. They trained their model for the sample data of 25 days and predicted the cases for the next 7 days.[44]

Moreover, Goshvarpour and Goshvarpour (2020) utilized an autoregressive (AR) model and the autoregressive moving average model on Iran's Covid-19 daily death data provided by Iran's Ministry of Health and Medical Education from March to July 2020 to predict future mortality rate of Covid-19 in Iran.[45] Khakaria et al. (2021) made use of nine different machine learning algorithm to predict the coronavirus disease cases for the 10 most highest densely populated countries and forecast the count of daily cases. Their set of models predicted new cases with high accuracy, and among them, Auto-Regressive Moving Average (ARMA) gives highest accuracy of 99.93% for the population of Ethiopia.[46]

Like different models, the autoregressive integrated moving average (ARIMA) and autoregressive integrated moving average with exogenous variables (ARIMAX) models are widely used by the research community for predictions about the Covid-19 effects. Shmueli and Burkom designed the time series ARIMA models for daily new cases and new deaths trained on the data of Hubie, China, and used these models in prediction of new cases and deaths in Italy because both countries had the same conditions of population and lockdown restrictions measures.[28] Researchers acquired the data of Covid-19 daily confirmed cases and death cases in Nigeria from 21 March to 5 May 2020 by the Nigerian Centre for Disease Control (NDCD), and they developed an ARIMA model using this data and forecasted the daily Covid-19 cases and deaths for 239 days (6th May 2020 to 31 December 2020).[47–49]

In similar research, Triacca and Triacca (2021) forecasted the spread of Covid-19 from May 19 to June 2 2020 and predicted the number of new daily confirmed cases. For their study, the authors combined a log-polynomial model and first-order integer-valued autoregressive model and then used the ARIMA model to compare the results of the two proposed methodologies.[50, 51] In February 2021, Yang et al., during the third wave of Covid-19 in South Korea, published their research in which they predicted the cumulative number of Covid-19 cases using the ARIMA model and compared the results with the actual number of confirmed cases.[52] Aslam et al. (2021) published an article in which they trained an ARIMA model on the Covid-19 data of Pakistan, India, and Bangladesh, separately, and predicted Covid-19 cases for 14 days starting from 1 July 2020 and ending on 14 July 2020.[53]

Several miscellaneous researches have been carried out in literature [54] have performed sentiment analysis of general people about Covid-19. Effectiveness of forecasting is based upon the quality of data source used for forecasting. Forecasting results may vary based on the impurities in the data sources.[55] Data mining and big data techniques always play a vital role in healthcare systems.[56–59] A geographic pandemic framework for estimating the number of deaths has been presented by Zhu et al. (2020) [3]. By taking into account the current dynamics of COVID, this research aims to propose a forecasting model that will examine the disease' expansion throughout the course of the upcoming month. There were three distinct scenarios taken into account: residents, residents with a history of travel to Wuhan, and individuals exposed to a local outbreak. The degradation rate was also included in the study to recognize the efforts made by various localities to stop the virus's growth. The statistics of city-savvy citizens who had returned from Wuhan were gathered via phone data, and the city-based model was trained using the existing records and confirmed against the cases reported as of 11 February 2019. In compliance with the applicable three scenarios, the same framework was utilized to forecast cases through 12 March 2020. Sameni (2020) suggested a virus pattern using mathematical modeling.[60] The easily susceptible infected-recovered (SIR) framework, a member of the well-known compartmental model class, is the framework used in this investigation. According to a survey, the actions taken by nations are having a favorable impact on the mortality rate. Additionally, the buildings built to accommodate those who are ill have made a significant contribution to halting the disease's spread.

This statistical model was created for the dataset that was emphasized; therefore, it has shortcomings in terms of precision.

After the extensive literature review, it has been observed that research on employing the ARIMAX model for Covid-19 data is not extensive.[12] Hence, this chapter also presents the implication of the ARIMA model on the data of Covid-19 daily cases and deaths in a developing country.

5.3 POTENTIAL ISSUES WITH THE FORECASTING FRAMEWORKS

Owing to its advantages for resource recovery or economic growth, forecasting is crucial in every field. It does, however, have its difficulties. Because the Covid-19 infection period is so prolonged and there are so few datasets available for this objective, projecting the total death toll and dissemination rate also presents significant issues. Several of these difficulties with prediction approaches are stated as follows:[61, 62]

- **Tracking of individuals:** One of the most challenging duties is locating infectious employees and any persons who have touched each other.
- **Increased duration of incubation:** It is impossible to classify individuals in advance due to Covid-19's 14-day incubation period. Patients can contaminate everyone who comes into direct contact with them within the defined duration of incubation.
- **Insufficient relevant data:** Unorganized data can be found on occasion. Therefore, before data enters the training phase, it is first crucial to preserve both its type and effectiveness—i.e. quantity and quality—directly. Data accuracy is crucial for developing effective prediction techniques.
- **The data is overflowed:** It is likely that the algorithm in concern will not perform well on additional knowledge if overfitting of the data takes place.
- **Excessively pure data:** For analytical purposes, clean data is essential; yet sometimes data cleaning loses quality.
- **A lot of information:** There are a lot of data accessible, however providing the model for all of this information will not increase accuracy.
- **Inappropriate algorithm and attribute choice:** If the incorrect method is chosen, the outcome may be deceptive. The same is true if the incorrect characteristic is chosen.
- **Modeling complexity:** The accuracy of the algorithm as a whole may suffer from overly complicated models.

Other challenges that indirectly effects the modeling of forecasting framework includes:

- **Effective lockdown:** Implementing a lockdown is exceedingly difficult in any nation. Choosing the appropriate circumstances for a lockdown is a very difficult undertaking.[63]
- **The best time to go into lockdown:** This is a significant task that must be completed at the right time.

- **Inform but do not incite panic:** It's necessary to inform people, but it's also essential to keep this in mind when doing so.
- **Delivery of critical services:** Prior to a shutdown, any nation must quickly identify its important services. Lack of these amenities might spark a severe fear even during a lockdown.

5.4 PROPOSED METHODOLOGY

Figure 5.1 shows the methodology for the proposed work. It comprises various steps as discussed below. This includes a collection of data through webscrapping, preprocessing, training and validation, and final testing of the data.

5.4.1 Data Collection

The dataset was extracted from a renowned source called Worldometers.[5, 27] They offered country-specific, regularly updated statistical Covid-19 data. Out of all the countries, we picked our desired country. Fetching the data isn't possible directly from the website, i.e. data is not available in tabular format and is not directly downloadable. Instead, we had to perform scrapping on the particular webpage using Python in order to pull out the data behind it. In that way, we were able to get the data to model. While there are other web sources where country-specific Covid-19 data is available, it was limited until August 2021, and the method of extracting data from that source was very lengthy. Therefore, Worldometers was the optimum website with accurate data and with current date data conveniently available.

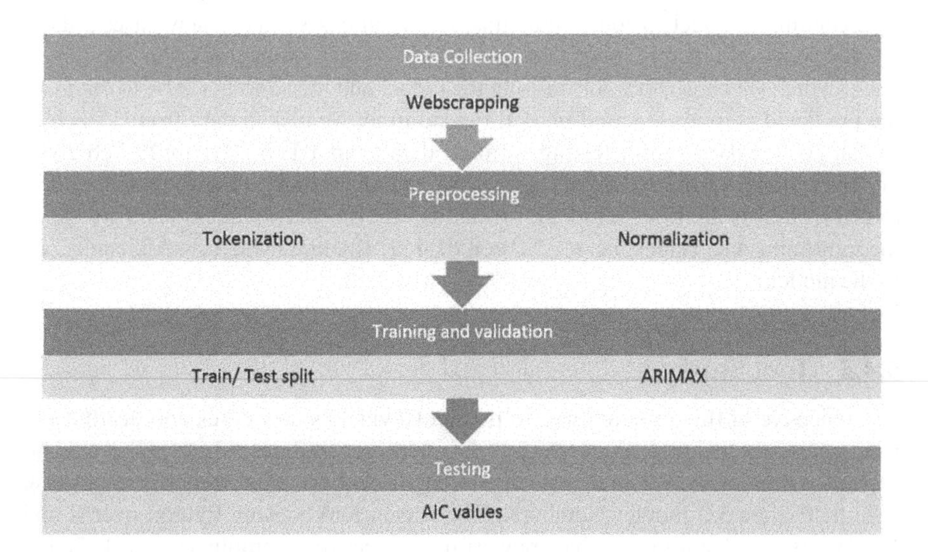

FIGURE 5.1 Methodology for the proposed work.

5.4.2 Preprocessing

After drawing out the Covid-19 data from our chosen source, we had to judge the nature of what the data said and where to store it in order to recall it when required to feed into the machine learning model. The nature of our dataset was time series data, i.e. *x* vs. *y* data. *x* is the date whereas *y* is the numbers of suspects reported. In the *y* we have 'new cases (daily)', 'deaths (daily)', 'deaths (total)', and 'total cases' columns. We stored the data in a column format in a comma separated values (csv) file. Each row of the csv file corresponded to each date, i.e. data fetched and stored date wise. Corresponding to each date, each *y* value is present.

5.4.3 Data Preparation

As discussed in the earlier section, the data was not directly available to pick from the selected source; instead we had to manually pick the data from the webpage source and use Python web scrapping to scrape the data for our use. Scrapped data was present with unwanted symbols and garbage texts which we removed, and we just kept the actual required suspects values of each *y* variable against each date. After finding the data, picking it, scrapping it, and cleansing it, we stored that in CSV file. Because it is temporal data, i.e. time series data, we also had to sort the tuples in the csv file time wise so that we could split the data into train and test parts.

5.4.4 Methodology

After getting the dataset ready, we made a code file (a Jupyter Notebook), and then we read the dataset CSV file into the Python. The data stored in Python was in the pandas table. After the data was imported in Python, we used the *x* column, i.e. date column, and picked four *y* columns, i.e. 'total cases', 'daily new cases', 'daily deaths', and 'total deaths'. Our data consisted of a total 380 records, i.e. 380 tuples, out of which we chose 80% for training the model and then another 20% to compare the predicted output vs. actual output. For training, we picked data from 15 Feb to 31 Jan 2021. Likewise, for testing we used data from 1 Feb 2021 to 26 Feb 2021. For the training of the ARIMAX model, we fed our training data to model, and in return it gave us multiple Akaike information criteria (AIC) values. Then we chose the minimum AIC values, i.e. (0, 1, 1) on (0, 1, 1, 12), and using this AIC value, we fit the model.

5.4.5 Tool Selection

The objective of this research was to predict Covid-19's new cases and death cases from across the country. We collected data and prepared according to the method explained previously. When it was time to begin. tool we chose to perform our task was the Python 3.0 Jupyter Notebook. The reason for choosing Python over R and MATLAB was that it gives robust powerful packages to accomplish our tasks in the most efficient and accurate way. There are basic advanced statistical machine learning packages available for both supervised and unsupervised learning in Python, which allows almost every type of data to be dealt with. In our case, the data is time series data and supervised learning. The Python packages we chose for our code file were NumPy, pandas, matplotlib, and statsmodel.

5.4.6 Model Selection

For our supervised time series data, various machine learning models can be used, like linear regression, lasso regression, support vector machines (SVM), multi-layer perceptron (also called artificial neural network), decision trees, Naïve Bayes, and random-forest. But the most efficient model for dataset suites is the ARIMAX model. Our data was time series data and had a sequential pattern. Hence, ARIMAX was expected to give better results.

5.4.7 ARIMAX Model

The ARIMAX model is one of the best models for predicting future events like coronavirus future cases predictions. This model requires data in a collection (pandas' data-frame is favourable) of data points where each point belongs to a specific time. In an autoregressive model (AR) model, the model predicts information about the next value by examining the previous values and uses following mathematical formula:

$$X_t = c + \Sigma b_i X_{t-I} + e_{t-i}, \text{ where } 1 < i < p$$

Here, c is constant, e is the noise term, and p is the order of AR model. The acronym MA in ARIMA stands for moving average model. This model does calculations that are based on noise that appears in data with the slope of the data. The formula for this model is represented as:

$$X_t = \mu + e_t + a_1 e_{t-1} + a_2 e_{t-2} + ------ + a_q e_{t-q}$$

Here, a_i are constants, μ is the mean, and q is the order of MA model. The I in ARIMAX model means integration, which is defined as taking the difference of time-series data by subtracting each value of data from the previous value. This works to make our data stationary because to apply the ARIMAX model we need data with trends or seasonality. The argument that is used for this is d, which represents how many times the data is to be differenced. The best practice is to try out different values of argument d and observe which value gives the minimum AIC value. AIC is used to measure how good a fit is of a statistical model. When comparing the two models, the model with the lower AIC value is generally assumed better.

Now let's talk about a model with extended form of ARIMA, univariate for single variable and ARIMAX (ARIMA model with exogeneous variable) for multivariate variable purposes. The general formula for the ARIMAX model constituted based on earlier discussion is:

$$X_t = (c + \mu + e_t) + b_1 X_{t-1} + b_2 X_{t-2} + ------ + b_p X_{t-p} + e_{t-i} + a_1 e_{t-1} + a_2 e_{t-2} + ------ + a_q e_{t-q}$$

```
ARIMA(0, 1, 0)x(1, 1, 0, 12)12 - AIC:5182.161988118738
ARIMA(0, 1, 0)x(1, 1, 1, 12)12 - AIC:5041.15812711268
ARIMA(0, 1, 1)x(0, 0, 0, 12)12 - AIC:5283.4536268298725
ARIMA(0, 1, 1)x(0, 0, 1, 12)12 - AIC:5121.694585874641
ARIMA(0, 1, 1)x(0, 1, 0, 12)12 - AIC:5362.412152634316
ARIMA(0, 1, 1)x(0, 1, 1, 12)12 - AIC:4983.913432035104
ARIMA(0, 1, 1)x(1, 0, 0, 12)12 - AIC:5148.947556905026
ARIMA(0, 1, 1)x(1, 0, 1, 12)12 - AIC:5123.683417944221
ARIMA(0, 1, 1)x(1, 1, 0, 12)12 - AIC:5138.3534341745435
ARIMA(0, 1, 1)x(1, 1, 1, 12)12 - AIC:4985.690850279028
ARIMA(1, 0, 0)x(0, 0, 0, 12)12 - AIC:5358.477748532646
ARIMA(1, 0, 0)x(0, 0, 1, 12)12 - AIC:5194.256833687073
```

FIGURE 5.2 Different AIC values (minimum one).

```
ARIMA(0, 1, 1)x(0, 1, 1, 12)12 - AIC:4983.913432035104
```

FIGURE 5.3 The best AIC value.

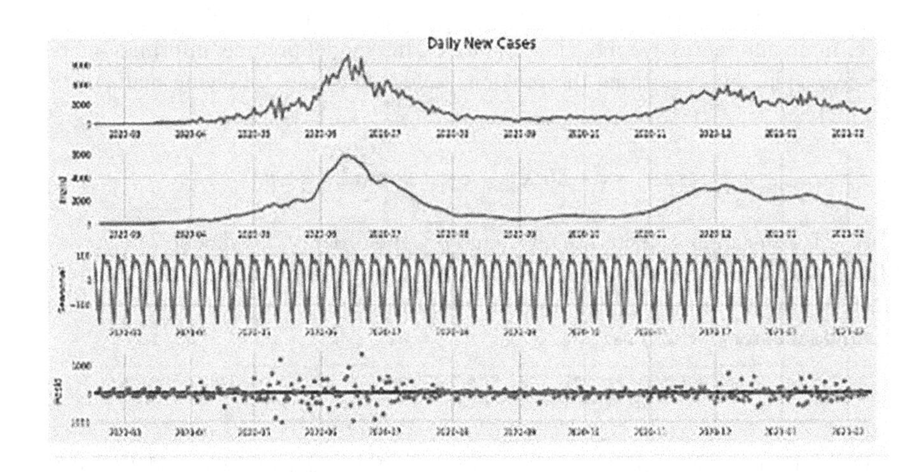

FIGURE 5.4 Seasonal daily new cases.

When it comes to time series dataset, the best model is ARIMAX. The factor which makes this model superior to Artificial neural network (ANN), regression, and SVM is that it gives a high precision of prediction. When using the ARIMAX model for our datasets, it gives multiple AIC values; among those AIC values we chose the one which was minimum; then we used that chosen minimum AIC values to fit the data to the model. We have decided to consider the initial AIC values coming out from ARIMAX model via training the model 50 times, so that we will able to produce the best AIC values (Figure 5.2 and Figure 5.3). We also viewed the data seasonal-wise to visualize the pattern (Figure 5.4 and Figure 5.5).

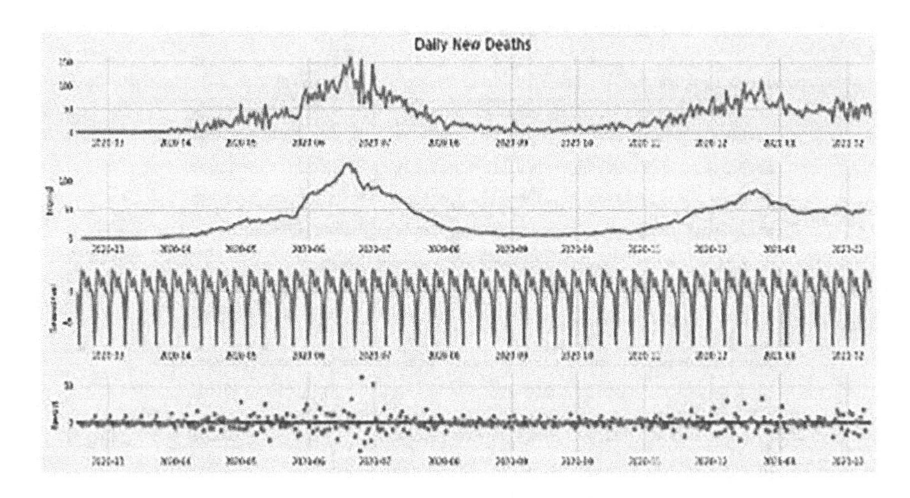

FIGURE 5.5 Seasonal death cases.

```
The Mean Squared Error of our forecasts is 118022.5        The Root Mean Squared Error of our forecasts is 343.54
```

FIGURE 5.6 Seasonal MSE and RMSE values of cases.

```
The Mean Squared Error of our forecasts is 195.16     The Root Mean Squared Error of our forecasts is 13.97
```

FIGURE 5.7 Seasonal MSE and RMSE values of deaths.

5.5 RESULTS AND DISCUSSION

After the model had been trained on training data, we plotted the predicted outputs, i.e. y-predicted vs. y-actual of test date. On comparing by visualizing both the predicted and actual outputs, we found that our model gives 90% accurate predicting values. The mean square error (MSE) and root mean square error (RMSE) for the cases reports in the country by the model are shown in Figure 5.6 and Figure 5.7.

From the model tuning step, we figured out the best model parameters having AIC value lower than the other is 4983.91. AIC is the factor that shows if the model is a good fit. We applied more than 50 parameters on the ARIMAX model to find the best AIC value, and the most optimum parameters we got were (0, 1, 1) on (0, 1, 1, 12); so we chose these parameters to find the fitness of the ARIMAX model. Figure 5.8 and Figure 5.9 shows the statistical results for the model.

Figure 5.8 and Figure 5.9 present coefficient tables of models by the ARIMA model. The *coef* column in Figures 5.8 and 5.9 contains the weights of the term defined in column 1, which are ma.L1, ma.S.L12, and sigma with their weights of -0.3716, -1.0001 and 1.7200 receptively. All model p-value using p>|z| were 0.0, which were less than 0.05; that shows this fitted model is highly significant. Figure 5.10 and Figure 5.11 show the multiple plot views of the daily new cases and death cases.

	coef	std err	z	P>|z|	[0.025	
ma.L1	-0.4003	0.031	-12.734	0.000	-0.462	
ma.S.L12	-1.0001	0.048	-20.634	0.000	-1.095	
sigma2	1.388e+05	3.49e-07	3.97e+11	0.000	1.39e+05	1

FIGURE 5.8 Seasonal statistical results of the model for daily new cases.

	coef	std err	z	P>|z|	[0.025
ar.L1	-0.1389	0.045	-3.060	0.002	-0.228
ma.L1	-0.6092	0.036	-17.129	0.000	-0.679
ma.S.L12	-0.9389	0.043	-22.061	0.000	-1.022
sigma2	195.3148	10.628	18.377	0.000	174.484

FIGURE 5.9 Seasonal statistical results of the model for daily new deaths.

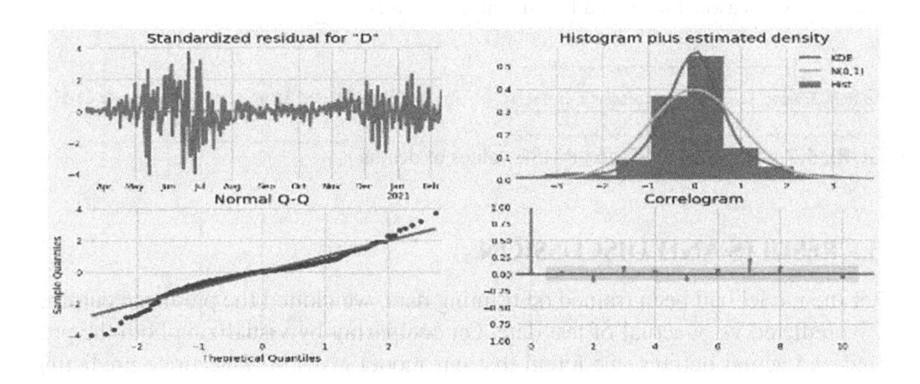

FIGURE 5.10 Seasonal multiple plot views of the daily new cases.

The normal Q-Q plot of blue dots that fall mostly on or near the red line means that if there are any changes in the model the distribution may be skewed. The last plot is the ACF plot, which explains that the residuals errors are not autocorrelated. Autocorrelation applies when some relationship exists in the model. Following are the visual and tabular representation of our final obtained predicted vs. actual outputs of the model, as shown in Figure 5.12.

Table 5.1 provides the comparison of actual and predicted new cases while Table 5.2 provide the actual and predicted death cases comparison. The following is the total cases visual and tabular representations of our final obtained predicted vs. actual outputs of the model. From the results, it estimated that by the end of January 2021, the expected cases were 546,592 and the actual case currently in the

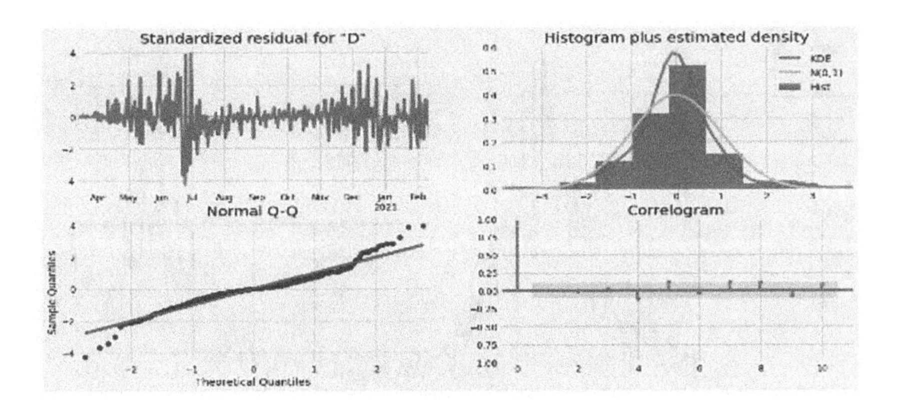

FIGURE 5.11 Seasonal multiple plot views of the daily death cases.

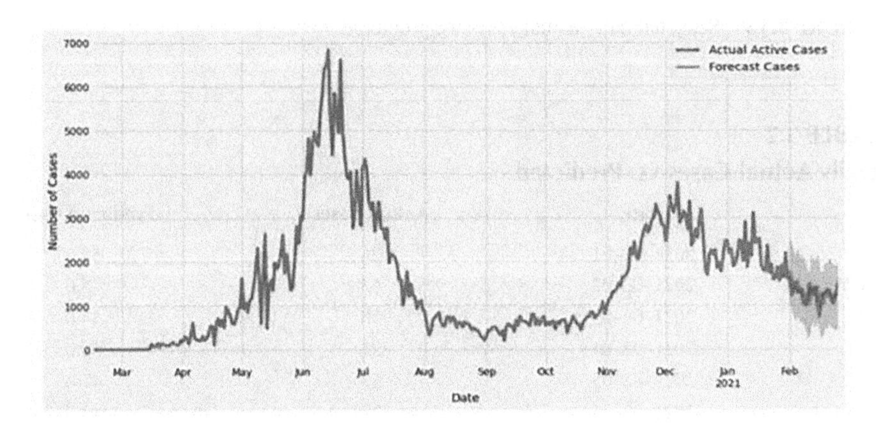

FIGURE 5.12 Seasonal daily actual cases vs. predicted.

TABLE 5.1
Daily Actual Cases vs. Predicted

	Date	Actual Cases	Predicted Cases
0	2021–02–01	1615	1787.0
1	2021–02–02	1220	1642.0
2	2021–02–03	1384	1464.0
3	2021–02–04	1508	1370.0
4	2021–02–05	1302	1540.0
5	2021–02–06	1286	1346.0
6	2021–02–07	1346	1372.0
7	2021–02–08	1037	1308.0
8	2021–02–09	1008	1189.0
9	2021–02–10	1072	1002.0
10	2021–02–11	1502	1017.0

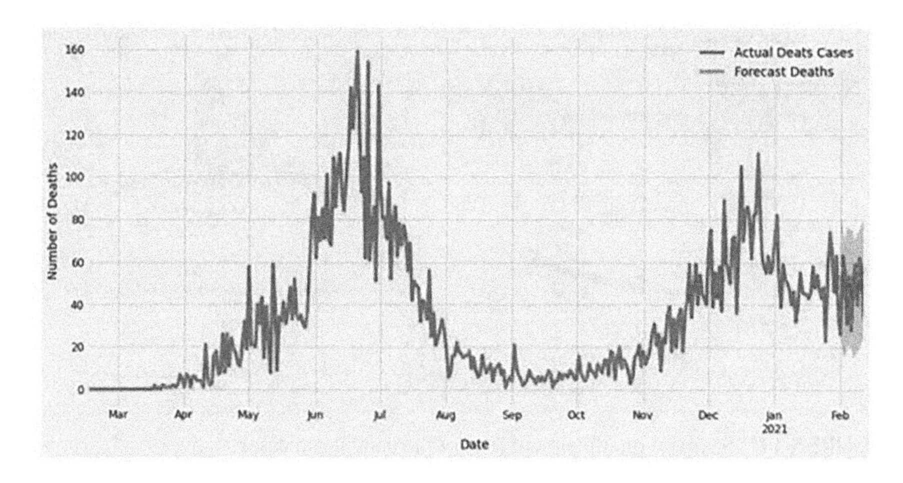

FIGURE 5.13 Seasonal daily actual cases vs. predicted.

TABLE 5.2
Daily Actual Cases vs. Predicted

	Date	Actual Cases	Predicted Cases
0	2021–02–01	26	54.0
1	2021–02–02	63	45.0
2	2021–02–03	56	44.0
3	2021–02–04	31	50.0
4	2021–02–05	53	47.0
5	2021–02–06	28	48.0
6	2021–02–07	53	43.0
7	2021–02–08	59	45.0
8	2021–02–09	40	46.0
9	2021–02–10	62	49.0
10	2021–02–11	57	51.0

country at the end of month January 2021 were 546,428. Figure 5.14 shows the actual vs. total predicted cases graph. The same is shown in Table 5.3.

The model predicted that there would be around 604,000 cases reported in the country with average daily deaths of 50, and the total deaths would reach approximately 13,500 due to Covid-19 effects by 15 March 2021 with an accuracy of around 90% by using the model's RMSE and MAE values during training and testing the model.

5.6 FUTURE WORKS AND CONCLUSION

The Covid-19 epidemic has emerged as the greatest threat to people in terms of health, the economy, and the financial industry. The Council for Mutual Economic Assistance (CMEA) is one of the most recent major banks or financial organizations

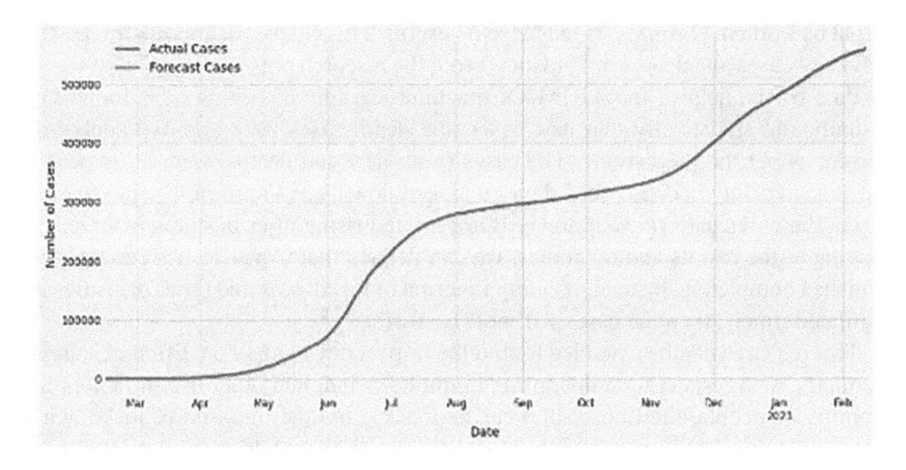

FIGURE 5.14 Seasonal total actual cases vs. total predicted output.

TABLE 5.3
Total Actual Cases vs. Total Predicted Output

	Date	Actual Cases	Predicted Cases
0	2021-02-02	1285	548064.0
1	2021-02-03	1509	549110.0
2	2021-02-04	1772	550403.0
3	2021-02-05	1684	552082.0
4	2021-02-06	1198	553191.0
5	2021-02-07	1542	554505.0
6	2021-02-08	1260	555787.0
7	2021-02-09	1441	556708.0
8	2021-02-10	2008	557531.0
9	2021-02-11	1732	558619.0
10	2021-02-12	1481	560486.0

to discontinue making predictions about the state of the world economy. The fear of Covid-19 has had a significant impact on the entire economy; in particular, markets around the world have been severely damaged, with sharp drops in investment and financial prices. The country's or city's lockdown regulations must be scrupulously adhered to by everyone. People wait to leave the house until it is absolutely required. Eliminating air conditioners is a smart idea because a regulated environment can easily have an impact on one's health. People's routines and ways of life may change during a lockdown. A total lockdown might make people overly nervous. TV entertainment and other forms of entertainment like Netflix, Amazon Prime, HotStar, etc. can offer some solace. The financial system is negatively impacted by the total lockdowns. In unforeseen circumstances like these, the work-from-home policies can be useful. Universities may provide online courses to students, so the academic loss might be limited. Online tests may also be useful. This chapter deals with forecasting Covid-19 cases and deaths in Pakistan. From the research presented earlier on Covid-19 data by the help of the ARIMAX machine learning algorithm, it is found both visually and statistically that new cases and death cases were approximately 90% correct as per the predictions. The cases (both new and deaths) were at its peak in the June and July because until then no locked down was imposed. During the lock down, the cases pattern went down. But it started rising after business resumed. So looking at the results and outcomes, we can deduce that Covid-19 has certainly not vanished completely. Instead, if safety in terms of lockdowns and other measures are regulated timely, then the cases will not rise further.

This research can be extended to find the impact of Covid-19 on different states in a country with respect to temperature, humidity, wind, and other useful factors like mobility of people, the impact of wearing masks, and the imposed of lockdown in different states. In future research, we will improve the models based on the aforementioned problems and continue to improve the generalizability of the models. This work can also be incorporated by using the same model proposed in this chapter or by using advanced deep neural networks with its different variations.

5.7 ACKNOWLEDGMENTS

The authors would like to thank Karachi Institute of Economics and Technology for their continuous support in the completion of this research. We would also like to thank NED University of Engineering and Technology, Pakistan.

REFERENCES

1. Lai, C.-C., et al. (2020). Severe acute respiratory syndrome coronavirus 2 (SARS-CoV-2) and coronavirus disease-2019 (COVID-19): The epidemic and the challenges. *International Journal of Antimicrobial Agents*, 55(3), 105924.
2. WHO director-general's opening remarks at the media briefing on COVID-19–11 March 2020. Retrieved 2021.
3. Zhu, N., et al. (2020). A novel coronavirus from patients with pneumonia in China, 2019. *New England Journal of Medicine*, 382, 727–733.
4. www.worldometers.info/. Retrieved June 2022.
5. Prabhakaran, S. (2019). Arima model–complete guide to time series forecasting in python. *Machine Learning Plus*, 18.
6. Tsay, R. S. (2000). Time series and forecasting: Brief history and future research. *Journal of the American Statistical Association*, 95(450), 638–643.
7. Hiironen, M.-L., et al. (2021). *Alakohtaiset lyhyet sairauspoissaolot LYHTY–Lyhyiden työpoissaolojen ennaltaehkäisy-hankkeen loppuraportti*. Centria University of Applied Sciences, 2021, Centria UAS. Raportteja ja selvityksiä 2342-933X, Nro 51.
8. Jain, A. (2016). *Complete guide to create a time series forecast (with codes in Python)*. Analytics Community. Analytics Discussions. Big Data Discussion. [online] Analytics Vidhya Available at: https://www.analyticsvidhya.com/blog/2016/02/time-series-forecasting-codes-python/ [Accessed 14 December 2022]
9. Zhang, J., et al. (2019). An end-to-end automatic cloud database tuning system using deep reinforcement learning. SIGMOD '19. In *Proceedings of the 2019 international conference on management of data*, Association for Computing Machinery, New York, NY, United States, pp. 415–432.

10. Statsmodels. *SARIMAX: Introduction*. Retrieved June 2022, from www.statsmodels.org/stable/examples/notebooks/generated/statespace_sarimax_stata.html
11. Ilyas, N., Azuine, R. E., & Tamiz, A. (2020). COVID-19 pandemic in Pakistan. *International Journal of Translational Medical Research and Public Health*, 4(1), 37–49.
12. Noreen, N., et al. (2020). COVID 19 pandemic & Pakistan: Limitations and gaps. *Global Biosecurity*, 2(1).
13. Yousaf, M., et al. (2020). Statistical analysis of forecasting COVID-19 for upcoming month in Pakistan. *Chaos, Solitons & Fractals*, 138, 109926.
14. Waris, A., et al. (2020). COVID-19 outbreak: Current scenario of Pakistan. *New Microbes and New Infections*, 35, 100681.
15. Abid, K., et al. (2020). progress of COVID-19 epidemic in Pakistan. *Asia Pacific Journal of Public Health*, 32(4), 154–156.
16. Qiu, H., et al. (2020). Clinical and epidemiological features of 36 children with coronavirus disease 2019 (COVID-19) in Zhejiang, China: an observational cohort study. *The Lancet Infectious Diseases*, 20(6), 689–696.
17. Lai, C.-C., et al. (2020). Global epidemiology of coronavirus disease 2019 (COVID-19): Disease incidence, daily cumulative index, mortality, and their association with country healthcare resources and economic status. *International Journal of Antimicrobial Agents*, 55(4), 105946.
18. Alasali, F., et al. (2021). Impact of the covid-19 pandemic on electricity demand and load forecasting. *Sustainability*, 13(3), 1435.
19. Le, T. T., et al. (2020). The COVID-19 vaccine development landscape. *Nature Reviews Drug Discovery*, 19(5), 305–306.
20. Mou, J. (2020). Research on the impact of COVID19 on global economy. *IOP Conference Series: Earth and Environmental Science*, 546(3), 032043.
21. Morgan, A. K., Awafo, B. A., & Quartey, T. (2021). The effects of COVID-19 on global economic output and sustainability: evidence from around the world and lessons for redress. *Sustainability: Science, Practice and Policy*, 17(1), 76–80.
22. Fattah, J., et al. (2018). Forecasting of demand using ARIMA model. *International Journal of Engineering Business Management*, 10, 1847979018808673.
23. www.voanews.com/science-health/coronavirus-outbreak/pakistan-detects-first-coronavirus-cases-links-iran-outbreak. Retrieved June 2022.
24. Li, Z., et al. (2020). Identification of drug-disease associations using information of molecular structures and clinical symptoms via deep convolutional neural network. *Frontiers in Chemistry*, 7, 924.
25. Aji, B. S., & Rohmawati, A. A. (2021). Forecasting number of COVID-19 cases in Indonesia with ARIMA and ARIMAX models. In *2021 9th international conference on information and communication technology (ICoICT)*. Conference Location: Yogyakarta, Indonesia, IEEE, pp. 71–75.
26. Covid-19 situation, national information technology board in collaboration with ECOM PK (Pvt.) Ltd, 2020. Retrieved June 2022, from https://covid.gov.pk/
27. Covid-19 pandemic in Pakistan, Wikipedia, 2020. Retrieved June 2022, from https://en.wikipedia.org/wiki/Covid-19_pandemic_in_Pakistan
28. Shmueli, G., & Burkom, H. (2010). Statistical challenges facing early outbreak detection in biosurveillance. *Technometrics*, 52(1), 39–51.
29. Rolka, H. R. (2011). 13th biennial CDC & ATSDR symposium on statistical methods info-fusion: Utilization of multi-source data preface. Wiley-Blackwell Commerce Place, 350 Main St, Malden 02148, MA, USA, pp. 401–402.
30. Punn, N. S., Sonbhadra, S. K., & Agarwal, S. (2020). COVID-19 epidemic analysis using machine learning and deep learning algorithms. MedRxiv.
31. Dandekar, R., & Barbastathis, G. (2020). Quantifying the effect of quarantine control in Covid-19 infectious spread using machine learning. MedRxiv.
32. World Health Organization (WHO). (2020). Coronavirus [cited April 13, 2021].

33. Islam, N., et al. (2021). Ternion: An autonomous model for fake news detection. *Applied Sciences*, 11(19), 9292.
34. Usmani, S. Z., et al. (2021). A comparative analysis of aPriori and FP-growth algorithms for frequent pattern mining using Apache spark. In *Proceedings of international scientific research conference*. Azerbaijan.
35. Muhammad, G., et al. (2022). GVDeepNet: Unsupervised deep learning techniques for effective genetic variant classification. *Pakistan Journal of Engineering and Technology*, 5(1), 16–22.
36. Frunza, O., Inkpen, D., & Tran, T. (2010). A machine learning approach for identifying disease-treatment relations in short texts. *IEEE Transactions on Knowledge and Data Engineering*, 23(6), 801–814.
37. Ben Abdessalem Karaa, W., Alkhammash, E. H., & Bchir, A. (2021). Drug disease relation extraction from biomedical literature using NLP and machine learning. *Mobile Information Systems*, 2021.
38. Tominaga, D. (2020). Logarithmic quadratic regression model for early periods of COVID-19 epidemic count data. *Archives of Clinical and Biomedical Research*, 5(5), 582–612.
39. Yang, Z., et al. (2020). Modified SEIR and AI prediction of the epidemics trend of COVID-19 in China under public health interventions. *Journal of Thoracic Disease*, 12(3), 165.
40. Pandey, G., et al. (2020). SEIR and Regression Model based COVID-19 outbreak predictions in India. arXiv preprint arXiv:2004.00958.
41. Al-Jameel, S. S., et al. (2021). Machine learning-based model to predict the disease severity and outcome in COVID-19 patients. *Scientific Programming*, 2021.
42. Ahmad, I., & Asad, S. M. (2020). Predictions of coronavirus COVID-19 distinct cases in Pakistan through an artificial neural network. *Epidemiology & Infection*, 148.
43. Khan, F., Saeed, A., & Ali, S. (2020). Modelling and forecasting of new cases, deaths and recover cases of COVID-19 by using vector autoregressive model in Pakistan. *Chaos, Solitons & Fractals*, 140, 110189.
44. Satu, M., et al. (2021). Short-term prediction of COVID-19 cases using machine learning models. *Applied Sciences*, 11(9), 4266.
45. Goshvarpour, A., & Goshvarpour, A. (2020). Estimation of Covid-19 mortality rate in Iran using the autoregressive model. *Journal of Critical Care Nursing*, *13*(4), 11–21.
46. Khakharia, A., et al. (2021). Outbreak prediction of COVID-19 for dense and populated countries using machine learning. *Annals of Data Science*, 8(1), 1–19.
47. Didi, E. I., Kingdom, N., & Harrison, E. E. (2021). ARIMA modelling and forecasting of COVID-19 daily confirmed/death cases: A case study of Nigeria. *Asian Journal of Probability and Statistics*, 12(3), 59–80. ISSN 2582-0230.
48. Ariyo, A. A., Adewumi, A. O., & Ayo, C. K. (2014). Stock price prediction using the ARIMA model. In *2014 UKSim-AMSS 16th international conference on computer modelling and simulation*. Cambridge, UK: IEEE, pp. 106–112.
49. Andreoni, A., & Postorino, M. N. (2006). A multivariate ARIMA model to forecast air transport demand. *Proceedings of the Association for European Transport and Contributors*, 1–14.
50. Yang, Q., et al. (2020). Research on COVID-19 based on ARIMA modelΔ—Taking Hubei, China as an example to see the epidemic in Italy. *Journal of Infection and Public Health*, 13(10), 1415–1418.
51. Triacca, M., & Triacca, U. (2021). Forecasting the number of confirmed new cases of COVID-19 in Italy for the period from 19 May to 2 June 2020. Infectious Disease Modelling, 6, 362–369.
52. Lee, D. H., et al. (2021). Forecasting COVID-19 confirmed cases using empirical data analysis in Korea. *Healthcare*, 9(3), 254. https://doi.org/10.3390/healthcare9030254

53. Aslam, F., et al. (2021). Prediction of COVID-19 confirmed cases in Indo-Pak sub-continent. *The Journal of Infection in Developing Countries*, 15(03), 382–388.
54. Mirza Hammad Baig, Danish Karim, & Noman Islam. (2021). Analysis of public sentiments about COVID-19 based on twitter's tweets. In *Proceedings of Euro Asia international conference*. Uzbekistan.
55. Petrevska, B. (2017). Predicting tourism demand by ARIMA models. *Economic Research-Ekonomska istraživanja*, 30(1), 939–950.
56. Bhatt, C., Dey, N., & Ashour, A. S. (2017). Internet of things and big data technologies for next generation healthcare. In Chintan Bhatt, Nilanjan Dey, Amira S. Ashour (Eds.), *Studies in big data book series (SBD)*, Electronic ISSN: Springer, vol. 23.
57. Hassanien, A. E., Dey, N., & Borra, S. (2018). *Medical big data and internet of medical things: Advances, challenges and applications*. Boca Raton: CRC Press.
58. Lan, K., et al. (2018). A survey of data mining and deep learning in bioinformatics. *Journal of Medical Systems*, 42(8), 1–20.
59. Jain, A., & Bhatnagar, V. (2017). Concoction of ambient intelligence and big data for better patient ministration services. *International Journal of Ambient Computing and Intelligence (IJACI)*, 8(4), 19–30.
60. Sameni, R. (2020). *Mathematical modeling of epidemic diseases; a case study of the COVID-19 coronavirus*. arXiv preprint arXiv:2003.11371.
61. Hu, S., et al. (2018). Forecasting China future MNP by deep learning. In Raymond Wong, Chi-Hung Chi and Prof. Patrick C. K. Hung (Eds.), *Behavior engineering and applications*. Cham: Springer, pp. 169–210.
62. Singh, N., & Mohanty, S. (2018). Short term price forecasting using adaptive generalized neuron model. *International Journal of Ambient Computing and Intelligence (IJACI)*, 9(3), 44–56.
63. Almasarweh, M., & Alwadi, S. (2018). ARIMA model in predicting banking stock market data. *Modern Applied Science*, 12(11), 309.

6 ML-Based Method for Detecting and Alerting to Cyber Attacks

D.V. Chandrashekar and K. Suneetha

CONTENTS

6.1 INTRODUCTION

As political and business entities increasingly utilize technology to disrupt, restrict, or degrade information on computer networks, cyber warfare is becoming an increasingly intricate phenomenon. When constructing a network, the protocols that are used must be dependable in order to ensure that they can withstand even severe attacks that can take control of a portion of the participants in the network. Both aggressive and passive attacks, such as eavesdropping and refusing to take part in the activity, might be launched by the party that is being controlled (e.g., jamming, message dropping, corruption, and forging). Because looking for telltale signals of an intrusion is essential to achieving the objective of preventing unauthorized access to computer systems and networks, regular scans are carried out on both. The vast majority of the time, this is accomplished by the automatic collection and examination of data from numerous systems and networks.[1] After that, an investigation is carried out to determine whether or not the data contains any potential vulnerability.

DOI: 10.1201/9781003310785-6

When it comes to securing networks and systems against assaults such as denial of service, traditional methods like firewalls, access control mechanisms, and encryption have their limitations. These traditional methods are used to discover and stop incursions. Also, the majority of the systems that use these approaches have a high rate of both false positives and false negatives, and they don't always update to keep up with how harmful people behave. This makes it difficult to determine whether or not a person has committed a crime. Over the course of the past decade, a wide variety of machine learning (ML) strategies have been implemented in an effort to accelerate the rate at which security breaches are discovered. They are frequently utilized in the process of keeping attack databases up to date and complete. Concerns over one's physical safety and one's protection from cybercriminals have risen to the top of the list of priorities in recent times. This is mostly the result of how rapidly computer networks have expanded and how many the apps are utilized by individuals or organizations for either private or professional reasons, particularly since the Internet of Things (IoT) has gained widespread acceptance.[2] When they target huge networks, cyber-attacks result in considerable physical damage as well as significant financial losses. The solutions that are currently available, such as hardware and software firewalls, user authentication, and data encryption methods, are insufficient to defend the computer network from the multiple cyber-threats that are present. Because of the more rapid and stringent evolution of intrusion methods, the traditional security mechanisms that were once acceptable safeguards are no longer adequate. A firewall must only control traffic going from one network to another in order to fulfill its duty of preventing unauthorized access to other networks. It does not give any warning, even in the event that an external attack takes place. It is recommended that machine learning-based intrusion detection systems (also known as IDS) be created in order to guarantee the system's safety. Systems or software known as IDS are used to monitor a network or system for potentially malicious activities or violations of established rules. During the normal operation of a network or system that is used to detect network security problems or assaults, such as denial of service (DoS) attacks, an IDS can identify inconsistencies and behaviors that are not intended to occur on a network.[3] Intrusion detection systems are also helpful in determining whether or not a system has been compromised, modified, or destroyed. This can be done by determining whether or not a breach has occurred. There are various distinct varieties of intrusion detection systems, each of which is designed to cater to a specific type of user. Intrusion detection systems that are based on the host computer as well as the network are one example.

6.2 LITERATURE SURVEY

The majority of the time, an IDS must contend with concerns such as enormous volumes of network traffic, very unequal data distribution, and other similar problems. It is difficult to differentiate between normal and aberrant behavior in a world that is continuously evolving, and constant adaptation is required in order to thrive in this environment. The majority of issues that arise in computer networks are caused by an inability to effectively distinguish various patterns of behavior. Misuse detection and anomaly detection are two prominent classification strategies for network behavior.

Common classification related to strategies for network behavior include: methods of signature matching are utilized in order to identify known instances of inappropriate behavior in system as well as network activity.[3] The most effective application of this technique is in the identification of previously reported assaults. The issue is that novel approaches are frequently disregarded, which leads to an increased number of false negatives. It's possible that the IDS will generate alarms, but it would be a waste of time and resources to respond to each and every one of them, which would eventually lead to the system becoming unstable. IDS should take its time collecting warnings and make judgments based on the correlation of those signals in order to tackle this problem. Instead of launching, an IDS should take its time collecting warnings and eradicate the first symptom immediately after recognizing and eliminating it. According to the findings of a number of studies that were conducted not too long ago, cybersecurity exerts a considerable amount of effect on businesses, organizations, and individuals.[4]

Cybercrime has, to this point, been responsible for around $400 billion worth of lost revenue and cleanup expenses owing to the theft of cash. It is anticipated that there would be a shortage of more than 1.8 million people working in cyber security by the year 2022. It is anticipated that businesses all over the world will spend a total of $100 billion per year on the protection of their digital infrastructure. Attacks using ransomware, such as WannaCry and CryptoWall, bring in more than one billion dollars annually for attackers.[5]

6.2.1 Cybersecurity Assurance Model Improvements

When carrying out an audit of compliance, information technology, or data security, it is not unusual for a large group of auditors to follow the same procedures. The process of designing an audit of the cybersecurity of a corporation is not dissimilar to the process of designing any other kind of audit. However, in order to accomplish this goal, a large investment of both time and resources is going to be required. Nevertheless, the vast majority of cyber capabilities are not covered within the purview of internal audits. This framework encompasses the entirety of the process of designing software, establishing a security program, and managing third parties, in addition to the management of information and asset security, access control, and threat and vulnerability management.[6] The requirement for assurance can be fulfilled through the use of risk analytics, information management and protection, management reviews, and cyber risk assessments. Aside from that, Deloitte's strategy is consistent with the industry standards that have been established by the National Institute of Standards and Technology (NIST), the Information Technology Infrastructure Library (ITIL), the Committee on Standards for the Information Technology Industry (COSO), and International Standard Organisation (ISO).[7]

6.3 BACKGROUND WORKS

Allowing harmful threats to operate anytime, anywhere, and in whatever shape they take is an intolerable condition that could result in major harm. Even in an ever-expanding subject like cybersecurity, it is difficult to pinpoint exactly why the topic has such a huge influence. The ever-evolving network of people who use the Internet and organizations that hold sensitive information is a challenge for cybersecurity groups who are finding it difficult to keep up with the pace of change. As part of their day-to-day activities on the global network, individuals, families, organizations, governments, and academic institutions all need to include considerations of cybersecurity into their routines. We will be able to provide a higher level of defense against cyberattacks by utilizing machine learning. The modern high-tech infrastructure, which consists of network and cybersecurity systems, is responsible for collecting and analyzing a substantial quantity of data and information regarding virtually all of the most significant aspects of mission-critical systems, despite the fact that human supervision and intelligence are still crucial to the operation of today's infrastructure. The majority of intrusion detection systems, beginning with your firewall, are designed to identify potential dangers on the outermost layers of your network. Your network does not care about east-to-west traffic, which is how many modern network threats penetrate your system. However, your network does not care about north-to-south traffic because it is protected from it. We are aware of the truth of this statement due to the fact that surveillance from north to south only uncovers 20% of all threats. It is standard practice for an IDS to send suspicious activity to a security information and event management system (SIEM) system.[8] This is done so that true threats can be detected among traffic irregularities and other false alarms. On the other hand, the longer it takes to recognize a threat, the more damage it is possible to inflict. IDSs are a lifesaver when it comes to network monitoring, but how effective they are is largely dependent on what you do with the information that they offer you. IDSs are a godsend. It is pointless to deploy detection tools as an additional layer of protection if you do not have the appropriate personnel and procedures in place to deal with potential dangers. An adversary can take advantage of the fact that an IDS is unable to read encrypted packets in order to obtain access to a network by exploiting this vulnerability. Your systems are still vulnerable to attack since an intrusion detection system, also known as an ID, won't notice these attacks until after the network has already been infiltrated. This is a significant cause for concern, particularly in view of the growing prevalence of the usage of encryption to protect sensitive personal data.[9] A significant limitation of IDSs is the occurrence of false positive warnings. The presence of genuine dangers is much more typical than the occurrence of false positives. The amount of effort spent by your engineers responding to these issues will not go away, despite the fact that false positives can be decreased by adjusting an IDS. It is possible to miss false positives, which would therefore allow actual attacks to get unreported

6.4 PROPOSED MODEL

In the event that a cyber assault has been carried out, approaches from machine learning can be utilized to identify it. End users or security engineers can be the recipients of an email alert that is sent as soon as the threat is found. Any one of several different classification methods could be used to categories a DoS or DDoS attack. An example of a classification technique that is based on supervised learning is known as a support vector machine (SVM). It examines the data in order to identify patterns. Because we are unable to control when, when, or how an attack will occur, and because we cannot yet guarantee that we will have complete security against these attacks, our greatest hope for avoiding an attack is early detection. In order to mitigate the effects of a cyberattack, organizations may choose to implement preexisting solutions or develop their own from scratch. The ideal system is one that does not require a significant amount of participation on the part of the user.

Problem Modeling:

- Administrators of the network
- Network administrators do the following tasks:
  - Network traffic can be snatched
  - Data packets should be read and stored
  - Keep an eye out for warnings and network statistics relating to recent cyberattacks

6.5 SYSTEMS CONCEPT DEVELOPMENT

Applications can utilize algorithms for machine learning to detect and stop cyberattacks before they can cause any damage to their systems. The majority of the time, this is accomplished by constructing a model based on security events and patterns of inappropriate behavior. When events of a similar nature occur, the necessary measures are taken care of on their own. Indicators of Compromise (IOC) are put to use in the construction of models and systems that are able to detect and respond to attacks in real time. This dataset frequently consists of IOC that have been identified and documented in the past. We also have the option of applying classification strategies derived from machine learning, based on IOC datasets.[10, 11] Figure 6.1 shows the system architecture.

a) The goal is to organize datasets in accordance with the characteristics of malicious software. Because of the patterns that have been uncovered, it is now possible to automatically locate and catagorize information. The complete original form related to the software is deadly dangerous and is adequate to security analyst on the job and other automated system, which can speed up the results and its potentiality. Locate and designate a novel category of threats, and after that, based on the available facts, take the appropriate actions.
b) In order to accomplish anything specific, a system is made up of a number of interdependent parts that are logically connected to one another in order to form a cohesive whole. Its major characteristics include being organized, interrelated, integrated, and centered on a central aim.
c) Problem-solving techniques that are assisted by computers are referred to as system analysis and design (or a combination of the two). The analyst needs to consider the system's outputs and inputs, as well as its processors, controls, feedback, and surroundings, before putting the system back together.

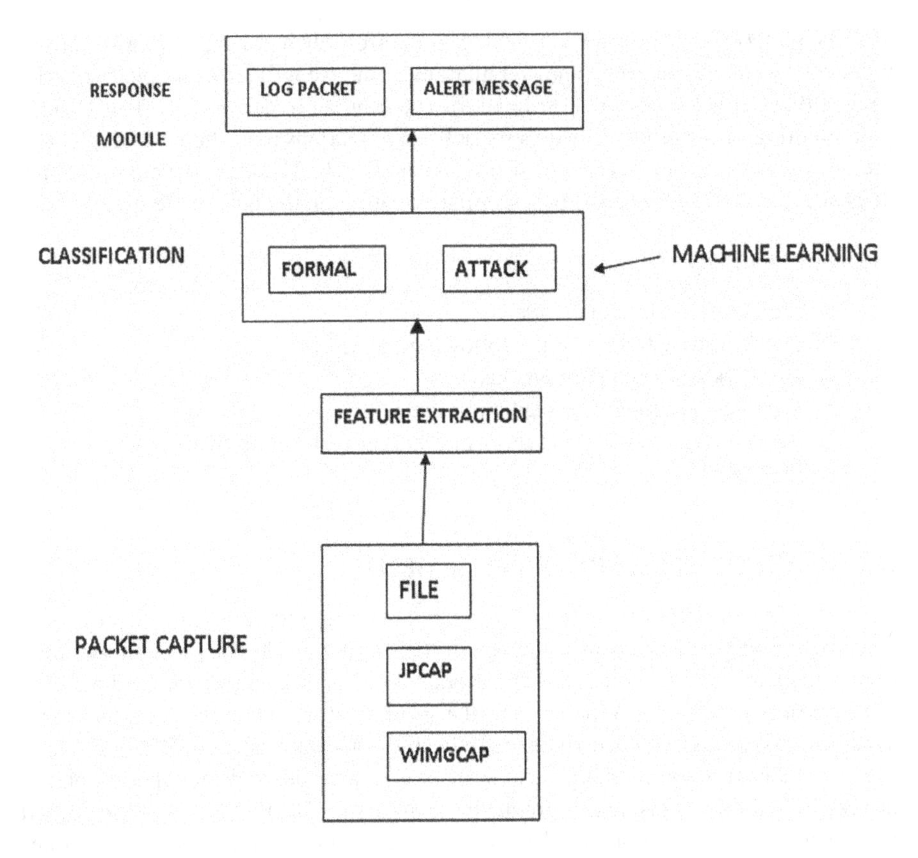

FIGURE 6.1 System architecture.

6.5.1 The Architecture of the Software

Applications can utilize algorithms for machine learning to detect and stop cyberattacks before they can cause any damage to their systems. The majority of the time, this is accomplished by reviewing data on security events in order to identify patterns of undesirable behavior. By making a look at the potential dangers, initially identify them, segregate them, block, and respond quickly based on the discovery of an attack. This is feasible due to the fact that we are able to categories malicious software by utilizing IOC datasets and machine learning classification approaches. It is now possible, using the patterns that have been learned, to automatically locate and categories newly discovered malicious software.[12]

This can be useful for automating processes, such as those used by security analysts. The identification and handling of a new type of threat both make use of decision-making processes that are driven by data.

6.5.2 Movement of Data across the Network

When we discuss network traffic, we are referring to the amount of data that is moving through a network at any particular moment. There is a significant amount of traffic on the network due to the fact that data transfers take place inside the network packets. The amount of network traffic that is present is used as the foundation for measurements, controls, and simulations. The standard of service in a network can be maintained if the traffic in that network is properly managed.[13]

When the traffic on a company's network is properly analyzed, the company can benefit in the following ways: traffic identity, protection of attack on traffic, and traffic monitoring. It is vital to identify users or programmers that use a considerable amount of bandwidth in order to be able to detect the bottlenecks in the network. Identifying these users and programmers is necessary. There are many different approaches that can be used in order to address these problems: initially, the protection of the network; second, the type of network initiated; and third, how attacks can be forbidden.

The presence of an attack on a network may be indicated by a significant increase in the amount of traffic. When attempting to thwart attacks of this nature, the information that is provided in network traffic reports is of great use. Engineering of the network—being aware of the utilization levels of the network—makes it possible to evaluate potential future needs.

6.6 TAKING IN DATA PACKETS

The act of seizing and keeping a data packet as it moves through a network node is referred to as "packet capture." The term "packet capture" is a networking term that is used to describe the operation. Before being downloaded or saved, packets are held for a predetermined amount of time so that they can be examined thoroughly for the purpose of locating and fixing network issues, such as

- Recognizing challenges to security
- Getting to the bottom of network issues
- Detecting congestion in a network
- Detecting data loss and reporting it
- Investigating forensics network packets

The task of organizing data into groups is frequently performed by supervisory machine learning algorithms. In the field of cybersecurity, classifiers that are derived from machine learning are utilized to differentiate between legitimate emails and spam. Spam filters are able to differentiate between communications that are legitimate and those that are spam.

Machine learning can be used to categories items in a variety of different ways, including logistic regression, k-nearest neighbors, support vector machine, Naive Bayes, decision tree, and random forest classification. Restricted Boltzmann machines (RBM), convolutional neural networks (CNN), recurrent neural networks (RNN), and long-short term memory (LSTM) cells are examples of the types of models that are used in deep learning categorization. Before deep learning, densely linked neural networks are significantly more effective than they were in the past at finding solutions to challenging situations. In order for the machine learning strategies mentioned above to be of any help, it is necessary to have huge volumes of data that has been labeled.[14,15]

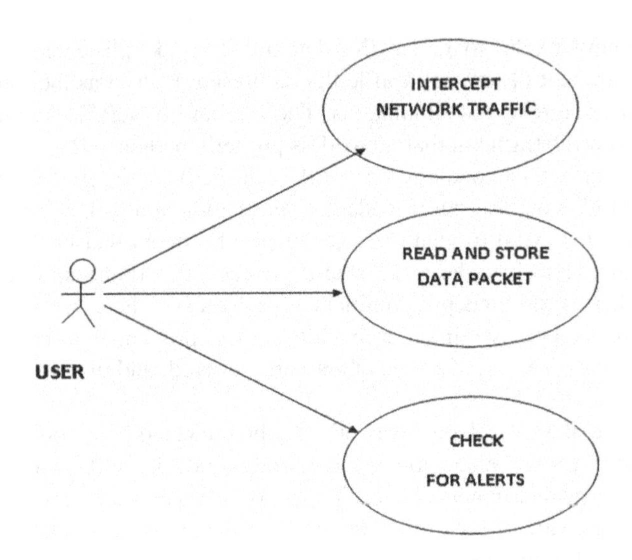

FIGURE 6.2 Use case diagram.

RNN describes how an organization reacts to incidents such as a data breach or a cyberattack and how it works to mitigate the harm that these incidents cause (the "incident"). At the end of the day, the objective is to cause the least amount of damage to other aspects, such as the reputation of a business, while simultaneously notifying security analysts of the breach. In this approach, both the amount of damage and the amount of time and money required to recover are kept to a minimum.

The flow of information from the beginning point to the final destination is depicted in Figure 6.2. As a result, it is the responsibility of the user to monitor and control all aspects (actors) of the system. These aspects include but are not limited to intercepting the network and acquiring packets from it, storing information about the network, and utilizing machine learning models to evaluate the collected data before looking for attack warnings.[16]

6.6.1 Details Regarding the Application of the Solution

The application program of Anaconda was utilized throughout the construction of the system. Anaconda is the most popular platform in the field of data science, and it serves as the foundation for the most advanced forms of machine learning. It was a first for us to use Python for data science, and we are pleased to show our support for the vibrant Python community and the open-source projects that will drive the scientific advancements of the next 10 years. Corporations and academic institutions are able to take benefit of open source's competitive advantages, ground-breaking research, and better world with the help of our enterprise-grade open-source solutions.

[17, 18] Powerful open-source software distributed in a manner that is centralized, collaborative, and version-controlled in an online library of software packages that include tools for auditing, tracing, and tracking that can be utilized to facilitate the monitoring of activities related to data science. There has been some testing and implementation of automated training and deployment of models, as well as various monitoring levels and scalable infrastructure based on containers.

1) The use of computers on both the system level and the user level—Sandboxed memory and real-time memory both use the size of the swap memory I/O and the disc space that is currently available.
2) The user's perspective—Different kinds of users with their privileges bestowed upon them to access the resources and directories which are inside their perimeter area or location within their log-in and log-out related to the activity of a program or software within an application to gain the required task for utilization and progress.
3) Keystroke pattern (for future use)—The average quantity of data that is sent and received with each packet and the amount of time that is spent online. The third level of structure shows the structure of organizational levels, operational levels, and the total number of levels required to connect and state organizational action and action taken.
4) Packet Level—The average quantity of data that is transmitted and received in each packet.

6.7 GRAPHICAL ANALYSIS

The result of proposed work is shown in Figures 6.3, 6.4, 6.5, and 6.7. Table 6.1 provides analysis of the cyber attacks.

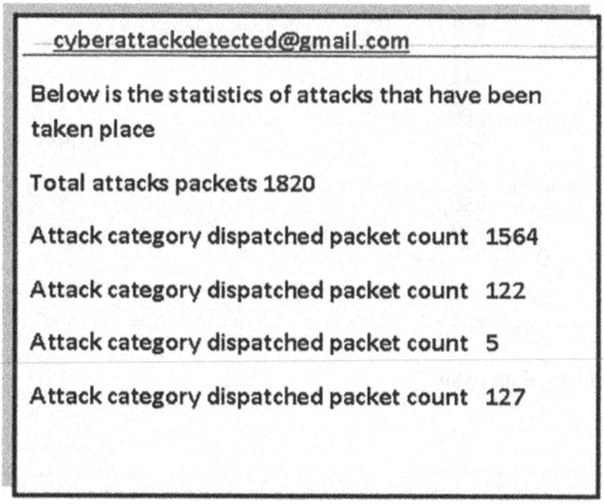

cyberattackdetected@gmail.com

Below is the statistics of attacks that have been taken place

Total attacks packets 1820

Attack category dispatched packet count 1564

Attack category dispatched packet count 122

Attack category dispatched packet count 5

Attack category dispatched packet count 127

FIGURE 6.3 By way of electronic mail.

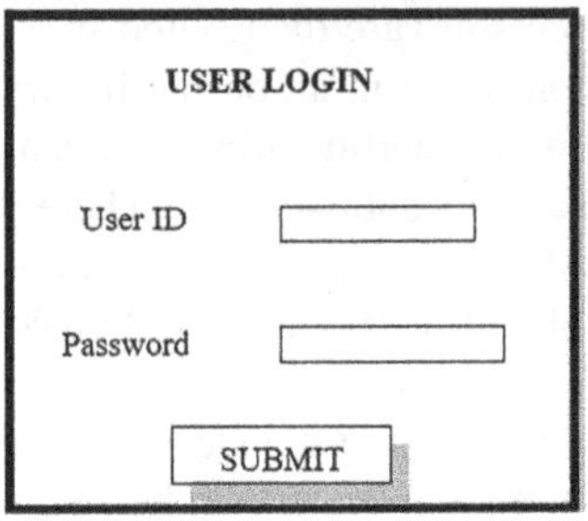

USER LOGIN
User ID
Password
SUBMIT

FIGURE 6.4 To access the log-in page.

TABLE 6.1
Cyber Attacks

Model	Type of Attacks
S	1820
U	122
KT	5
PL	127

2000
1500
1000
500
0
L
KT
U
S
Series1

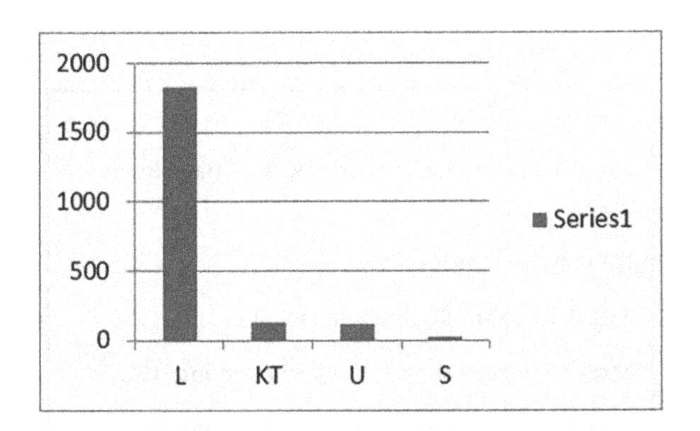

FIGURE 6.5 Statistical analysis.

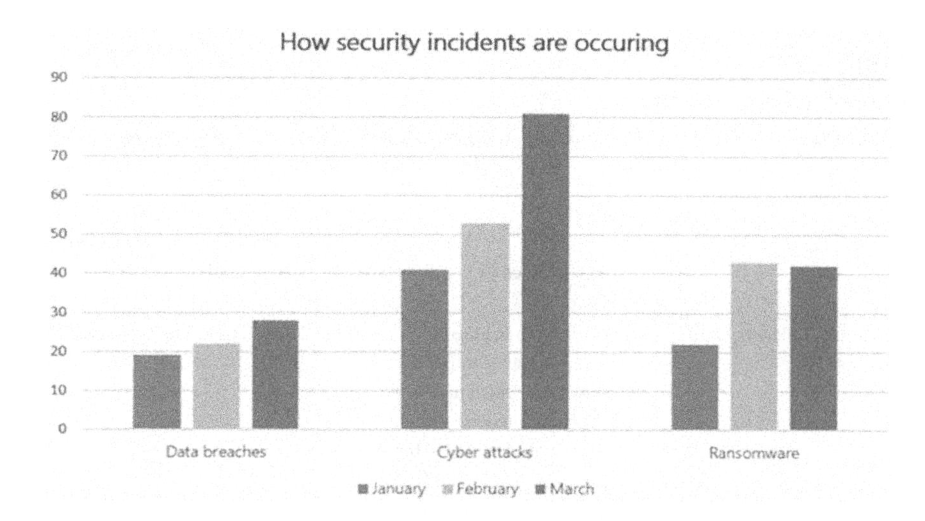

FIGURE 6.6 A graphical breakdown of the many sorts of assaults that were launched against the process in relation to the total number of packets targeted.

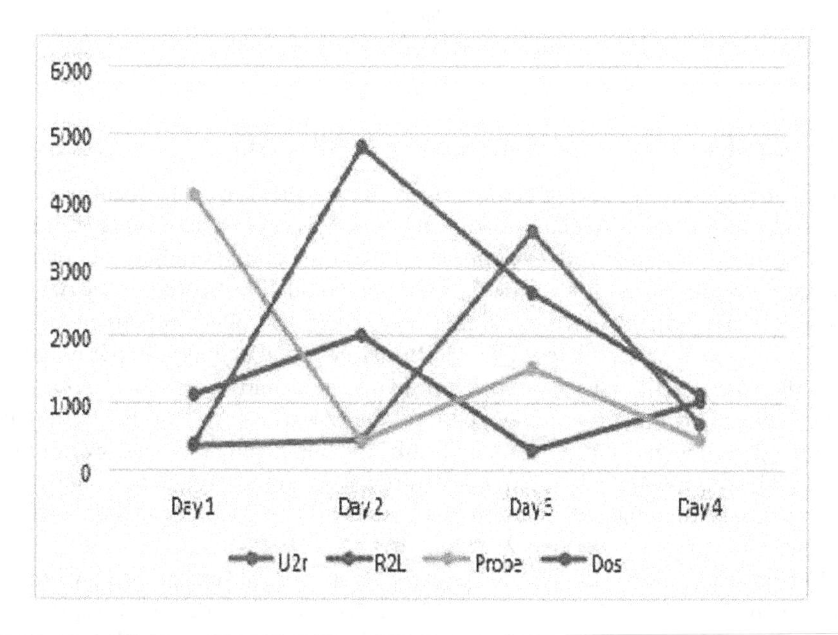

FIGURE 6.7 Each day of the week is shown by a bar graph, and weekly assaults are analyzed using a statistical model.

TABLE 6.2
Brief of Existing Works

Researcher	Research Work Done	Comments
M. Zamani; December 2013	Dynamic anomalies detection system	Intrusion detection of anomalies
H. Alqahtani & H. Minhaz; July 2020	Devised various machine learning methods such as DT, RF, Bayes Naïve	Machine learning techniques were used for learning
M. Shasi & Lakshmanrao; January 2020	It was used to manage large amount of traffic on network in which data was evenly distributed, which was a challenging task for identification of boundaries in between distant variant	Large range of network traffic
P. Malik, R Devakunchari & Sourabh; May 2019	Proposed a time-based detection system	IDS
K. Blanch, Raymond & Manjeet; 2020	Develop a method which was very easy to identify unsecured data, which can be planned future to reduce attacks on it	Extended security system on data solutions

6.8 CYBER ATTACKS USE MACHINE LEARNING

Social engineering is a sort of artificial intelligence that is used by identity thieves as a technique to deceive and influence users into providing sensitive data or performing functions, such as performing a wire transfer and clicking on anything that could be hazardous. The operations of criminals are facilitated by machine learning (ML), which makes it simple and quick to collect intelligence about firms, workers, and associates. This facilitates criminal activity. To put it another way, the use of artificial technology mitigates the negative effects of perception management.[19] Phishing, malware, and ransomware are all forms of computer hacking that rely on exploiting human fallibility. Spam, phishing, and spear phishing are all subsets of phishing. To put it another way, every single person needs to be duped. In these kinds of situations, ML is also utilized to educate nanotechnology by producing real-world scenarios.

Impersonation, also known as spoofing, is a form of deception in which an attacker pretends to be someone else, typically a businessperson, a representative of a well-known brand, or an expert in their field. Hackers will do in-depth research on a variety of target aspects using a variety of techniques. Imagine for a moment that an adversary wants to pretend to be the CEO in order to send phishing malware to you. He makes use of technology in order to understand how the CEO operates by reading messages and blogs on social media websites. Fake emails, photos, and even voices can be used to help machine learning or artificial intelligence. Ransomware, Trojan horses, spyware, and other forms of malware are unidentified and can cause damage to the datasets and information. The vast majority of online threats, on some level, utilize ransomware in the form of hostage, adware, or malware. Email transmission of unauthorized content and connections is responsible for a significant number of Structured Query Language (SQL) infections. The hostile actors are employing AI and ML for learning-based generation of their attacks. The use of malware in defense strategies is appropriate.[20]

Discovering vulnerabilities: The process of finding defects in software applications and computer systems relies heavily on artificial intelligence and machine learning, including the structures of these technologies. Flaws are the same thing as weaknesses, and weaknesses are what allow software to be hacked. AI and ML make it possible to quickly and effectively locate errors and bugs of this kind. In the coming weeks, months, or even years, there is a possibility that an error will be discovered.

Capuches and passwords: In the event of an infringement of spammers and codes, criminal hackers may utilize machine learning models to bypass security measures like as capuches and passwords. In the case of capuches, ML makes it possible for convicted criminals to train the robot (in addition to the automata) to overcome such security challenges.

Bots and automation: Machine learning has the potential to improve a number of different aspects and procedures of a strike. Imagine a programmer is responsible for the creation of an email containing identity theft. You are required to send a predetermined number of emails to these individuals on a regular basis. Machines might be able to be of assistance. Formulas are often necessary in order to organize and counteract distributed denial of service attacks (DDoS) that make use of ransom ware or skeleton computers.

6.9 CONCLUSION

The constantly shifting and intricate nature of assaults on computer networks renders the IDS tactics that are currently being used to defend against them unsuitable. As a result of this, it is possible to achieve higher detection rates, reduced false alarm rates, and provide cheaper costs for calculation and transmission if you make use of efficient adaptive approaches such as machine learning. We took a look at a variety of significant algorithms for the identification of intrusions based on machine learning. Because of their one-of-a-kind characteristics, ML approaches are an excellent choice for this application. These techniques are able to provide IDS with high detection rates and low rates of false positives, and they are also able to react swiftly to changes in harmful behavior. IDS uses a variety of different machine learning strategies, including random forest, decision tree, and logistic regression, to name just a few of them. The most effective solutions should be provided by the IDS in accordance with the requirements. Any company or organization that does not implement these strategies immediately or in the near future puts the safety of its data in jeopardy.

REFERENCES

1. Dipankar Dasgupta. (1999). Immunity-based intrusion detection system: A general frame-work. In *Proceedings of the 22nd national information systems security conference (NISSC)*. Arlington: ACM Online Publications.
2. Jonatan Gomez & Dipankar Dasgupta. (2002). Evolving fuzzy classi_ers for intrusion detection. In *Proceedings of the 2002 IEEE workshop on information assurance*. West Point, NY: IEEE Explore.
3. Steven, A. H., Forrest, S., & Somayaji, A. (1998, August). Intrusion detection using sequences of system calls. *Journal of Computer Security*, 6(3), 151–180.
4. Peter Mell Karen Scarfone. (2007). Guide to intrusion detection and prevention systems (IDPS). *National Institute of Standards and Technology, NIST SP*, 800–894.
5. Mohammadi, S., Mirvaziri, H., Ghazizadeh-Ahsaee, M., & Karimipour, H. (2019). Cyber intrusion detection by combined feature selection algorithm. *Journal of Information Security and Applications*, 44, 80–88.
6. Tapiador, J. E., Orfla, A., Ribagorda, A., & Ramos, B. (2013). Key-recovery attacks on kids, a keyed anomaly detection system. *IEEE Transactions on Dependable and Secure Computing*, 12(3), 312–325.
7. Tavallaee, M., Stakhanova, N., & Ghorbani, A. A. (2010). Toward credible evaluation of anomaly-based intrusion-detection methods. *IEEE Transactions on Systems, Man, and Cybernetics, Part C (Applications and Reviews)*, 40(5), 516–524.
8. Foroughi, F., & Luksch, P. (2018). Data science methodology for cybersecurity projects. arXiv preprint arXiv:1803.04219.
9. Rainie, L., Anderson, J., & Connolly, J. (2014). Cyber attacks likely to increase. *Digital Life in.*, 2025.
10. Fischer, E. A. (2005). Creating a national framework for cybersecurity: An analysis of issues and options. In *Library of congress*. Washington, DC: Congressional Research Service.
11. Craigen, D., Diakun-Thibault, N., & Purse, R. (2014). Defning cybersecurity. *Technology Innovation Management Review*, 4(10), pp. 13–21.
12. Jang-Jaccard, J., & Nepal, S. (2014). A survey of emerging threats in cybersecurity. *Journal of Computer and System Sciences*, 80(5), 973–993.
13. Mukkamala, S., Sung, A., & Abraham, A. (2005). Cyber security challenges: Designing efficient intrusion detection systems and antivirus tools. In V. Rao Vemuri (Ed.), *Enhancing computer security with smart technology*. Auerbach Publications, Taylor & Francis Group, pp. 125–163. (Auerbach, 2006).
14. Bilge, L., & Dumitraş, T. (2012). Before we knew it: An empirical study of zero-day attacks in the real world. In *Proceedings of the 2012 ACM conference on computer and communications security*. ACM Digitial Online, New York, pp. 833–844.
15. Davi, L., Dmitrienko, A., Sadeghi, A-R., & Winandy, M. (2010). Privilege escalation attacks on android. In *International conference on information security*. New York: Springer, pp. 346–360.
16. Alazab, A., Hobbs, M., Abawajy, J., & Alazab, M. (2012). Using feature selection for intrusion detection system. In *2012 international symposium on communications and information technologies (ISCIT)*. IEEE Explore, USA, pp. 296–301.
17. Viegas, E., Santin, A. O., Franca, A., Jasinski, R., Pedroni, V. A., & Oliveira, L. S. (2016). Towards an energy-efficient anomaly-based intrusion detection engine for embedded systems. *IEEE Transactions on Computers*, 66(1), 163–177.
18. Xin, Y., Kong, L., Liu, Z., Chen, Y., Li, Y., Zhu, H., Gao, M., Hou, H., & Wang, C. (2018). Machine learning and deep learning methods for cybersecurity. *IEEE Access*, 6, 35365–35381.

19. Dutt, I., Borah, S., Maitra, I. K., Bhowmik, K., Maity, A., & Das, S. (2018). Real-time hybrid intrusion detection system using machine learning techniques. In Rabindranath Bera, Subir Kumar Sarkar, Swastika Chakraborty (Eds.), *Advances in communication, devices and networking*, Springer, Singapore, pp. 885–894.
20. Ragsdale, D. J., Carver, C., Humphries, J. W., & Pooch, U. W. (2000). Adaptation techniques for intrusion detection and intrusion response systems. In *SMC 2000 conference proceedings. 2000 IEEE international conference on systems, man and cybernetics. 'cybernetics evolving to systems, humans, organizations, and their complex interactions' (cat. No. 0)*, vol. 4. Greece: IEEE, Nashville, TN, pp. 2344–2349.

7 Machine Learning in Natural Language Processing—Emerging Trends and Challenges

Chhaya Suhas Patil and Amit Prakashrao Patil

CONTENTS

7.1 INTRODUCTION

Natural language processing (NLP) and machine learning (ML) are major subfields in artificial intelligence (AI) that have recently received wider appeal and application in several fields. ML and NLP are critical components in converting an artificial agent into an AI-powered agent. As a result of advances in NLP, AI systems can process information more intuitively from the environment. Using machine learning techniques, an AI system may evaluate received data and make better predictions for its actions.

DOI: 10.1201/9781003310785-7

Learning from examples and previous experiences is the capability of machine learning. Genetic algorithms (GA) carry out a predetermined set of operations in accordance with their programming, and therefore are unable to address unforeseen issues. The majority of problems encountered in the real world have a large number of unknown variables, making standard algorithms ineffective. Here is where machine learning comes into play; it is significantly more capable to handle such unknown issues with the help of previous examples.

The classical example is the detection of spam mail. There are many unknown variables to consider in order to detect spam or a genuine email. Spam filters can be circumvented in a variety of ways. Traditional algorithms rely on every variable and feature being hardcoded, which is nearly impossible. The ability of a machine learning algorithm to learn and formulate a general rule will allow it to work well in this kind of environment.

NLP uses the ability of a system to comprehend and process natural languages. Human languages like Hindi or Marathi are incomprehensible by computers, but binary code can be comprehended. It was achieved through NLP, which makes the system capable of understanding Hindi or Marathi languages. The ease of use provided by NLP has led to its growing adoption in recent years. While natural language processing has made it easier for consumers to communicate with sophisticated electronics, behind the scenes much processing takes place to make this possible. Apart from assisting in the appropriate processing of natural language, machine learning has also contributed to the construction and for improved performance of major natural language processing applications like information retrieval (IR) systems, question answering systems, machine translation, etc.

7.2 NLP AND ML TECHNIQUES

It takes a lot of processing to translate human language into a machine. Tasks included in translating natural language into a machine include recognizing words and sentences from a character stream, confirming that the identification of the sentence is consistent with the language rules, identifying the sentence meaning, determining whether the sentence refers to something that is not part of it, finding the intended sentence meaning. Analysis of morphology, analysis of syntax, analysis of semantics, analysis of discourse, and analysis of pragmatics are the names given to the tasks. Machine learning techniques play a vital role in improving the NLP application accuracy and efficiency.

In this section, we give an overview of the work done in NLP using the ML technique, which highlights the importance of machine learning in the field of NLP.

Machine learning techniques such as support vector machines, naive Bayes, random forest, and decision trees as well as deep learning techniques such as convolution neural network and recurrent neural network have been found to contribute significantly to almost all analysis stages of NLP.[1]

Since data is received by the computing system in the form of 0s and 1s, with the help of American Standard Code for Information Interchange (ASCII), these 0s and 1s can be converted into alphabets. As a result, when a paragraph or sentence is given to a machine, it will receive a bunch of characters. It is important to recognize the words and phrases in the text through a process known as tokenization. Tokenization has been accomplished using a variety of machine learning and techniques of deep learning, including recurrent neural network (RNN) and support vector machine.

A machine that has completed tokenization will have a collection of sentences and words. Affixes are found in the majority of sentences. These affixes make things more difficult for machines because it's nearly impossible to build a dictionary with every possible affix of every word. The morphological analysis level's next purpose is to eliminate these affixes. Stemming or lemmatization can be used to eliminate these affixes.

Researchers have been experimenting with how machine learning algorithms can increase tokenization effectiveness. Anand Kumar et al. (2009) used tokenizing morphological information in Tamil texts by sequence labelling classification algorithms.[2] They found that support vector machine-based machine learning tools gave good results compared to other machine learning-based tools. Nisha et al. (2015) described the classification problem used for morphological analysis of Malayalam language words.[3] They used the memory-based language processing (MBLP) algorithm. The goal of the proposed system is to write meaningful parts of words instead of giving detailed morphology of words.

Every natural language by nature is ambiguous. In natural language, a single word has many meanings and identifying which meaning should be taken for the word that occurs in a sentence is decided by the word context. This technique is known as word sense disambiguation (WSD), and it's very challenging to find the correct meaning of the word. WSD is an important research area and work is still being carried out by many researchers. The problem of WSD has been viewed as a classification problem in machine learning. The authors Borah et al. (2014) proposed Assamese as an automatic WSD system using a naive Bayes classifier.[4] They used unigram co-occurrences (UCO), POS of target word (POST), POS of next word feature (POSN), and local collocation (LC) features of their system. The proposed system has performed best giving when combined all features by giving an F1 measure of 86%.

The authors Walia et al. (2018) developed WSD system using a supervised k-nearest neighbors (K-NN) algorithm for the Gurumukhi (Punjabi) language.[5] They used the Punjabi corpora and two sets of features; the first feature was a group of words that regularly appeared alongside ambiguous words, and the second was a collection of words that surround an ambiguous word. A five-fold cross-validation technique was used to divide the dataset into training datasets and testing datasets.

Mahmoodvand and Hourali (2017) experimented for the Persian language WSD system with a minimal supervision machine learning approach.[6] This system was developed using a news dataset. They got 88% recall, 95% precision, and 93% accuracy rate for 5368 documents. The authors presented an approach to unsupervised learning for developing WSD systems. They used the non-aligned bilingual Portuguese and Chinese bilingual corpus. They found a 6% improvement in the proposed system compared to baseline methods. The proposed system goal was to translate the sentence along with handling word sense disambiguation.[7]

The authors Shree and Shambhavi (2015) evaluated WSD systems developed for Indian languages using different techniques including ML techniques, knowledge-based techniques, and hybrid techniques.[8] They concluded that machine learning approaches outperform compared to other methods.

When trying to make sense of a voluminous text, reference resolution is crucial. ML approaches were also employed to resolve coreference. Veena et al. (2017) concluded that the support vector machine outperforms compared to other ML algorithms for the used dataset.[9] The authors experimented to resolve coreference on a noun phrase in unrestricted text using a decision tree machine learning approach. They evaluated the proposed system on the MUC-6 coreference corpora.[10]

The sarcastic reorganization approach is extremely useful for improving automatic sentiment analysis obtained from various microblogging sites and social media. Sentiment analysis refers to the expressed opinions and attitudes of Internet users in a specific community, as well as the aggregation of those opinions and attitudes. Pawar and Bhingarkar (2020) proposed the sarcasm detection system for English as well as Hindi tweets on Twitter.[11] They used neural network (NN), support vector machine (SVM), and a random forest classifier. This approach shows that the random forest classifier gave higher accuracy as compared to other classifiers.

Sharma et al. (2021) discussed the many levels of language processing as well as the difficulties associated with natural language understanding (NLU).[12] In addition, we've discussed ML approaches that can be used for NLP sub-problems including text categorization, disambiguation, syntactic analysis, part-of-speech tagging (POS) and semantic analysis, etc.

This study makes use of four cutting-edge machine learning classifiers, namely BFTree, OneR, naive Bayes and J48 for sentiment analysis optimization. Three hand crafted datasets are used in the trials, two of which were obtained from Amazon and one from IMDB film reviews. The effectiveness of these four classification methods was examined and evaluated. The naive Bayes was discovered to be quite quick at learning, while OneR appears more potential in producing accuracy. The OneR has 91.3% of precision, 97% of F-measure, and 92.34% in correct classifications.[13]

Soumya and Pramod (2020) explained classification of tweets as negative or positive comments using ML classifications algorithms like RF, SVM and NB. Unigram having SentiWordNet with negations words and Unigram having SentiWordNet are used as input for classification. Three thousand one hundred eighty-four tweets were extracted by the author using the tweeter application programming interface (API). Three thousand eight hundred forty-four tweets have been analyzed with positive 954, negative 1,018, and stop words 145 identified using the classifications algorithms. TF vs IDF and BOW were used as features for sentiment analysis. The best accuracy achieved by the random forest technique using Unigram and SentiWordNet was 95.6%.[14]

The authors attempted to employ ML algorithms. They researched meta and voting classifier combinations, which are two distinct types of methodologies. Tweepy data was collected using API 17. There were many neutral and sarcastic tweets with the negative and positive tweets. There were 438,931 total tweets gathered, of which 75,774 were negative and 75,774 were positive. Preprocessing was done on dataset by removing noisy data such as hashtags, emotions, pictures, retweets, removing non-Arebic letters, normalizing Arabic analogue letters, and tokenization. The classifiers like SVM, RR, NB, BNB, MNB, PA, SGD, LR, Ada boost, and ME were used to extrapolate from given tweets to determine their polarity. The best level of accuracy attained by RR and PA was 99.96%. BNB, Ada boost and LR had accuracy that was at its lowest, which was less than 60%.[15]

Vanaja and Belwal (2018) compared the ML algorithms, support vector machine, and Naïve Bayes algorithm performance for finding the neutrality, positivity, and negativity on Amazon product customer reviews. The input dataset was processed using the Apriori technique to extract the commonly utilized features. SentiWordNet calculates the neutrality, negativity, and positivity score before applying the classifier. Utilizing the F-1 measure, recall, precision, and accuracy of each classification, it was possible to compare the algorithm's performance. By the results of the comparison, naive Bayes classifier algorithm had a higher level of accuracy than a support vector machine. FN samples, FP samples, TN samples, and TP samples were used in the calculations.[16]

One of the main concerns affecting users is the prevalence of spam mails marketing. In order to examine the email messages, the author combines personality recognition and sentimental analysis. Validating the proposed method, two different datasets were used. CSDMC 2010 is spam dataset consist of 2,949 messages from emails, and TREC 2007 is public corpus having 75,419 emails (25,220 legitimate and 50,199 spam mails). CSDMC 2010 was used for training the algorithm, and TREC 2007 was used for validation of model. This technique was tested on two separate datasets, and it increased accuracy in both scenarios from 98.98% to 99.18% and from 99.15% to 99.24%. Additionally, this technique is used for many forms of validation, including social media and short messages service validation.[17]

In this paper, the authors Chakraborty et al. (2020) explained the analysis of public sentiment using social media during the Covid-19 pandemic. The author illustrates how crazy individuals are acting in context of the Covid-19 outbreak. Social media is a huge platform where you may express yourself in any situation. For the victim, gathering organized information via social media would be simpler. 226,668 tweets from dataset one were utilized as a starting point, and dataset two has the most retweeted tweets. Data were divided into train, validation, and test sets in order to train the model. Trigram, bigram, and unigram were used to demonstration the performance accuracy. Using different classifiers, dataset 1's accuracy was 81% while dataset 2's accuracy was 75%. As a result of the analysis, the author concluded that social media is not sufficient to assist people.[18]

Authors Oscar et al. (2017) use machine learning to analyze the stigma associated with Alzheimer's disease on Twitter. A machine learning technique modelled stigmatization exhibited in 31,150 tweets linked to Alzheimer's disease that were gathered using the tweeter API. The preprocessing was done on collected dataset by removing the tweets which not relevant to Alzheimer's disease, removal of retweets, and removal of topic name present in username. The sample of analysis was defined through the keywords: "dementia", "senility", "Alzheimer", "alz", and "memory loss". The results of two researchers' hand coding were 24.50% ridicule, 23.79% humor, 19.29% organization, 21.22% metaphors, and 43.41% informational content.[19]

ML may be a class of artificial intelligence which consists of all the techniques and methods that allow computers to automatically learn by utilizing mathematical formulas to extract useful information from a provided information. The SVM, naive Bayes, CRF, TnT, ANN, Brill, HMM, and neural network ML algorithms are the most frequently used for POS tagging. There can be statistical relationships between system variables in particular cases. Nevertheless, it could be challenging to convey these links in a conclusive manner.[20] Naïve Bayes (NB) networks are a kind of probability network model that can be utilized to take advantage of these arbitrary connections or correlations between a problem's variables. The question of "What is the chance of a certain word appearing before the other words in a sentence?" is answered by the probabilistic model using the conditional probability.[21]

Hirpssa and Lehal (2020) solve the POS tagging issue; an automated identification of POS tags for word in the Amharic language was put forth.[22] Comparison of the statistical POS taggers was done. With the identical training and testing datasets, HMM-based tagger, Trigrams'n'Tags tagger, naive bayes, and conditional random field are all compared for performance. The experimental finding demonstrates that the CRF-based tagger outperformed others' performance. The experiment's highest accuracy, 94.08%, was attained by the CRF-based tagger.

Vapnik [23, 24] is the person who first proposed support vector machines in 1998. SVM is a ML technique used in binary classification-required applications, which has been adapted for usage in a variety of domain challenges, including NLP. Using the support vector machine learning algorithm, the authors have attempted to create a predictive model for Sindhi POS. Along the same dataset, RBA and SVM are tested. According to the evaluations, SVM has outperformed RBA tagging methods in terms of detection performance. In essence, an SVM algorithm creates a linear hyperplane which divides the positive collection set from the negative collection set with the maximum boundaries.[25]

A technique for creating discriminative probability models that separate and label a set of sequential data is called a conditional random field data.[26–30] An undirected x, y graphical model known as a conditional random field is one in which each vertex (yi) denotes a random variable whose distribution is dependent on a particular observation variable (X), and each margin (xi) denotes a dependency between the two (yi) random variables. A collection of f functions defines the relationship between Yi and Xi (Yi-1,Yi,X,i).[31] An Urdu part of speech tagger with both independent feature and language-dependent sets was proposed by Khan et al. (2019) [32] and is based on CRF. To compare the efficiency of machine learning and DL methods, it employed both DL and ML methodologies using the language-dependent feature set.[33]

The most popular model for part-of-speech tagging appropriateness is the HMM. HMM works well when one thing is hidden and another is being seen. In this scenario, the ones being examined are words, while the one being covered is labeled.[34–37] Soon et al. (2001) presented an HMM-based POS for the Awngi language. The dataset used for developing this system was 23 manually tagged sets and 94,000 sentences were collected.

Efficacy of the Awngi HMM POS tagger was assessed using a tenfold cross-validation technique. The experimental finding demonstrates that bigram and unigram taggers achieve tagging accuracy of 94.77% and 93.64%, correspondingly[38]

Ware and Mullett (2012) described noun phrase coreference resolution in unrestricted text using a learning strategy. The method learns on a limited, annotated corpus, and the task involves addressing broad noun phrases rather than merely a specific sort of noun phrase, for example, pronouns. The entity kinds of the noun phrases are also not restricted by it; hence, coreference is given regardless over whether these are of "person," "organization", or other sorts. We test our method using widely available data sets, the MUC-6 and MUC-7 coreference corpora, and the findings are promising. They show that for the broad noun phrase coreference task the learning technique shows promise and achieves accuracy that is on par with nonlearning methods. On these data sets, our system would be the first learning-based system to provide performance on par with cutting-edge nonlearning systems.[39]

Despite being a very interesting area in the NLP community, coreference resolution of ideas has not yet been widely used in clinical papers. As a result, the 2011 i2b2 contest that focused on this topic was a necessary and important task. The goal of the study was to compile various conceptual chains from corpora of clinical data. These ideas fall under the "testing", "treatment methods", "person", and "issues" categories. To group related coreferent concepts, a ML method based on graphical representations was used. The chosen features were split into sets that were both domain-independent and domain-specific. The chosen features were split into sets that were both domain-specific and domain-independent. The 6,949 chains from 489 documents as a training set provided by i2b2 were used for training. A total of 322 documents were tested. When no domain-specific features were present in the feature set, the learning engine produced a f1 measure of 84.23%, and when both domain-specific features and domain-independent were present, it produced a f1 measure of 84.83%. This ML method is an effective way to identify various ideas, which is helpful for real-world uses like the compilation of issue and prescription lists from clinical documentation.[40]

Based on findings from three coreference resolution data sets, we demonstrated that, when analyzing learning techniques according to current practice, we cannot accurately draw many conclusions about each method's fitness for a particular task. We demonstrate that the preliminary distinctions among learning techniques are easily overcome when considering things like algorithm parameter optimization, their interactions, sample selection, and feature selection on the task of coreference resolution in an empirical study of the behaviour of representatives of two machine learning techniques, namely rule induction and lazy learning.[41]

In order to determine if two references are coreferent or not, conventional learning-based coreference resolvers work by building a mention-pair classifiers. Two separate boundaries of recent studies have tried to enhance these mention-pair classifiers: one by training an entity-mention classifier to evaluate whether a prior group is distinct from a given reference and the other by having learned a mention-ranking model to rank preceding mentions for a provided anaphor. We suggested a coreference resolution method based on cluster-ranking that uses a combination of mention rankers and entity-mention models. Furthermore, we demonstrated how our cluster-ranking approach intuitively enables the learning of discourse-new entity recognition along with coreference resolution. Its greater performance versus competing techniques is demonstrated by experimental findings on the ACE datasets.[42]

One of the first preprocessing phases in NLP is known as part-of-speech tagging. Additionally, a number of models with tagging accuracy closer to 97% have indeed been designed for POS Tagging. The POS labeling task entails determining which part of speech each token belongs to and then tagging it appropriately. Many ML techniques have been used extensively for tagging on English language texts. It was demonstrated that a ME model with probability techniques can achieve labeling accuracy of more than 96%.[43] The Trigrams'n'Tags tagger utilizes statistical approaches, and it has been shown to be an effective means of tagging when applied through HMMs. SVMs are frequently utilized for POS tasks as well. Nakagawa et al. (2001) concluded that SVM performed very well for POS tagging as compared to HMM. [44] They employ contextual and substring data from the keywords that come before and after, which may be efficiently combined in SVM to create educated assumptions about the POS tags of unidentified words and to create right to assume about the POS tags of unidentified words. HMM can be used for POS tagging as well. Schmid and Laws (2008) concluded that conventional techniques like TnT and SVMs cannot effectively handle morphologically complex languages with a large number of part-of-speech tags.[45] A technique named ensemble learning involves layering different machine learning approaches. Marquez et al. (1999) observed that a technique named ensemble learning involve layering different machine learning approaches.[46]

The part-of-speech tagging stage is typically followed by the NLP subtask of chunking. Using regular expressions, it entails grouping numerous distinct tokens into cohesive units, such as noun phrases. It is particularly helpful for obtaining information on names of things, such as, places, corporations, people etc. Kudo and Matsumoto (2001) have discovered that SVM method exhibit excellent generalization performance, making them ideally suited for the chunking process.[47] Support vector machine algorithm is excellent for dimension reduction because NLP datasets are known to have a huge number of features. Zhang et al. (2002) used a generalized version of the winnow algorithm to perform text chunking with high accuracy by treating it as a sequential prediction model. The literature provides evidence that semi-supervised learning techniques are useful for text chunking.[48] Collobert and Weston (2008) explained utilization DL approach to handle the majority of the natural language processing tasks such as POS, chunking, NER, etc. The authors employ the idea of multitask learning, in which a deep NN model learns to tackle several tasks at once. This strategy has led to the majority of the tasks being completed with decent precision.[49]

Natural language is said to be very ambiguous, indicating that a term or word may take on numerous meanings depending on the context, which makes it challenging to accurately determine its conceptual and syntactic meaning. The utilization of existing resources like MRD, thesaurus, collocation resources, annotated corpora, and ontologies, as well as considerable lexical and syntactic understanding of a language, are requirements for WSD.[50–55] The supervised approaches used most often for WSD include SVMs, decision trees, neural networks, instance-based learning, decision lists, and evolutionary algorithms. Unsupervised techniques include employing word-based clustering, context-based clustering and co-occurrence graphs. The WSD using a corpus has also been successfully accomplished using graph-based algorithms.[51]

SRL is a sophisticated NLP challenge that entails attempting to determine the answers to questions like "who?," "what?," "where?," "when?," etc. Applications for SRL include summarization, translation, information extraction, and other processes. In this paper, the authors Màrquez et al. (2008) and Carreras and Màrquez (2005) described that the most often utilized approaches for SRL are conditional random fields, SVM with kernels, memory-based learning, ME methods, and DT techniques.[52–53]

Named entities are specified as statements containing names of individuals, organizations, places, etc. The ME model was the approach for NER that was used the most frequently. Use of HMM, algorithms like AdaBoost, SVM, and CRFs are additional techniques that produce the best results. Parsing, stemming, syntax analysis, and tokenization, etc. are further NLP activities. For practically all of these tasks, ML methods are applied, particularly for languages like English that have a sizable number of labelled data that are freely accessible online.[54]

7.3 EMERGING TRENDS IN NLP USING ML

NLP is continuously evolving, and the number of applications for NLP is increasing by the minute. NLP will grow in popularity in the coming years as a result of pre-trained models, low-code, ready-to-use, and no-code technologies that are available to everyone.

Here we explore some of the main natural language processing trends.

7.3.1 Transfer Learning

Transfer learning is a ML technique that involves training a model for one activity, and then using it in another related activity. Instead of creating and training a model from scratch, which is time-consuming, costly, and requires a large amount of data, we simply fine-tune one that has already been built. Transfer learning is very popular in computer vision applications along with other NLP tasks such as NER, sentiment analysis, and intent classification.

7.3.2 BERT and ELMo Transformer

Generative Pre-trained Transformer 3 makes it now possible to comprehend word context in a way previously not possible. The Embedding Language Models (ELMo) and Bidirectional Encoder Representations Transformers (BERT) will be the focus of the NLP community in the near future. These models have indeed been trained on massive quantities of data and can significantly increase the performance of a variety of NLP tasks.

7.3.3 No-Code or Low-Code Tools

The NLP activities previously restricted to technology experts like data scientists and developers are now available to non-technical individuals, thanks to SaaS enterprises that strive to generalize machine learning technology and NLP.

7.3.4 Integrating Supervised and Unsupervised ML Techniques

Integrating unsupervised and supervised ML techniques improve the performance of ML models while training the models for NLP, particularly for text analysis.

7.3.5 Automating Service to Customer (Chatbots and Ticket Tagging)

The Covid-19 pandemic has resulted in a massive surge in support tickets across many businesses, including finance and travel. Companies are discovering that they need to automate simple customer care operations because getting a consumer's inquiry answered faster and more effectively is more important than ever.

At the front lines of dealing with customers, using natural language processing (NLP) techniques, a critical role will be played by chatbots and support desk software together.

7.3.6 Cyberbullying and Detecting Fake News

NLP has evolved into a vital tool for detecting fake news and preventing cyberbullying by reducing human effort and time. The transformer technique aids in the development of a high-performance fake news detection system, and classifiers are also important in the development of a cyberbullying detection system.

7.4 IMPLEMENTATION CHALLENGES WITH NLP

7.4.1 Data Challenges

The fundamental issue is information overload, which makes it difficult to find a particular, crucial piece of information among massive data. Due to qualitative and usefulness difficulties, lexical and contextual comprehension is crucial but difficult for summarization technologies. Another important task is determining the contexts of interactions between things and entities, particularly when dealing with low-quality data, complex data, heterogeneous, and high dimensional.

Uncertainties in the data make it more difficult to interpret the context. Finding the relationships between objects and entities depends on semantic. Without knowing the semantic and context of interactions, entities and objects extracted from textual and visual cannot deliver reliable information. Additionally, instead of using keywords, the contemporary search engines may look for entities or objects.

7.4.2 Challenges Related to Text

ML and NLP have become the most powerful and popular technologies used for textual analysis and text classification. Understanding multi-label data preprocessing is necessary for multi-level problem solving in big data analysis. The hardest part of language translation isn't interpreting words; it's understanding sentences' meanings in terms of giving an accurate interpretation. Comparing NLP models to empirical ML models, NLP models are bigger and use more memory.

7.4.3 Languages that Lack Resources

It is a well-known fact that while there is a tonne of information available for widely used languages like English or Chinese, thousands of lesser-used languages get much less focus. Africa alone has between 1,250 and 2,100 languages, yet there is little information available about them. Additionally, it is still exceedingly difficult to transfer activities that require real-world language comprehension from largest resource to smallest resource languages.

7.4.4 Evaluation

We need detailed research that explains why some methods work and others don't in order to establish assessment measures based on it. The issue of evaluating language technology, especially something as complicated as dialogue, is frequently overlooked; but it is a crucial issue. To determine whether our methods are truly generalizable to the whole range of human language, we require a new set of testing tasks and datasets.

7.4.5 Massive or Voluminous Documents

Another major unresolved issue is coping with huge or numerous documents, as existing models are primarily built on recurrent neural networks, which have been incapable of representing extended context well. The second issue is that supervision is expensive and hard to obtain when applied to large or numerous documents.

The primary motivation for incorporating NLP and ML into the system is to obtain a significant advantage, but the use of ML techniques in NLP tasks has the challenges because ML is not 100% reliable and requires time-consuming training. There is still the chance of inaccuracies in predictions and outcomes, which must be considered. It can take several weeks to get a high degree of performance when a new model is constructed without the help of a pre-trained model.

7.5 CONCLUSION

NLP and ML are very important fields of AI, which have gained fame over the past 20 years. NLP entails many tasks that are crucial including sarcasm detection, word sense disambiguation, stemming, syntactic parsers, named entity recognition and reference resolution, and part-of-speech tagging. This chapter highlighted the machine learning techniques that are successfully used to perform these NLP tasks to gain high performance and accuracy. The chapter also covered the current trends in NLP applications using machine learning techniques and the challenges faced by ML-based NLP applications.

Although ML methods have yielded better results than other methods for carrying out various NLP tasks, it was found that they are suitable for implementation in a real-world scenario to adapt to diverse situations and datasets. So there is an abundance of research opportunities for studying the various states of NLP as well as specific tasks like sarcasm detection, WSD, stemming, syntactic parsers, NER systems, resolution of reference, and part-of-speech tagging.

REFERENCES

1. Tatwadarshi, P., Nagarhalli, Dr. V. V., & Rana Dr., N. K. (2021). Impact of machine learning in natural language processing: A review. In *Third international conference on intelligent communication technologies and virtual mobile networks (ICICV 2021)*. IEEE, Conference Location: Tirunelveli, India, DOI: 10.1109/ICICV50876.2021.9388380, Date of Conference: 04-06 February 2021.
2. Anand Kumar, M., Arun Kumar, C., Dhanalakshmi, V., Rekha, R. U., Soman, K. P., & Rajendran, S. (2009). Morphological analyzer for agglutinative languages using machine learning approaches. In *IEEE international conference on advances in recent technologies in communication and computing*, pp. 433–435.
3. Nisha, M., Reji, R. K., Rekha, R. C. T., & Reghu Raj, P. C. (2015). Malayalam morphological analysis using MBLP approach. In *IEEE international conference on soft-computing and networks security (ICSNS)*, pp. 1–5.
4. Borah, P. P., Talukdar, G., & Baruah, A. (2014). Assamese word sense disambiguation using supervised learning. In *IEEE international conference on contemporary computing and informatics (IC3I)*, pp. 946–950.
5. Walia, H., Rana, A., & Kansal, V. (2018). A supervised approach on Gurmukhi word sense disambiguation using K-NN method. In *IEEE 8th international conference on cloud computing, data science & engineering (confluence)*, pp. 743–746.
6. Mahmoodvand, M., & Hourali, M. (2017). Semi-supervised approach for Persian word sense disambiguation. In *IEEE 7th international conference on computer and knowledge engineering (ICCKE)*, pp. 104–110.
7. Oliveira, F., Wong, F., & Li, Y. P. (2005). An unsupervised & statistical word sense tagging using bilingual sources. In *IEEE international conference on machine learning and cybernetics*, vol. 6, pp. 3749–3754.
8. Shree, M. R., & Shambhavi, B. R. (2015). Performance comparison of Word sense disambiguation approaches for Indian languages. In *IEEE international advance computing conference (IACC)*, pp. 166–169.
9. Veena, G., Gupta, D., Daniel, A. N., & Roshny, S. (2017). A learning method for coreference resolution using semantic role labeling features. In *IEEE international conference on advances in computing, communications and informatics (ICACCI)*, pp. 67–72.
10. Dzunic, Z., Momcilovic, S., Todorovic, B., & Stankovic, M. (2006). Coreference resolution using decision trees. In *IEEE 8th seminar on neural network applications in electrical engineering*, pp. 109–114.
11. Pawar, N., & Bhingarkar, S. (2020). Machine learning based sarcasm detection on Twitter data. In *IEEE 5th international conference on communication and electronics systems (ICCES)*, pp. 957–961.
12. Sharma, H., et al. (2021, October). Improving natural language processing tasks by using machine learning techniques. In *2021 5th international conference on information systems and computer networks (ISCON)*.
13. Singh, J., Singh, G., & Singh, R. (2017). Optimization of sentiment analysis using machine learning classifiers. *Human-centric Computing and Information Sciences*, 7, 32.
14. Soumya, S., & Pramod, K. J. I. E. (2020). Sentiment analysis of Malayalam tweets using machine learning techniques. *ICT Express*, Volume 6, Issue 4, December 2020, Pages 300–305, https://doi.org/10.1016/j.icte.2020.04.003
15. Gamal, D., Alfonse, M., El-Horbaty, E.-S. M., & Salem, A.-B. M. J. P. C. S. (2019, January). Implementation of machine learning algorithms in Arabic sentiment analysis using N-gram features. In *Proceedings of the 9th international conference of information and communication technology [ICICT-2019]*, vol. 154. Nanning, Guangxi, China, Science Direct Available online at www.sciencedirect.comProcedia Computer Science 154 (2019) 332 -3401877-0509 © 2019 The Authors. Published by Elsevier Ltd. pp. 332–340.
16. Vanaja, S., & Belwal, M. (2018). Aspect-level sentiment analysis on e-commerce data. In *2018 international conference on inventive research in computing applications (ICIRCA)*. Coimbatore, India: IEEE, pp. 1275–1279.

17. Ezpeleta, E., Velez de Mendizabal, I., Hidalgo, J. M. G., & Zurutuza, U. J. L. J. O. T. I. (2020). Novel email spam detection method using sentiment analysis and personality recognition. *Logic Journal of the IGPL*, 28(1), 83–94.
18. Chakraborty, K., Bhatia, S., Bhattacharyya, S., Platos, J., Bag, R., & Hassanien, A. E. J. A. S. C. (2020). Sentiment analysis of COVID-19 tweets by deep learning classifiers—A study to show how popularity is affecting accuracy in social media, 97, 106754.
19. Oscar, N., et al. (2017, March). Machine learning, sentiment analysis, and tweets: An examination of Alzheimer's disease stigma on Twitter. *The Journals of Gerontology Series B Psychological Sciences and Social Sciences*, 72(5).
20. Tseng, C., Patel, N., Paranjape, H., Lin, T. Y., & Teoh, S. (2012). Classifying Twitter data with naive bayes classifier. In *2012 IEEE international conference on granular computing classifying*, pp. 1–6.
21. Kumar, S., & Nezhurina, M. I. (2019). An ensemble classification approach for prediction of user's next location based on Twitter data. *Journal of Ambient Intelligence and Humanized Computing*, 10(11), 4503–4513.
22. Hirpssa, S., & Lehal, G. S. (2020). POS tagging for Amharic text: A machine learning approach. *INFOCOMP*, 19(1), 1–8.
23. Antony, P. J., Mohan, S. P., & Soman, K. P. (2010). *SVM based part of speech tagger for Malayalam.* In *ITC 2010–2010 international conference recent trends information telecommunication computer*, pp. 339–341.
24. Surahio, F. A., & Mahar, J. A. (2018). Prediction system for Sindhi parts of speech tags by using support vector machine. In *2018 International conference on computing, mathematics and engineering technologies (iCoMET)*, pp. 1–6
25. Surahio, F. A., & Mahar, J. A. (2018). Prediction system for Sindhi parts of speech tags by using support vector machine. In *2018 International conference on computing, mathematics and engineering technologies (iCoMET).*
26. Pisceldo, F., Adriani, M., & Manurung, R. (2009). Probabilistic part of speech tagging for Bahasa Indonesia. In *Proceedings 3rd international MALINDO work. Coloca.* Singapore: Event ACL-IJCNLP.
27. Khan, W., et al. (2019). Part of speech tagging in Urdu: Comparison of machine and deep learning approaches. *IEEE Access*, 7, 38918–3836.
28. Gashaw, I., & Shashirekha, H. (2018, December). Machine learning approaches for Amharic parts-of-speech tagging. In *Proceedings of ICON-2018*. Patiala, India, 2018 NLP Association of India (NLPAI). pp. 69–74.
29. Suraksha, N. M., Reshma, K., & Kumar, K. S. (2017). Part-of-speech tagging and parsing of Kannada text using conditional random fields (CRFs). In *2017 International conference on intelligent computing and control (I2C2).*
30. Sutton, C., & McCallum, A. (2011). An introduction to conditional random fields. *Foundations and Trends in Machine Learning*, 4(4), 267–373.
31. Hall, J. (2003). A probabilistic part-of-speech tagger with suffix probabilities a probabilistic part-of-speech tagger with suffix probabilities. MSc: Thesis, Växjö University.
32. Khan, W., et al. (2019). Part of speech tagging in Urdu: Comparison of machine and deep learning approaches. *IEEE Access*, 7, 38918–38936.
33. Khorjuvenkar, D. N., Ainapurkar, M., & Chagas, S. (2018). Parts of speech tagging for Konkani language. In *Proceedings 2nd international conference computer methodologies communication*. Erode, India: ICCMC, pp. 605–607.
34. Ankita, A., & Nazeer, K. A. (2018). Part-of-speech tagging and named entity recognition using improved hidden Markov model and bloom filter. In *2018 international conference on control, power, communication and computing technologies*. USA: GUCON, Greater Noida, India. pp. 1072–1077.
35. Mohammed, S. (2020). Using machine learning to build POS tagger for under-resourced language: the case of Somali. *International Journal of Information Technology*, 12(3), 717–729.
36. Mathew, W., Raposo, R., & Martins, B. (2012). Predicting future locations with hidden Markov models. In *Proceedings of the 2012 ACM conference on ubiquitous computing*, pp. 911–918.

37. Demilie, W. B. (2019). Parts of speech tagger for Awngi language. *International Journal of Engineering Science Computer*, 9(1).
38. Soon, W. M., Ng, H. T., & Yong Lim, D. C. (2001). *A machine learning approach to coreference resolution of noun phrases*. Cambridge, MA: Association for Computational Linguistics.
39. Ware, H., & Mullett, C. J. (2012, May 12). *Vasudevan Jagannathan, corresponding author and Oussama El-Rawas, machine learning-based coreference resolution of concepts in clinical documents*. doi:10.1136/amiajnl-2011-000774
40. Hoste, V., & Daelemans, W. (2005). Comparing learning approaches to coreference resolution. There is more to it than 'bias'. In *Proceedings of the ICML-2005 workshop on meta-learning*. Appearing in Proceedings of the ICML-2005 Workshop on Meta-learning, Bonn, Germany, 2005. Copyright 2005 by the author(s)/owner(s).
41. Vincent N. (2017, February 4). *Machine learning for entity coreference resolution: A retrospective look at two decades of research*. Association for the Advancement of Artificial Intelligence press, Hilton San Francisco, San Francisco, California, USA: AAAI.
42. Ratnaparkhi, A. (1996). A maximum entropy model for part-of-speech tagging. In *Conference on empirical methods in natural language processing*. Philadelphia, PA, USA.
43. Brants, T. (2000). TnT: A statistical part-of-speech tagger. In *Proceedings of the sixth conference on applied natural language processing*. Seattle, Washington, USA: Association for Computational Linguistics, pp. 224–231.
44. Nakagawa, T., Kudo, T., & Matsumoto, Y. (2001). Unknown word guessing and part-of-speech tagging using support vector machines. In *NLPRS*, pp. 325–331.
45. Schmid, H., & Laws, F. (2008). Estimation of conditional probabilities with decision trees and an application to fine-grained POS tagging. In *Proceedings of the 22nd international conference on computational linguistics*, vol. 1. Manchester, UK: Association for Computational Linguistics, pp. 777–784.
46. Marquez, L., Rodriguez, H., Carmona, J., & Montolio, J. (1999). Improving POS tagging using machine-learning techniques. In *1999 Joint SIGDAT conference on empirical methods in natural language processing and very large corpora*.
47. Kudo, T., & Matsumoto, Y. (2001). Chunking with support vector machines. In *Proceedings of the second meeting of the North American chapter of the association for computational linguistics on language technologies*. Association for Computational Linguistics, pp. 1–8.
48. Zhang, T., Damerau, F., & Johnson, D. (2002, March). Text chunking based on a generalization of winnow. *Journal of Machine Learning Research*, 2, 615–637.
49. Collobert, R., & Weston, J. (2008). A unified architecture for natural language processing: Deep neural networks with multitask learning. In *Proceedings of the 25th international conference on machine learning*. ACM, pp. 160–167.
50. Navigli, R. (2009). Word sense disambiguation: A survey. *ACM Computing Surveys (CSUR)*, 41(2), 10.
51. Bender, O., Agirre, E., & Edmonds, P. (Eds.). (2007). Word sense disambiguation: Algorithms and applications. *Springer Science & Business Media*, 33.
52. Màrquez, L., Carreras, X., Litkowski, K. C., & Stevenson, S. (2008). Semantic role labeling: An introduction to the special issue. *Computational Linguistics*, 34, 145–159.
53. Carreras, X., & Màrquez, L. (2005). Introduction to the CoNLL-2005 shared task: Semantic role labeling. In *Proceedings of the ninth conference on computational natural language learning*. Stroudsburg, PA, USA: Association for Computational Linguistics, pp. 152–164.
54. Josef Och, F., & Ney, H. (2003). Maximum entropy models for named entity recognition. In *Proceedings of the seventh conference on Natural language learning at HLT-NAACL 2003*, vol. 4. Edmonton, Canada: Association for Computational Linguistics, pp. 148–151.

8 Machine Learning and Future Directions

Koteswara Rao Makkena, Anjali Gautam and Karthika Nataranjan

CONTENTS

DOI: 10.1201/9781003310785-8

8.1 MACHINE LEARNING ON 5G NETWORK

8.1.1 Approach

From medical image diagnosis to self-driving cars, machine learning (ML) is everywhere.[1]

Individuals in a wireless network can use machine learning to find and gather information through communicating on content. Engineers and researchers from all over the world could be led to the practical application of 5G standards with the multiple machine learning protocols.[2–4]

Traditional approaches have a lot of flaws, and machine learning is a wonderful way to solve them. These problems might be solved by using information to analyze as well as by substituting programs with machine learning approaches that automatically learn from old data. An ML task can tackle complex problems involving enormous amounts of data, find patterns that humans overlook, forecast future situations, respond to new information, and discover anomalies.[5]

8.1.2 Types of Learning and its Applications

The categories of machine learning algorithms are often used to define the essential concepts of ML algorithms and related applications in 5G. The prototypical concept of learning can also be used to analyze the differences between these three learning methods. "If a computer program's performance at tasks in T, as measured by P, improves with experience E, it is said to learn from experience E with respect to some class of tasks T and performance measure P."[6]

8.1.2.1 Supervised Learning on 5G

In supervised learning, every knowledge management should be provided with its associated label. The purpose is to use problem instances to train the learning model, which will later be used to identify appropriate results. A collection of samples x (things) and a label y for each one will be present in a high-level synthesis.

Supervised learning can be divided into two types, i.e., classification and regression. Classification is the process of predicating based on discrete values and regression is the process of predicating based on continuous values. The most difficult aspect of supervised learning algorithms is generalization,[7] which refers to how well they function with additional, unknown inputs rather than merely the information they were learned on. A training error is another type of error and is also a big issue. Training error is when training a supervised model an error occurs on training. It is possible to decrease training error by describing a simple optimization issue.

The challenges of supervised learning are underfitting and overfitting. A model which is too "simple" to reflect all relevant class characteristics is said to be underfitting. Overfitting tends to make a model too "complex" and allows it to fit irrelevant data characteristics (noise).[8] When the capacity of machine learning algorithms is proportionate to the job complexity, they perform better.

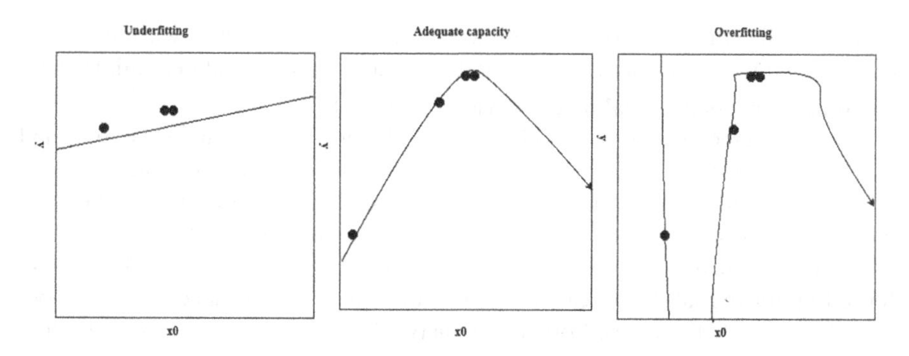

FIGURE 8.1 a) a linear function; b) a quadratic function; c) a polynomial [7]

TABLE 8.1
Highlights the Supervised Learning-Based Ideas for 5G

ML Technique	Learning Model	Applications
Supervised learning	Linear regression	Determine the scheduling strategies for the harvesting nodes, forecast, and design the energy supplies. It will allow the harvesting nodes to adapt to resource availability.[10]
	Statistical linear regression	Assigning channel capacity and frequency distribution dynamically for auto LTE compact cell developments.[9]
	Supervised classifier	Determine the connection need, flexibly allocate the available resources, and established the architecture and data speed in accordance with the connectivity characteristics, such as capacity, latency, and instability.[11]
	Support vector machine	In order to choose the best antenna indices in MIMO,[12] the propagation loss forecasting model for urban contexts[13] classifies the training channel state information.
	Neural network based approximation	Channel learning to infer unobservable channel state information from an observable channel.[14]
	Probabilistic learning	Adjustment of the TDD Uplink–Downlink in XG-PON-LTE system to maximize the network performance based on the ongoing traffic conditions in the hybrid optical-wireless network.[15]
	Artificial neural networks (ANN), multi-layer perceptrons (MLPs)	Modeling and approximation of objective functions for link budget and propagation loss for next-generation wireless networks.[16–20]
	Deep neural networks	Forecasting and synchronization of beam forming matrices at the BSs utilizing uplink pilot signals,[21] channel capacity, and direction of arrival (DOA) estimation in MIMO,[22] as well as training embedding linked to the type with some conditions.

The architecture that follows outlines the adaptive supervised learning methods for original dataset produced by random x value samples and deterministically choosing y for a polynomial function with minimum loss.

The parameters of a model could be updated based on the testing, validation, and implementation stages, and the algorithm could be retained using feedback.

To cope with potential bandwidth requirements, long-term evolution (LTE) tiny units are progressively being created in mobile communications. As a result of the interference, structures of these micro cells are unknown and variable, and the desire for self-optimized substitutes can result in smaller failures, faster speeds, and cheaper costs for users. SONs are self-organizing networks which can learn and adapt to a range of situations.[9]

8.1.2.2 Unsupervised Learning on 5G

The data shown to train the machine learning system in unsupervised learning is an unstructured set of characteristics. Without any guidance, the algorithm is able to group variables with similar characteristics. When we wish to find trends and regularities in a data set, we use this technique. Clustering is a common machine learning technique that produces outstanding results when clustering hardware in a cell network. Unsupervised learning is demonstrated in Figure 8.2.

8.1.2.2.1 Techniques Specified in Unsupervised Learning

8.1.2.2.1.1 Autoencoders One type of ANN used to learn unsupervised data encodings is the autoencoder (AE).

An autoencoder's goal is to prepare its structure that collect the much more essential bits of an incoming picture in order that acquire the lesser form (encoded) for greater data, which is typically used for data reduction.[23] Autoencoders are made up of three parts.

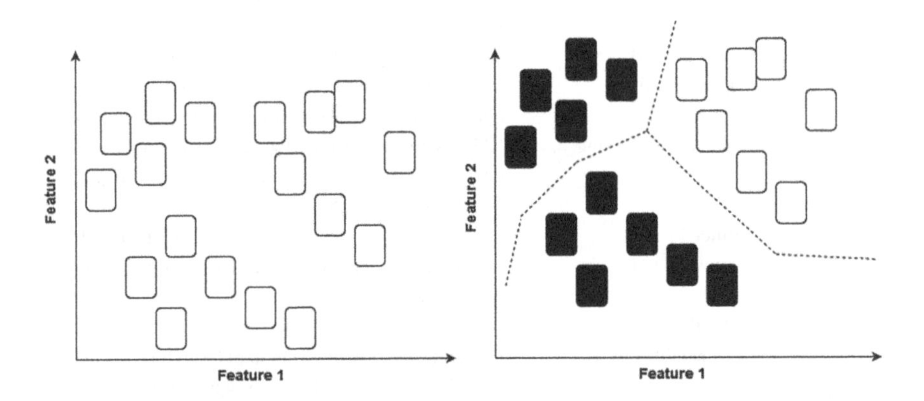

FIGURE 8.2 a) Shows data samples along with features that can be represented in two dimensional axes; b) the occurrences are fed.[1]

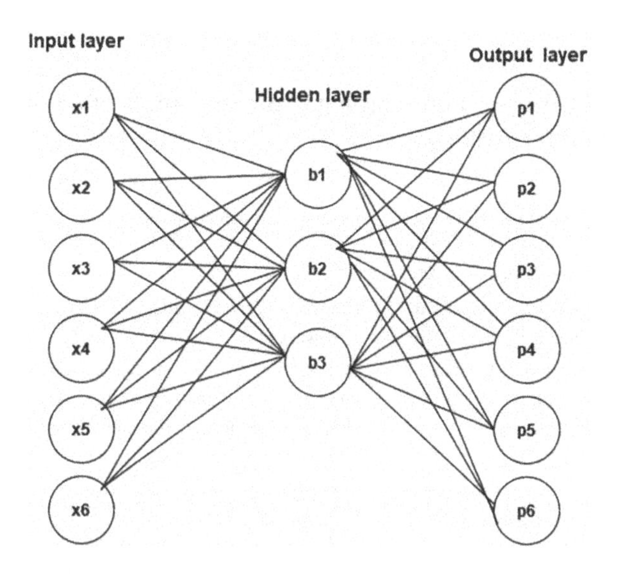

FIGURE 8.3 Decoder architecture.[24]

1) Encoder: An encoded representation of the input data that is typically several orders of magnitude less for the input data.
2) Bottleneck: The compressed knowledge representations are stored inside this module, providing it the most essential part of the network.
3) Decoder: A component that assists the network in "offloading" cognitive concepts and compiles input from its encrypted context. The result is then checked with the reference point.

The architecture as a whole look something like Figure 8.3.

AEs are a sort of neural network that has been utilized for image compression and knowledge discovery for quite some time. As a final efficient ML function, the concept of AE might be applied to higher layers of a data transmission. In a specific situation, the input restoration inside a neural net could be studied. We sought to optimize a huge nonlinear neural network with parameter for a basic high-level goal.

8.1.2.2.1.2 K-Means Clustering Clustering is an unsupervised learning technique in which data points are defined into categories based on their degree of similarity.[25]

8.1.2.3 Reinforcement Learning on 5G

Wireless networks operate in surprisingly stochastic environments. In order to optimize the intended objective, a separate stochastic control process for mathematical framework modeling could be simulated system dynamics in an uncertain situation.[38]

TABLE 8.2
A Summary of Unsupervised Learning's Potential Applications in 5G Wireless Communication Technologies

ML Technique	Learning Model	Applications
Unsupervised learning	Gaussian Mixture Model (GMM), Expectation Maximization (EM)	Mutual band detection.[26]
	Hierarchical clustering analysis	Detection of irregularities, defects, and invasions in mobile networks [27]
	Vector quantization method	Storing the data center contents in clusters to reduce the data travel among distributed storage system.[28] Optimal handover estimation by clustering the UEs according to their mobility patterns.[29] Relay node selection in vehicular networks.[30]
	Unsupervised soft-clustering	Latency reduction by clustering fog nodes by automatically decide which low power node (LPN) is upgraded to high power node (HPN) in heterogeneous cellular networks.[31]
	Self-organizing map (SOM) learning	Planning the coverage of HetNets with dynamic clusters.[32]
	Autoencoders (AE)	Channel characterization by interpreting a communication system design as an end-to-end reconstruction task, in order to jointly optimize transmitter and receiver components in a single process.[33]
	Adversarial autoencoders (AAE)	Detection anomalous behaviour in wireless spectrum by using power spectral density (PDS) data in an unsupervised learning setting.[34]
	Affinity propagation clustering	Data-driven resource management for ultra-dense small cells.[35]
	Non-parametric Bayesian learning	Traffic reduction in a wireless network by proactively serving predictable user demand via caching at BSs and users' devices.[36]
	Generative deep neural networks (GDNN)	Capture the presence of traffic correlations that impact the reading of multiple sensors deployed in the same geographical area.[37]

A learning creature known as an agent engages with its environment. Actors choose an action from options present in the following scenario while making each one as big as possible.[39]

R produces a corresponding incentive or punishment (negative reward) as a reaction from the service and moves towards states for where the action can be executed, as specified in Figure 8.4(a).

While sensing an observation, the model utilizes the stochastic mathematical framework model, which is a generic model and also has to directly examine the environment, but instead it has a poor experience, as shown in Figure 8.4(b).

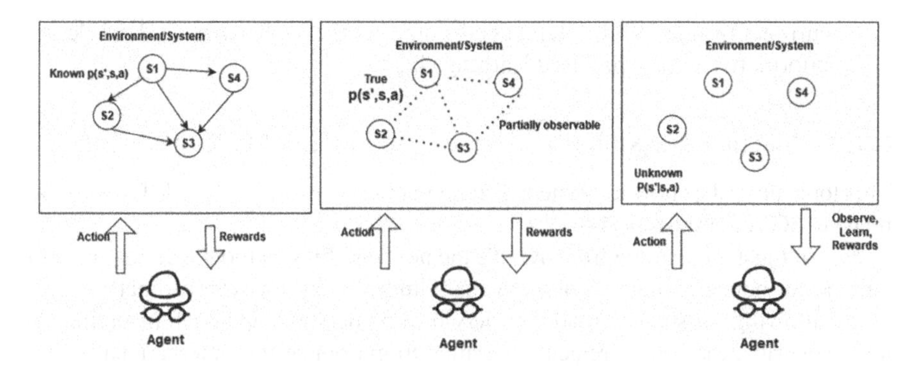

FIGURE 8.4 a) Markov decision process; b) partially observed Markov decision process; c) Q-learning.[40]

Q-learning, as seen in Figure 8.4(c) is a prominent prototype RL method.

$$V(s) = \max R(s) + \Sigma P(S^1 | s,a) R(S^1) \quad V(s) = max\ R(s) + \Sigma O(s^1 | s,a) R(s^1) \quad \text{Q}$$
$$\text{(a)} \qquad\qquad \text{(b)} \qquad\qquad \text{(c)}$$

For optimality, we are using a dynamic programming equation.

$$V^{\pi^*}(S) = \max_{a \in A} \sum_{s' \in S} \mathrm{P}_{(s'|(s,a))} \left(\mathrm{R}(s,a) + \gamma V^{\pi^*}(S') \right)$$

Where γ is a sale price that decreases the combined value of the incentives to a fixed number.

The RL algorithm could be used to discover an optimum approach π * in MDP issues.

Design RL should be applied when an MDP framework is provided. Otherwise, the model-free RL method may also be employed.[41]

The previous equation can be rewritten as follows:

As seen here, the Q-learning process turns on above equation into an iterative approximation procedure.

$$V^{\pi^*} = \max_{a \in A} Q^{\pi}(s,a)$$

$$Q(s,a) \leftarrow (1-\alpha) + \alpha[R(s) + \gamma \max_{a \in A} Q(s',a)]$$

Where α denotes the learning rate, with a greater value denoting a faster pace, the more the agent relies on the benefit and minimizes it's got, the lower the current Q-value. While RL takes advantage of the feedback reward, it is less informative than supervised learning.[42, 43]

8.1.3 Machine Learning is an Approach to Support 5G Requirements

This topic describes how to create a different network as well as how ML algorithms relate to 5G requirements.[44]

5G isn't just an upgrade from 4G; it's the next big thing in mobile communication with performance advantages of many magnitudes over parameter estimates.

By allowing for such a broader variety of providers and approach involving various scenarios, these requirements are aims to guarantee that international mobile telecommunication-2020 offers more adaptability, safety, and dependability than previous technologies.[45, 46]

The following sections outline the basic services.

8.1.3.1 Enhanced Mobile Broadband (eMBB)

Recent advances with higher data rate demands will be able to operate more than a consistent service area by altering the present MBB service (Figure 8.5).

8.1.3.1.1 Optimum Data Rate

To handle highly sought data-driven use cases, the maximum average throughput will be enhanced. From change of bandwidth in 5G, IMT-2020 equipment will be needed to achieve a 20-fold increase in data rate over the preceding technological specification.

This is the maximum achievable bandwidth in optimal error-free condition given to a one base station if all of the addressable radio spectrum for the relevant connection is utilized.

A rise in maximum bandwidth for 5G should be used as an upgrade that makes use of all available distributed resources.

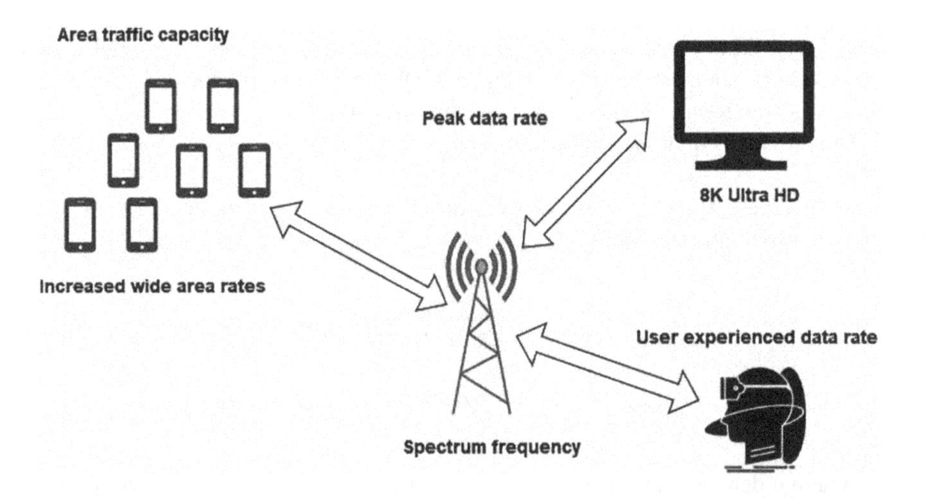

FIGURE 8.5 eMBB applications provide greater energy bit rate and greater broad throughput.[47]

As advancements emerge, the need for additional frequency has increased. As telecommunication providers evaluate their options for rolling out 5G networks, they will need a sizable amount of channel bandwidth to rich resource the efficiency value of the proposed technology.

The key to releasing the full potential of eMBB is believed to be centimetre waves (cmWave) and millimetre waves (mmWave).[47]

8.1.3.1.2 User Experienced Data Rate

The five percent point of the cumulative distribution function (CDF) of the user throughput over a period of active time is the description of the user perceived data rate in a congested urban environment. IMT-2020 aims to deliver a 10-fold increase in user data rate over 4G LTE.

It can be utilized as a 5G measure of quality in practical applications due to its tight association to other requirements like maximum transmission rate and latency.

For instance, wireless network virtualization (WNV) would improve system speed and expand system capacity, providing end users with a better user experience.[48]

8.1.3.1.3 Potential of the Region's Transport

Area traffic capacity, also known as the number of successfully bit stream contained in a service data unit (SDU) above a particular amount of time, is the massive traffic output provided by territory per Mb/s.

Area traffic capacity was enhanced from 0.1 to 10 Mbits/s/m2 in 5G. Cell densification has been proposed to handle this region traffic capacity need with promising results in 5G scenarios.[49, 50]

Greater geographical resources utilization is made possible by the installation of a lot of small base stations (SBSs) of different cell sizes, such as micro, pico, and femto cells.

Additional tiers of tiny cells boost the spectrum reuse factor dramatically and allow more bandwidth per UE to be allocated.

Network coordination deals with inter-cell interference, reducing the interference especially at the cell edge, which results in much needed additional capacity and increased UE throughput.[13]

8.1.3.1.4 Spectrum Efficiency

Peak spectral efficiency for downlink and uplink in IMT-2020 must be at least 30 bit/s/Hz and 15 bit/s/Hz, respectively.

Maximum efficiency, normalized by carrier frequency band (in bit/s/Hz), is the peak data throughput achievable under ideal conditions.

The frequency accessible will range up to 30 GHz in 5G. Access to flexible strategies that maximize spectrum efficiency is also required to meet these criteria.

We now see frequency usage in terms of multiple access rather than exclusive ownership in our quest for the electromagnetic spectrum's most effective use.

The negative implications of sharing licences in the process of dynamic spectrum access has generated doubts among scholars.

Reinforcement learning has been utilized in the new 3.5 GHz Citizens Broadband Radio Service (CBRS) band to access shared spectrum opportunistically, reducing the detrimental effects of spectrum sharing on priority access licences (PAL) nodes.[51]

To alter the access of the secondary general authorized access (GAA) nodes, a Q-learning algorithm was employed to learn an ideal energy-detection threshold (EDT) for carrier sensing.[52]

The local learning framework used in this work can be developed to an intelligence gathering that uses inter-learning to cooperatively optimize inside and outside of various shared-spectrum installations.

Detection of anomalous behaviour in the mobile networks has remained a serious task because of the complex electromagnetic use of the spectrum.

Since mobile network irregularities can change significantly based on the unwanted signal in a frequency spectrum, manual identification is a challenge.[53]

Adversarial autoencoders (AAE), a machine learning technique that analyses frequency data in an unsupervised learning situation to find abnormalities in the wireless spectrum, have now been used to solve this issue.

8.1.3.2 Massive Machine-Type Communications (mMTC)

The capacity of 5G telecom services to scale connection demand to accommodate the expanding range of wireless network nodes is another aspect that sets them apart by concentrating on efficient tiny load delivery above a large distribution area.[54, 55]

Body-area networks, smart homes, the Internet of Things (IoT), and drone delivery will all generate occasional traffic between a large number of geographically dispersed devices and will need massive machine-type communications (mMTC) to serve novel but unanticipated use cases.

The two most important prerequisites for mMTC are

1) Frequency of links:
   Connecting a huge number of devices to the internet and advancing IOT smart cities, homes, and buildings from 100,000 connections per km^2 in 4G to 1,000,000 connections per km^2 in 5G is a huge challenge for 5G systems. For low speed and transmitting sites, this requirement should be satisfied.

   Due to the Internet-of-Things, a substantial portion of data is collected from users and devices, which future wireless system BSs must manage.[56, 57]
2) Network energy efficiency:
   For eMBB, system efficiency is critical, with estimates of annual uses ranging from 1x on 4G to 100x by 2020.

   Due to new power multimedia applications, energy consumption from devices connected under massive machine type communication uses must be considered in future network architecture.

   Energy-efficient networks have become a significant and difficult problem that both business and academia have been seeking to solve with the proliferation of mobile terminals in 5G deployments.[58] It will continue to be a subject of growing interest for a long period of time.[59]

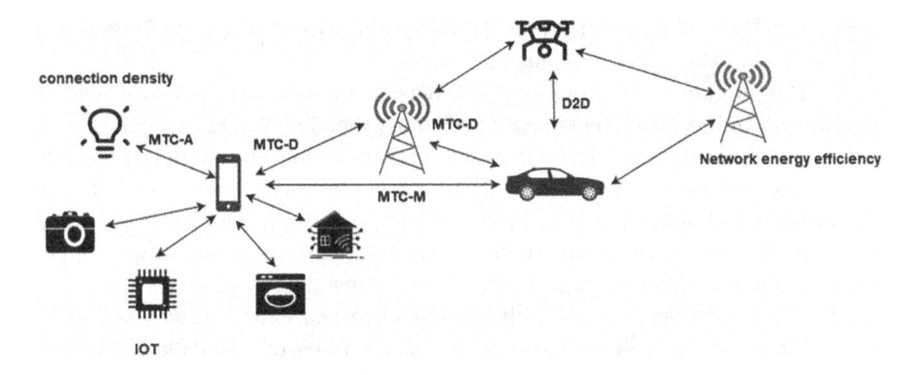

FIGURE 8.6 mMTC provides efficient connectivity for the deployment of a massive number of geographically spread devices (e.g., sensors and smart devices).[54, 55]

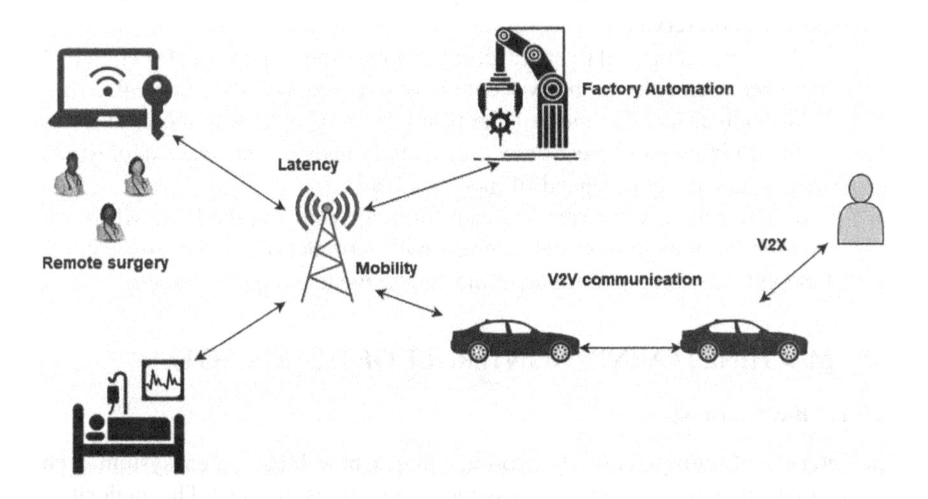

FIGURE 8.7 Ultra-reliable low-latency (URLLC) for demanding applications.[60]

8.1.3.3 Ultra-Reliable Low-Latency Communications (URLLC)

Future Internet applications will prioritize extreme trustworthiness, low bandwidth, and agility over data rates (Figure 8.7), including connected healthcare, telemedicine, mission-critical applications, automated vehicles, vehicular (V2V) communications, rising railroad interconnection, and intelligent applications (Figure 8.7). The following are the essential prerequisites for URLL communication:

1) Delay:
   One of the most crucial evaluation criteria for 5G is latency. Low latency is necessary for a functional 5G system, and even a few milliseconds (ms) will have a major impact, which makes it a critical field for 5G researchers and engineers.

The IMT-2020 specifications, typically range from 10 ms in 4G to 1 ms in 5G, do not allow for unbounded latency.

The success of URLLC will also depend on predictive network services, which can foresee system needs in relation properly.[60, 61]

Authentic links between driverless cars, e-health, remote object tracking, augmented and virtual reality (AR/VR), and other technologies will be made possible by URLLC.

A self-driving car on the road, for example, must recognize other vehicles, pedestrians, motorcycles, and other things in real time, not later.

In a perfect world, the local BS would always have the desired material.

If not, the user will need to get this from a very far-off cloud platform, which will correspondingly increase latency.[62, 63]

2) Mobility:

Mobility is the highest mobile unit speed (in km/h) that may be achieved for a specific good service.

The user is supposed to be travelling at a maximum speed of 500 km/h in the high-speed vehicular mobility scenario, as opposed to 350 km/h on 4G.

An optimal identification of the beam forming vectors is required to enable these highly mobile use cases, particularly in dense mmWave deployments where users must often hand-off between BSs.

To anticipate and coordinate beam forming vectors at the BSs, ML models can use the uplink pilot signal received at the terminal BSs and learn the implicit mapping function pertaining to the environment configuration.[64]

8.2 MACHINE LEARNING: INTERNET OF THINGS (IOT)

8.2.1 Background

The Internet of Things (IoT) is evolving into a new large, open system architecture that enables dispersed, accessible applications.[65, 66] The majority of new gadgets, including detectors, mobile phones, and other connected devices, are linked. These interconnected gadgets can transfer data and communicate each other. Complex systems can be created to enhance quality of life, including machine diagnostics, human body functions, monitoring systems, geolocation, and architectural tracking. Massive data created by huge sensors is generated as IoT gets more popular and widely used, and numerous applications were designed to give increasingly precise and perfectly good services. The Internet of Things' massive data can be studied and processed further to deliver intellectual ability to IoT customers and suppliers. Many IoT applications require a lot of data-driven procedures and the efficient the use IoT sensing data. AI algorithms have recently been introduced into IoT data analytics.

The Internet of Things' massive data can be studied and processed further to deliver intelligence to IoT customers and suppliers.[67] Many IoT applications require a lot of data-driven procedures and the efficient the use IoT sensing data. AI

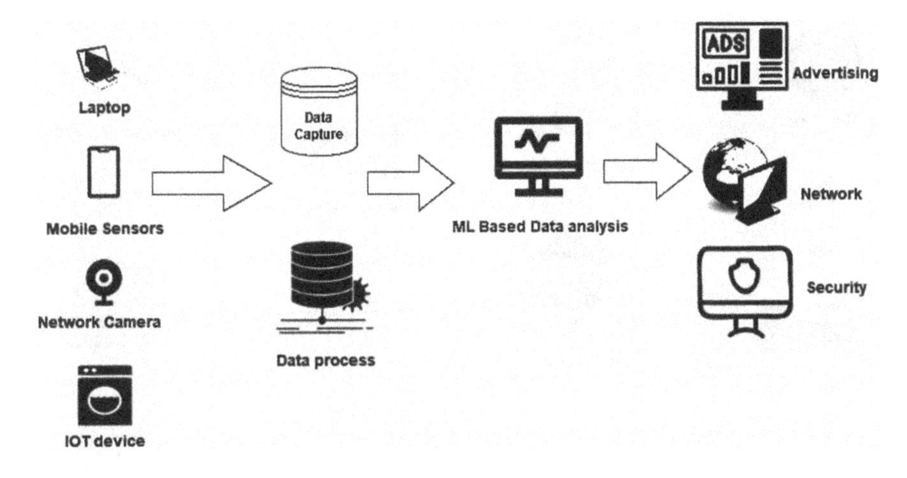

FIGURE 8.8 IoT device identification.[65]

algorithms have recently been introduced into IoT data analytics.[68, 69] Machine learning is a vital part of autonomous smart/intelligent network management. The majority of Internet of Things technologies are becoming more active, diverse, and complex as time passes on.

8.2.2 IoT Device Identification

Device identification is a procedure for detecting the sort of Internet of Things machine based on its features.[70] For service providers and infrastructure managers, ability to identify IoT devices is critical for business reasons and security. The model for device identification can be seen in the Figure 8.8.

The IoT identification problem was defined as follows: Various data is collected from the device for input. The output is an IoT device label that describes the device type.

8.2.3 Security

IoT security issues are becoming more critical due to higher threats. With the characteristics of IoT devices and communication protocols, the IoT networks are more vulnerable than traditional networks.[71] For example,

1) IoT devices are usually equipped with low battery and micro-controller
2) IoT communicates with each through Bluetooth, and various technologies are more vulnerable to attacks.

In an IoT network, there are usually three components, including device, gateway, and controller.[72] These components become targets for attackers. Figure 8.9 shows the IoT security problem model.

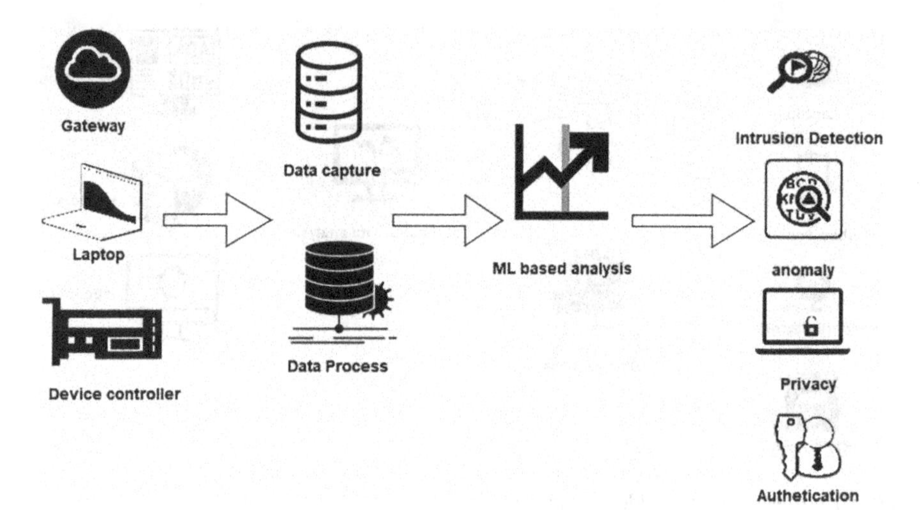

FIGURE 8.9 IoT security.[65]

8.3 MACHINE LEARNING: HYPERAUTOMATION

8.3.1 Overview

Hyperautomation is the application of a more successful method approach to the commencement of automation that combines intelligence. The method addresses the importance of finding the correct equilibrium among manual work substitution, as well as robotics and improving difficult phases.

Business intelligence specialists are good at identifying opportunities for automation that can be managed by a wide group of individuals. Users can easily use the capabilities of hyperautomation systems.

With these systems and different types of data platforms, hyperautomation lets the organization easily convey the value of integrations.[73–75] The main advantages of hyperautomation in organizations are increasing productivity, reducing errors, enhancing quality, saving time, and reducing costs. While introduction of hyperautomation, the majority of automation has been researched. Furthermore, hyperautomation surveillance tools may make knowledge workers scared to exploit this data.

Through preconfigured components purchased from a mobile app, hyperautomation enables the integration of ML capabilities into automation.

Automated tools must be easy to incorporate into existing technology without needing a significant amount of IT experience. The system has to be able to incorporate and match with a broad variety of equipment in order to accomplish hyperautomation.

Hyperautomation systems can be immediately implemented on top of current software companies. The best way to apply automation to a particular program component is through conventional business automation techniques. Due to business hyperautomation, there are numerous approaches to enhance company processes.

While hyperautomation is a technological advance, it does not match the way robotic process automation completes tasks.

8.3.2 Definition

The term "hyperautomation" is identified by Gartner's on different technologies as [76] the application of advanced technologies such as AI, machine learning, and robotic process automation to replace previously manual operations. Hyperautomation refers not only to the tasks and procedures that can be automated but also to the extent to which they'll be automated. The next value of digital transformation is hyperautomation.

Humans will not be totally eclipsed by hyperautomation. With hyperautomation, users focus on high-value tasks for the organization rather than repetitive and low-value tasks. Organizations that combine robotics and social interaction can enhance consumer experiences while lowering administrative costs and increasing revenue.[77]

Hyperautomation depends heavily on the ability to integrate individuals in the digitization process. Robotic process automation was a big part of the first wave of automation technology (RPA).

The employment of robotics to do repeated human tasks is known as robotic process automation (RPA). This is a rule-based process that uses structured data to execute tasks. RPA is solely concerned about human actions. Hyperautomation allows digital workers to work alongside humans to provide unmatched efficiency.

8.3.3 Requirement of a Hyperautomation

To digitize greater skilled work, which includes everyone in an organization, hyperautomation is required. Different tasks can be performed using hyperautomation, which combines a variety of machine learning and automation tools. Becoming more automated allows a greater number of people to benefit from advanced technologies. Hyperautomation refers not just to a lopping technology but also to a broad range of instruments. Hyperautomation has the ability to integrate corporate strategy with the end objective of generating and optimizing final procedures that enable new enterprise concepts to emerge.[78–80]

By minimizing user's involvement in repetitive and time taking operations, hyperautomation can improve the efficiency, productivity, and moral standards. Hyperautomation has the potential to improve the healthcare industry by producing a great user health, more dependable results, and more accurate data.[81, 82]

8.3.4 Technologies Associated with Hyperautomation

Figure 8.10 shows various technologies associated with hyperautomation.[83]

Some of the technology involved includes process mining, RPA, AI/ML, digital twins of organizations (DTO), optical character recognition (OCR), and NLP. Hyperautomation technologies are created with the purpose of automated processes and planning and scheduling goals in mind. Hyperautomation improves a

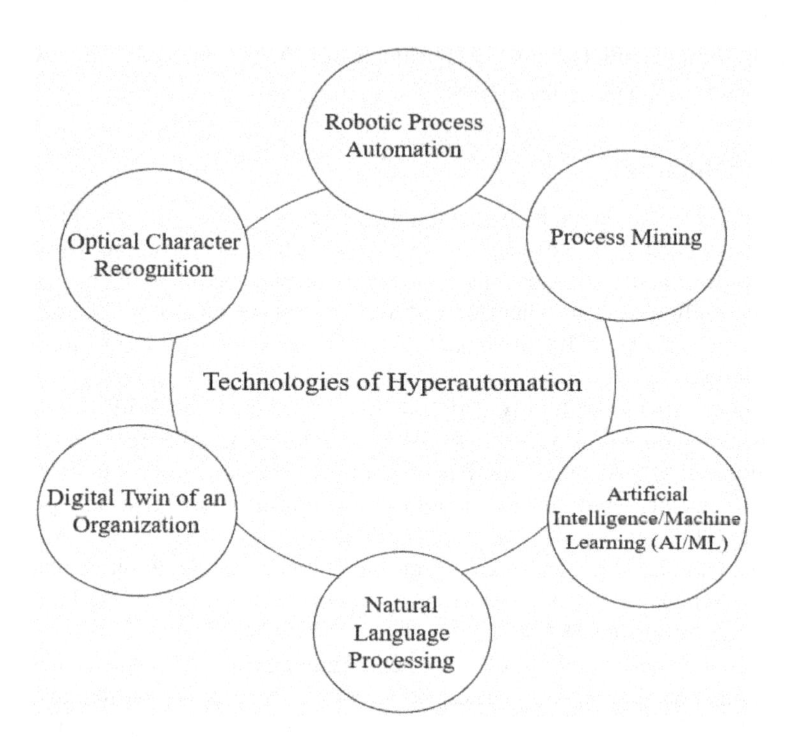

FIGURE 8.10 Various Technologies Associated with Hyperautomation.[83]

product's production and quality control. Low-value work can be carried out using automation techniques. The unique design allows efficient autonomous manufacturing with little human participation. Data can be used to create a more flexible, dynamic place where employees can make quick and efficient decisions.[84–86] To accommodate different departments in the organization, basic operations automation, data recovery solutions, and some more automation solutions are needed. Hyperautomation is a technique to increase human expertise in an organization by leveraging the advanced automation technology ecosystem. By integrating information and knowledge, hypererauotmation can automate corporate processes and boost learning by making decisions simpler. Previously, the organizations utilized automation to recognize the effects of positive robotic processes and activities.[87, 88] ML is a broad field that encompasses a large range of methods for constructing systems in ways. Before it can predict self, a supervised algorithm develops inputs and outputs. Pattern recognition insights are produced by unmonitored algorithms which monitor structured input.

Robotic process automation is known as RPA. The term "robot" refers to a program that can be configured to do tasks similarly to those that a computer and computer applications can. The term "process" represents the process you intend to do. Making work happen on its own through "automated processes."[89]

Hyperautomation can handle both labeled and unlabeled data, while RPA can specify structured data. RPA relates to the automation of services that mimic human activity. RPA is a grab term for tools with other computer systems' interfaces.[90]

RPA can enable the collaboration of many digital instruments in order to increase efficiency. By creating extra impacts for business analysts, experts, and remaining business users, the transformation process has become an important element. Robots and humans can work together more successfully thanks to hyper-automation technologies.[91]

Hyper-automation integrates higher knowledge and a broader piece of equipment to encourage firms to embrace core methods for establishing and automating workflow processes across the organization, as well as enhancing, discovering, organizing, measuring, and analyzing them.

Protocols for all automation phases are also elements of hyperautomation. This system involves process discovery, process optimization, process discovery, design, planning, use, and monitoring. A digital technology ecosystem is needed for the next production stage. Novel process ecology is made by mixing fully automated techniques and hyperautomation. The hyperautomation system ensures that multiple data systems in the organization are connected, adaptable, and technologically flexible. Hyperautomation provides businesses with higher efficiency—which increases profits and controls mechanisms—as well as fresh perspectives, improved data, and automated operations.

8.3.5 Hyperautomation Workflow Process

The process of hyperautomation approach is depicted in the Figure 8.11. At the beginning of the problem statement step, this is followed by extracting information by RPA robots, validation, and the development of enriched machine-readable data.

Hyper automation helps automate various elements of business decision-making. Rapid process automation, robust statistical applications, better worker happiness, as well as engagement, value-added workforce labor, better precise views, controlling mechanism and lower risk, enhanced efficiency, and better collaboration are just a few of the features.[92–94]

Domains in hyperautomation include claim handling (CH), travel and expense handling, order management, customer service operations, fully digital process, and anti-money laundering (AML). As technology changes on a regular basis, adopting hyperautomation has become essential. The modern concept in robotics began with hyperautomation, and clever businesses all over the world are going that way. Customer and industry benefits from hyperautomation are inversely correlated.[95–97]

8.4 MACHINE LEARNING: AUTOML

8.4.1 Summary

There has been a rush of machine learning initiatives in recent decades, with a variety of methods presented and effectively used in a variety of applications. Multiple

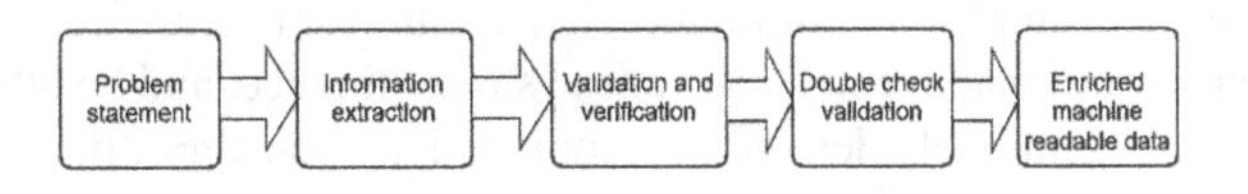

FIGURE 8.11 Workflow progressive steps for hyperautomation.[75]

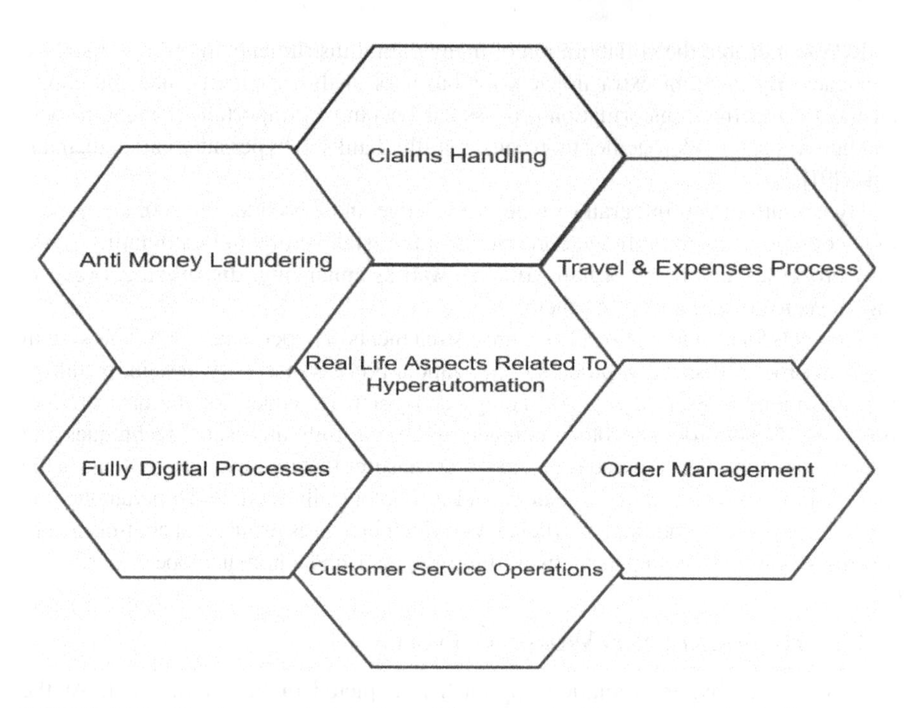

FIGURE 8.12 The diagram depicts the real-time aspects of hyperautomation.[75]

machine learning algorithms' performances are highly dependent on a variety of factors, and good results require a significant amount of human effort. For such a well-designed machine learning system, domain expertise and highly skilled researchers are necessary. Automated machine learning (AutoML) focuses on making the application of machine learning algorithms accessible to people with varied degrees of competence.[98]

Artificial intelligence is a field of computer science that aims to create machine intelligence that can do tasks that are similar to human intelligence.

Machine learning methods can be both expensive and difficult to implement in the context of computer resources. Algorithms have a number of hyper-parameters that, when chosen before learning, directly affect how well the model performs. Subject expertise and experience is required for manual hyper-parameter tuning, which is a difficult and time-consuming operation for both professional and non-professional users.

Under the machine learning framework, techniques for automatic search have been proposed (AutoML). This streamlines a method of choosing the optimal model and determining the hyper-parameters that go with it, as well as the application of machine learning algorithms. The AutoML focuses mainly on supervised learning, which is where a lot of the methods have been proposed.

Anomaly detection is a well-studied subject in machine learning and data mining. It entails locating examples that dramatically deviate from the majority of a dataset. Network security and management, storage systems, finance, and healthcare are just a few of the uses for anomaly detection. Anomaly detection algorithms, for example, are critical elements in monitoring network breaches, financial transaction fraud, data leaks, and existence situations in health information systems, which often contain unlabeled data.

8.4.2 AutoML

The efficiency of a ML determines how well it operates on a given data. It is based on a number of characteristics, including modeling, efficiency, storage capacity, and computation. A data scientist is continuously confronted with numerous possibilities. For example, a large variety of different methods are selected depending on the ML tasks. While choosing an approach with several hyper-parameters, a top value option is provided. These can be done based on assumptions and experience. To specify these concerns, conceptual and AutoML algorithms are employed. To increase the overall forecasting power with its system, AutoML is an autonomous system that chooses methods and sets hyper-parameters.

8.4.3 Model Generation

AutoML can be regarded as a technique for making machine learning tasks easier. It also saves time and effort for both non-machine learning specialists and machine learning professionals by eliminating the need for human model selection and hyper-parameter optimization.

Model generation consist of two main components: a learning algorithm and hyper-parameter optimization techniques. The AutoML process can be supplemented with other components, such as feature engineering prior to learning. This component employs a classification algorithm or extraction of features to decrease the input feature space.[99] Principal component analysis can be used to hold the relevant features of a model to improve its performance.

8.4.4 Meta-Learning

Meta-learning is a branch of research to study how various ML methods perform on different datasets. When given a fresh unfamiliar ML problem, AutoML approaches often begin building a machine learning pipeline. There are various definitions for meta-learning, but they all include concepts about the effectiveness of using training methodologies to enhance performance.[100–103]

This is based on the evaluation of the presentation of various ML algorithms in various configurations on various databases. Meta-learning selects the most promising configurations the with top results for a fresh unidentified database by comparing new and training data.

8.4.5 Automated Approaches

Machine learning methods including such AutoML and meta-learning have been developed. The performance of traditional machine learning is a common goal for various approaches.

Figure 8.13 shows different automated approaches of AutoML.

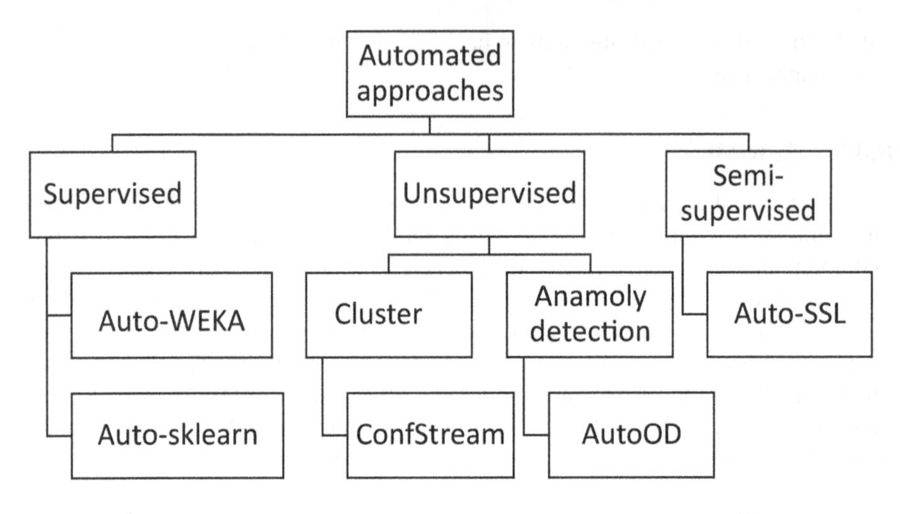

FIGURE 8.13 Classification of AutoML for various tools and approaches.[99]

8.4.5.1 Supervised AutoML

The first solution is based on mechine learning, so it focuses on finding the training datasets that are closest to a given sample data base based on features. The algorithm is then evaluated based on the skill of the surrounding datasets, and the best model is chosen.[103]

8.4.5.1.1 Auto-WEKA

Written in Java and based on the machine learning software, Auto-WEKA [104] was one of the earliest free software supervised AutoML systems. Sequential model-based optimization for general algorithm configuration (SMAC) [105] is a technique that Auto-WEKA uses to refine algorithm hyper-parameters throughout many occurrences. SMAC is a tool for analyzing supervised learning algorithms as well as their hyper-parameter settings.

8.4.5.1.2 Auto-Sklearn

Auto-sklearn [106] has been implemented in Python. Recently, auto-sklearn 2.0 was presented, which employs meta-learning that can maximize the productivity of auto-sklearn utilizing inventories.

8.4.5.2 Unsupervised AutoML

Machines that use unlabeled data and learn without supervision are known as unsupervised learning. Machine attempts to reveal a pattern in unlabeled data and reacts.

8.4.5.2.1 Clustering

AutoML is currently solely concerned with supervised learning. For unsupervised tasks, some approaches may be applicable. Clustering job proposal hyper-parameter settings and hypothesis choices must be performed without prior experience in data labels.[107]

To improve cognitive or indeed any qualitative metric for the clustering process, external and internal validity measures can be used. The comparison of how previous instances in same group were to one another is used to evaluate a clustering algorithm.

A novel method named confStream [108] has been introduced in the stream framework for autonomous method design algorithms for developing data feeds.

8.4.5.2.2 Anomaly Detection

Anomaly detection, often known as feature extraction, is one of the areas of data mining. It entails recognizing cases that deviate from the norm and exhibit normal behavior. As a result, there seem to be a variety of algorithms with varied degrees of efficiency.[109, 110] This function is critical in a range of applications, including fraud prevention, penetration testing, and surveillance of medical systems.

In an environment such as a time series period, a situational outlier is an occurrence that deviates greatly from the other examples.

Group outliers are a group of instances that deviates greatly from the entire database; single examples of set may not even be anomalies; however, the subset as a whole is.

8.4.5.2.2.1 Automated Outlier Detection (AutoOD) The AutoOD [111] framework seeks to identify the best NN for unsupervised feature extraction within a predetermined search space. It is based on a continuous learning technique that chooses blocks from the search space with the help of a controller for a recurrent neural network (RNN). For the object tracking job, a model with the greatest result on the validation set is used.

8.4.5.3 Semi-Supervised AutoML

Until now, the majority of AutoML research has been on supervised machine learning tasks that require only labelled data for classification or regression.

8.4.5.3.1 AutoSSL

Automated learning systems employed graded conceptual and augmented concepts to help them select the examples which are most likely to do well in a semi-supervised learning system. To improve fine hyper-parameters, AutoSSL [112] uses a big margin separation method.

8.5 CONCLUSION

We've demonstrated that machine learning holds enough promise that we can test out a vision where machine learning is a fundamental component of wireless communication systems. The expenses, duration, and delay introduced by some machine learning approaches are too great for some authentic services, thus it's crucial to keep this in mind while discussing fifth-generation networks. As a discipline, machine learning and fifth-generation networks still have more potential to grow. Whereas the major telecom industry has total trust in ML, due to the need to be careful and preserve existing systems, the rate of growth in the area will be severely limited. Our focus must be balanced with great caution because ML has the potential to increase ambiguity and difficulty in any network.

Machine learning has a great deal of potential as a key IoT technology. Machine learning has become much more popular as a way to offer analytics for IoT applications. This chapter attempts to fill the gap in the machine learning literature about its uses for IoT services despite the recent success of machine learning in networking.

Emerging technologies such as artificial intelligence (AI) are used with automation to solve complex problems and improve productivity. Hyperautomation has the ability to bring people together by letting technology and people work together side by side. As significant policymaker, it exploits technology to analyze big data and apply insights to its company. Hyperautomation changes enterprises by optimizing corporate processes by reducing repetitive operations and automating manual ones. With hyperautomation companies can do operations with consistency, precision, and speed.

In recent years, we've seen a continuous increase in the amount of available data, and the creation of new ML models to study, extract, and search it. As a result, a number of tasks, including choosing the optimal ML model and directly adjusting its hyper-parameters, are very dependent on specialist advice. Among various unsupervised tasks, methods attempting to automatically detect anomalies are given special focus.

REFERENCES

1. Morocho-Cayamcela, M. E., Lee, H., & Lim, W. (2019). Machine learning for 5G/B5G mobile and wireless communications: Potential, limitations, and future directions. *IEEE Access*, 7, 137184–137206. doi:10.1109/ACCESS.2019.2942390
2. Morocho-Cayamcela, M., & Lim, W. (2018). Artificial intelligence in 5G technology: A survey. In *2018 international conference on information and communication technology convergence (ICTC)*. IEEE, Jeju, Korea (South), pp. 860–865.
3. Zhang, C., Patras, P., & Haddadi, H. (2019). Deep learning in mobile and wireless networking: A survey. *IEEE Communications Surveys & Tutorials*, 21(3), 2224–2287, 3rd Quart.
4. Fu, Y., Wang, S., Wang, C., Hong, X., & McLaughlin, S. (2018, November/December). Artificial intelligence to manage network traffic of 5G wireless networks. *IEEE/ACM Transactions on Networking*, 32(6), 58–64.
5. Geron, A. (2017). *Hands-on machine learning with scikit-learn and tensor-flow: Concepts, tools, and techniques to build intelligent systems* (N. Tache, Ed., First Edition). Sebastopol, CA: O'Reilly Media, p. 543. http://shop.oreilly.com/product/0636920052289.do
6. Mitchell, T. M. (1997). *Machine learning* (First Edition). New York: McGraw-Hill.
7. Caruana, R., Lawrence, S., & Giles, C. L. (2001). Overfitting in neural nets: Back propagation, conjugate gradient, and early stopping. *Processing Advances in Neural Information Processing Systems*, 402–408.
8. Huang, G.-B., Ding, X., & Zhou, H. (2010, December). Optimization method based extreme learning machine for classification. *Neurocomputing*, 74, 155–163.
9. Bojović, B., Meshkova, E., Baldo, N., Riihijärvi, J., & Petrova, M. (2016). Machine learning-based dynamic frequency and bandwidth allocation in self-organized LTE dense small cell deployments. *Journal of Wireless Communications and Networking*, 2016(1), Art. no. 183.

10. Azmat, F., Chen, Y., & Stocks, N. (2016, January). Predictive modelling of RF energy for wireless powered communications. *IEEE Communications Letters*, 20(1), 173–176.
11. Martin, A., Egaña, J., Flórez, J., Montalbán, J., Olaizola, I. G., Quartulli, M., Viola, R., & Zorrilla, M. (2018, June). Network resource allocation system for QoE-aware delivery of media services in 5G networks. *IEEE Transactions on Broadcasting*, 64(2), 561–574.
12. Timoteo, R. D. A., Cunha, D., & Cavalcanti, G. D. C. (2014). A proposal for path loss prediction in urban environments using support vector regression. In *Proceedings advanced international conference on telecommunications*, vol. 10. Paris, France, pp. 119–124.
13. Bassoy, S., Farooq, H., Imran, M. A., & Imran, A. (2017). Coordinated multi-point clustering schemes: A survey. *IEEE Communications Surveys & Tutorials*, 19(2), 743–764, 2nd Quart.
14. Sanchez-Fernandez, M., de-Prado-Cumplido, M., Arenas-Garcia, J., & Perez-Cruz, F. (2004, August). SVM multiregression for nonlinear channel estimation in multiple-input multiple-output systems. *IEEE Transactions on Signal Processing*, 52(8), 2298–2307.
15. Liu, J., Deng, R., Zhou, S., & Niu, Z. (2014). Seeing the unobservable: Channel learning for wireless communication networks. In *2015 IEEE global communications conference (GLOBECOM)*. IEEE, San Diego, CA, USA, pp. 1–6.
16. Sarigiannidis, P., Sarigiannidis, A., Moscholios, I., & Zwierzykowski, P. (2017, July). DIANA: A machine learning mechanism for adjusting the TDD uplink-downlink configuration in XG-PON-LTE systems. *Mobile Information Systems*, 2017, Art. no. 8198017.
17. Ayadi, M., Zineb, A. B., & Tabbane, S. (2017, July). A UHF path loss model using learning machine for heterogeneous networks. *IEEE Transactions on Antennas and Propagation*, 65(7), 3675–3683.
18. Sotiroudis, S. P., Goudos, S. K., Gotsis, K. A., Siakavara, K., & Sahalos, J. N. (2013). Application of a composite differential evolution algorithm in optimal neural network design for propagation path-loss prediction in mobile communication systems. *IEEE Antennas and Wireless Propagation Letters*, 12, 364–367.
19. Mom, J. M., Mgbe, C. O., & Igwue, G. A. (2014). Application of artificial neural network for path loss prediction in urban macrocellular environment. *American Journal of Engineering Research*, 3(2), 270–275.
20. Popescu, I., Nikitopoulos, D., Constantinou, P., & Nafornita, I. (2006, September). ANN prediction models for outdoor environment. In *2006 IEEE 17th international symposium on personal, indoor and mobile radio communications*. Helsinki, Finland: IEEE, pp. 1–5.
21. Sotiroudis, S. P., Siakavara, K., & Sahalos, J. N. (2007). A neural network approach to the prediction of the propagation path-loss for mobile communications systems in urban environments. *PIERS Online*, 3(8), 1175–1179.
22. Huang, H., Yang, J., Huang, H., Song, Y., & Gui, G. (2018, September). Deep learning for super-resolution channel estimation and DOA estimation based massive MIMO system. *IEEE Transactions on Vehicular Technology*, 67(9), 8549–8560.
23. Bourlard, H., & Kamp, Y. (1988). Auto-association by multilayer perceptrons and singular value decomposition. *Biological Cybernetics*, 59(4–5), 291–294.
24. www.jeremyjordan.me/content/images/2018/03/Screen-Shot-2018-03-06-at-3.17.13-PM.png
25. Song, W., Zeng, F., Hu, J., Wang, Z., & Mao, X. (2017, June). An unsupervised-learning-based method for multi-hop wireless broadcast relay selection in urban vehicular networks. In *2017 IEEE 85th vehicular technology conference (VTC Spring)*. Sydney, NSW, Australia: IEEE, pp. 1–5.

26. Sobabe, G. C., Song, Y., Bai, X., & Guo, B. (2017, October). A cooperative spectrum sensing algorithm based on unsupervised learning. In *2017 10th international congress on image and signal processing, biomedical engineering and informatics (CISP-BMEI)*. Shanghai, China: IEEE, pp. 1–6.
27. Parwez, M. S., Rawat, D., & Garuba, M. (2017, August). Big data analytics for user-activity analysis and user-anomaly detection in mobile wireless network. *IEEE Transactions on Industrial Informatics*, 13(4), 2058–2065.
28. Liao, Z., Zhang, R., He, S., Zeng, D., Wang, J., & Kim, H.-J. (2019). Deep learning-based data storage for low latency in data center networks. *IEEE Access*, 7, 26411–26417.
29. Wang, Z., Li, L., Xu, Y., Tian, H., & Cui, S. (2018, December). Handover control in wireless systems via asynchronous multiuser deep reinforcement learning. *IEEE Internet of Things Journal*, 5(6), 4296–4307.
30. Song, W., Zeng, F., Hu, J., Wang, Z., & Mao, X. (2017, June). An unsupervised-learning-based method for multi-hop wireless broadcast relay selection in urban vehicular networks. In *2017 IEEE 85th vehicular technology conference (VTC Spring)*. Sydney, NSW, Australia: IEEE, pp. 1–5.
31. Balevi, E., & Gitlin, R. D. (2017, December). Unsupervised machine learning in 5G networks for low latency communications. In *2017 IEEE 36th international performance computing and communications conference (IPCCC)*. San Diego, CA, USA: IEEE, pp. 1–2.
32. Gazda, J., Šlapak, E., Bugár, G., Horváth, D., Maksymyuk, T., & Jo, M. (2018). Unsupervised learning algorithm for intelligent coverage planning and performance optimization of multitier heterogeneous network. *IEEE Access*, 6, 39807–39819.
33. O'Shea, T., & Hoydis, J. An introduction to deep learning for the physical layer. *IEEE Transactions on Cognitive Communications and Networking*, 3(4), 563–5752017, December 2017.
34. Rajendran, S., Meert, W., Lenders, V., & Pollin, S. (2018, October). SAIFE: Unsupervised wireless spectrum anomaly detection with interpretable features. *IEEE International Symposium on Dynamic Spectrum Access Networks (DySPAN)*, 1–9.
35. Wang, L.-C., & Cheng, S. H. (2018, July/September). Data-driven resource management for ultra-dense small cells: An affinity propagation clustering approach. *IEEE Transactions on Network Science and Engineering*, 6(3), 267–279.
36. Bastug, E., Bennis, M., & Debbah, M. (2014, August). Living on the edge: The role of proactive caching in 5G wireless networks. *IEEE Communications Magazine*, 52(8), 82–89.
37. Zorzi, M., Zanella, A., Testolin, A., De Grazia, M. D. F., & Zorzi, M. (2015, August). Cognition-based networks: A new perspective on network optimiza-tion using learning and distributed intelligence. *IEEE Access*, 3, 1512–1530.
38. Abu Alsheikh, M., Hoang, D. T., Niyato, D., Tan, H.-P., & Lin, S. (2015). Markov decision processes with applications in wireless sensor networks: A survey. *IEEE Communications Surveys & Tutorials*, 17(3), 1239–1267, 3rd Quart.
39. Jaakkola, T., Singh, S. P., & Jordan, M. I. (1995). Reinforcement learning algorithm for partially observable Markov decision problems. *Advances in Neural Information Processing Systems*, 7, 345–352.
40. Li, R., et al. (2017). Intelligent 5G: When cellular networks meet artificial intelligence. *IEEE Wireless Communications*, 24, 175–183.
41. Sutton, R. S., & Barto, A. G. (2018). *Reinforcement learning: An introduction* (Second Edition). London: MIT Press.
42. Cherkassky, V., & Mulier, F. M. (2007). *Learning from data: Concepts, theory, and methods*. Hoboken, NJ: Wiley.

43. Busoniu, L., Babuska, R., & De Schutter, B. (2008, March). A comprehensive survey of multiagent reinforcement learning. *IEEE Transactions on Systems, Man, and Cybernetics—Part C: Applications and Reviews*, 38(2), 156–172.
44. El Hattachi, R., & Erfanian, J. (2015). Next generation mobile networks alliance 5G initiative. Berkshire, U.K., 5G White Paper 1.0.
45. Andrews, J. G., Buzzi, S., Choi, W., Hanly, S. V., Lozano, A., Soong, A. C. K., & Zhang, J. C. (2014, June). What will 5G be? *IEEE Journal on Selected Areas in Communications*, 32(6), 1065–1082.
46. Agyapong, P. K., Iwamura, M., Staehle, D., Kiess, W., & Benjebbour, A. (2014, November). Design considerations for a 5G network architecture. *IEEE Communications Magazine*, 52(11), 65–75.
47. Shafi, M., Molisch, A. F., Smith, P. J., Haustein, T., Zhu, P., De Silva, P., Tufvesson, F., Benjebbour, A., & Wunder, G. (2017, June). 5G: A tutorial overview of standards, trials, challenges, deployment, and practice. *IEEE Journal on Selected Areas in Communications*, 35(6), 1201–1221.
48. Hossain, E., & Hasan, M. (2015, June). 5G cellular: Key enabling technologies and research challenges. *IEEE Instrumentation and Measurement Magazine*, 18(3), 11–21.
49. Bhushan, N., Li, J., Malladi, D., Gilmore, R., Brenner, D., Damnjanovic, A., Sukhavasi, R. T., Patel, C., & Geirhofer, S. (2014, February). Network densification: Thedominant theme for wireless evolution into 5G. *IEEE Communications Magazine*, 52(2), 82–89.
50. Kamel, M., Hamouda, W., & Youssef, A. (2016). Ultra-dense networks: A survey. *IEEE Communications Surveys & Tutorials*, 18(4), 2522–2545, 4th Quart.
51. Bassoy, S., et al. (2017). Coordinated multi-point clustering schemes: A survey. *IEEE Communications Surveys & Tutorials*, 19(2), 743–764.
52. Tonnemacher, M., et al. (2018). Opportunistic channel access using reinforcement learning in tiered CBRS networks. In *2018 IEEE international symposium on dynamic spectrum access networks (DySPAN)*. Seoul, Korea (South): IEEE.
53. Noam, E. M. (1995, December). Taking the next step beyond spectrum auctions: Open spectrum access. *IEEE Communications Magazine*, 33(12), 66–73.
54. Javaid, N., Sher, A., Nasir, H., & Guizani, N. (2018, October). Intelligence in IoT-based 5G networks: Opportunities and challenges. *IEEE Communications Magazine*, 56(10), 94–100.
55. Ejaz, W., Naeem, M., Shahid, A., Anpalagan, A., & Jo, M. (2017, January). Efficient energy management for the internet of things in smart cities. *IEEE Communications Magazine*, 55(1), 84–91.
56. Sauter, M. (2014). *From GSM to LTE-advanced: An introduction to mobile networks and mobile broadband*. Hoboken, NJ: Wiley.
57. Beyranvand, H., Lim, W., Maier, M., Verikoukis, C., & Salehi, J. A. (2015, June). Backhaul-aware user association in FiWi enhanced LTE—A heterogeneous networks. *IEEE Transactions on Wireless Communications*, 14(6), 2992–3003.
58. Ejaz, W., Naeem, M., Shahid, A., Anpalagan, A., & Jo, M. (2017, January). Efficient energy management for the internet of things in smart cities. *IEEE Communications Magazine*, 55(1), 84–91.
59. Cao, X., Liu, L., Cheng, Y., & Shen, X. S. (2018). Towards energy-efficient wireless networking in the big data era: A survey. *IEEE Communications Surveys & Tutorials*, 20(1), 303–332, 1st Quart.
60. Chen, K.-C., Zhang, T., Gitlin, R. D., & Fettweis, G. (2019, March/April). Ultra-low latency mobile networking. *IEEE Network*, 33(2), 181–187.
61. Azari, A., Ozger, M., & Cavdar, C. (2019, March). Risk-aware resource allocation for URLLC: Challenges and strategies with machine learning. *IEEE Communications Magazine*, 57(3), 42–48.

62. Mao, Y., You, C., Zhang, J., Huang, K., & Letaief, K. B. (2017). A survey on mobile edge computing: The communication perspective. *IEEE Communications Surveys & Tutorials*, 19(4), 2322–2358, 4th Quart.
63. Wang, S., Zhang, X., Zhang, Y., Wang, L., Yang, J., & Wang, W. (2017). A survey on mobile edge networks: Convergence of computing, caching and communications. *IEEE Access*, 5, 6757–6779.
64. Alkhateeb, A., Alex, S., Varkey, P., Li, Y., Qu, Q., & Tujkovic, D. (2018). Deep learning coordinated beamforming for highly-mobile millimeter wave systems. *IEEE Access*, 6, 37328–37348.
65. Cui, L., Yang, S., Chen, F., Ming, Z., Lu, N., & Qin, J. (2018). A survey on application of machine learning for internet of things. *International Journal of Machine Learning and Cybernetics*, 9. doi:10.1007/s13042-018-0834-5
66. IoT Analytics. Why the internet of things is called internet of things: Definition, history, disambiguation. https://iot-analytics.com/internet-of-things-definition/
67. Sharma, S. K., & Wang, X. (2017). Live data analytics with collaborative edge and cloud processing in wireless IoT networks. *IEEE Access*, 5, 4621–4635.
68. Chau, D. H., Kittur, A., Hong, J. I., & Faloutsos, C. (2011). Apolo: Making sense of large network data by combining rich user interaction and machine learning. In *Proceedings of the SIGCHI conference on human factors in computing systems*. Vancouver, BC: CHI '11, pp. 167–176.
69. Suthaharan, S. (2014). Big data classification: Problems and challenges in network intrusion prediction with machine learning. *SIGMETRICS Performance Evaluation Review*, 41(4), 70–73.
70. Stöber, T., Frank, M., Schmitt, J., & Martinovic, I. (2013). Who do you sync you are? smartphone fingerprinting via application behav-iour. In *Proceedings of the sixth ACM conference on security and privacy in wireless and mobile networks*. Budapest, Hungary: ACM, pp. 7–12.
71. Kotenko, I., Saenko, I., Skorik, F., & Bushuev, S. (2015, May). Neural network approach to forecast the state of the internet of things elements. In *2015 XVIII international conference on soft computing and measurements (SCM)*. St. Petersburg, Russia: IEEE, pp. 133–135.
72. Baldini, G., Giuliani, R., Steri, G., & Neisse, R. (2017, June). Physical layer authentication of Internet of Things wireless devices through permutation and dispersion entropy. In *2017 global internet of things summit (GIoTS)*. Geneva: IEEE, pp. 1–6.
73. Machado, J. P., Notare, M. R., Costa, S. A., Diverio, T. A., & Menezes, P. B. (2001, February 19). Hyper-automation system applied to geometry demonstration environment. In *International conference on computer-aided systems theory*. Berlin, Heidelberg: Springer, pp. 457–468.
74. Awan, U., Sroufe, R., & Shahbaz, M. (2021, May). Industry 4.0 and the circular economy: A literature review and recommendations for future research. *Business Strategy and the Environment*, 30(4), 2038–2060.
75. Abid Haleem, Mohd Javaid, Ravi Pratap Singh, Shanay Rab, & Rajiv Suman. (2021). Hyperautomation for the enhancement of automation in industries. *Sensors International*, 2, 100124. ISSN 2666–3511.
76. https://www.iberdrola.com/innovation/hyperautomation#:~:text=Hyperautomation%20consists%20of%20increasing%20the,Robotic%20Process%20Automation%20(RPA).
77. Gartner Top 10 Strategic Technology Trends, 2021. Retrieved July 2, 2021, from www.gartner.com/smarterwithgartner/gartner-top-10-strategic-technology-trends-for-2020/ Access
78. Trbovich, A. S., Vuckovic, A., & Draskovic, B. (2020). Industry 4.0 as a lever for innovation: Review of Serbia's potential and research opportunities. *Ekonomika preduzeca*, 68(1–2), 105–120.

79. Ivanov, S. H. (2021, March 14). Robonomics: The rise of the automated economy. *Journal of the Automated Economy*, 1(11).
80. Poux, F., Mattes, C., & Kobbelt, L. (2020, September 3). Unsupervised segmentation of indoor 3D point cloud: Application to object-based classification. *International Archives of the Photogrammetry, Remote Sensing and Spatial Information Sciences*, 44(W1–2020), 111–118.
81. Jaekel, F. (2019). Fundamentals of interfirm networks and networked horizontal LSPcooperation. In *Cloud logistics*. Wiesbaden: Springer Gabler, pp. 53–144.
82. Silva, S. C., Corbo, L., Vlačić, B., & Fernandes, M. (2021). Marketing accountability and marketing automation: Evidence from Portugal. *EuroMed Journal of Business*, 18(1), 145–164.
83. *Sensors International*, 2(2021), 100124. ISSN 2666–3511.
84. Haleem, A., Javaid, M., Singh, R. P., Rab, S., & Suman, R. (2021). Hyperautomation for the enhancement of automation in industries. *Sensors International*, 2, 100124.
85. Chih-Yi, S., & Bou-Wen, L. (2021, June 1). Attack and defense in patent-based competition: A new paradigm of strategic decision-making in the era of the fourth industrial revolution. *Technological Forecasting and Social Change*, 167, 120670.
86. Soni, N., Sharma, E. K., Singh, N., Kapoor, A. (2020, January 1). Artificial intelligence in business: From research and innovation to market deployment. *Procedia Computer Science*, 167, 2200–2210.
87. Taylor, M. P., Boxall, P., Chen, J. J., Xu, X., Liew, A., & Adeniji, A. (2020, January 1). Operator 4.0 or maker 1.0? Exploring the implications of Industrie 4.0 for innovation, safety and quality of work in small economies and enterprises. *Computers & Industrial Engineering*, 139, 105486.
88. Machado, J. P., Notare, M. R., Costa, S. D., Diverio, T. A., & Menezes, P. B. (2001). 1 computer aided systems theory-1.1 mathematical and logic formalisms-hyper-auto mation system Applied to geometry demonstration environment. *Lecture Notes in Computer Science*, 2178, 457–468.
89. Goettl, C. (2021, April 1). Prioritising risk for better efficiency and collaboration. *Computer Fraud & Security*, 2021(4), 13–16.
90. Javaid, M., & Haleem, A. (2019, May 1). Industry 4.0 applications in medical field: A brief review. *Current Medicine Research and Practice*, 9(3), 102–109.
91. Saxena, M., Bagga, T., & Gupta, S. (2021, May 4). Fearless path for human resource personnel's through analytics: A study of recent tools and techniques of human resource analytics and its implication. *International Journal of Information Technology*, 1–9.
92. Jacoby, M., & Uslander, T. (2020, January). Digital twin and internet of things—current standards landscape. *Applied Sciences*, 10(18), 6519.
93. Kronz, A., & Thiel, T. (2021, May 10). 19 Digitization applied to automate freight paper processing. In *Robotic process automation*. De Gruyter Oldenbourg, pp. 393–402.
94. Goher, G., Mason, M., Amrin, A., & Abd Rahim, N. (2021). Disruptive technologies for labor market information system implementation enhancement in the UAE: A conceptual perspective. *International Journal of Advanced Computer Science and Applications*, 370–379.
95. Jankowska, B., Di Maria, E., & Cygler, J. (2021, May 1). Do clusters matter for foreign subsidiaries in the Era of industry 4.0? The case of the aviation valley in Poland. *European Research on Management and Business Economics*, 27(2), 100150.
96. Chen, Y. (2021, February 1). Research on convolutional neural network image recognition algorithm based on computer big data. *Journal of Physics: Conference Series*, 1744(2), 022096 (IOP Publishing).
97. Lemieux, V. L., Manhattan, A., Safavi-Naini, R., & Clark, J. (2021). A cross-pollination of ideas about distributed ledger technological innovation through a multidisciplinary and multisectoral lens: Insights from the blockchain technology Symposium'21. *Technology Innovation Management Review*, 11(6).

98. Bughin, J., Hazan, E., Ramaswamy, S., Chui, M., Allas, T., Dahlstrom, P., Henke, N., & Trench, M. (2017). *Artificial intelligence: The next digital frontier?* McKinsey Global Institute. *https://apo.org.au/node/210501*
99. Bahri, M., Salutari, F., Putina, A., et al. (2022). AutoML: state of the art with a focus on anomaly detection, challenges, and research directions. *International Journal of Data Science and Analytics*. https://doi.org/10.1007/s41060-022-00309-0
100. Sun, Q., & Pfahringer, B. (2013). Pairwise meta-rules for better meta-learning-based algorithm ranking. *Machine Learning*, 93(1), 141–161.
101. Vanschoren, J. (2018). Meta-learning: A survey. arXiv preprint arXiv:1810.03548.
102. Zhao, Y., Rossi, R. A., & Akoglu, L. (2020). Automating outlier detection via meta-learning. arXiv preprint arXiv:2009.10606.
103. Elshawi, R., Maher, M., & Sakr, S. (2019). Automated machine learning: State-of-the-art and open challenges. arXiv preprint arXiv:1906. 02287.
104. Kotthoff, L., Thornton, C., Hoos, H. H., Hutter, F., & Leyton-Brown, K. (2019). Auto-WEKA: Automatic model selection and hyperparameter optimization in WEKA. In *Automated machine learning*. Cham: Springer, pp. 81–95.
105. Hutter, F., Hoos, H. H., & Leyton-Brown, K. (2011, January). Sequential model-based optimization for general algorithm configuration. In *International conference on learning and intelligent optimization*. Berlin, Heidelberg: Springer, pp. 507–523.
106. Feurer, M., Klein, A., Eggensperger, K., Springenberg, J., Blum, M., & Hutter, F. (2015). Efficient and robust automated machine learning. In *Advances in neural information processing systems*, vol. 28. NeurIPS Proceedings. https://papers.nips.cc/paper_files/paper/2015/hash/11d0e6287202fced83f79975ec59a3a6-Abstract.html
107. Speranza, E. A., Ciferri, R. R., & Ciferri, C. D. A. (2016). Clustering approaches and ensembles applied in the delineation of management classes in precision agriculture. In *Brazilian symposium on geoinformatics*, vol. 17. Campos do Jordão Proceedings. São José dos Campos: INPE.
108. Carnein, M., Trautmann, H., Bifet, A., & Pfahringer, B. (2020, May). Confstream: Automated algorithm selection and configuration of stream clustering algorithms. In *International conference on learning and intelligent optimization*. Cham: Springer, pp. 80–95.
109. Handola, V., Banerjee, A., & Kumar, V. (2009). Anomaly detection: A sur-vey. *ACM Computing Surveys (CSUR)*, 41(3), 1–58.
110. Hodge, V., & Austin, J. (2004). A survey of outlier detection methodologies. *Artificial Intelligence Review*, 22(2), 85–126.
111. Li, Y., Chen, Z., Zha, D., Zhou, K., Jin, H., Chen, H., & Hu, X. (2020). Autood: Automated outlier detection via curiosity-guided search and self-imitation learning. arXiv preprint arXiv:2006.11321.
112. Li, Y., Zha, D., Venugopal, P., Zou, N., & Hu, X. (2020, April). Pyodds: An end-to-end outlier detection system with automated machine learning. In *Companion proceedings of the web conference 2020*, ACM digital library, pp. 153–157.

9 Towards a Web Standard for Neuro-Symbolic Integration and Knowledge Representation Using Model Cards

Paola Di Maio

CONTENTS

9.1 INTRODUCTION

Despite half a century of computer science, during which the benefits of integrated and hybrid systems have been demonstrated, recent developments in machine learning (ML) have evolved by leveraging the extreme views of radical connectionism,[1]

DOI: 10.1201/9781003310785-9

which essentially envisaged a connectionist approach to artificial intelligence (AI) without knowledge and without explicit representation. This trend has contributed today to the absence of knowledge representation (KR) in ML, which increases known risks and vulnerabilities such as lack of explainability, reproducibility and accountability of systems automating so-called biological intelligence— intelligent responses not mediated by logic and cognition but on reaction to stimuli. However neural-symbolic integration that includes KR, intended as symbolic representation also of non-symbolic/connectionist constructs, has been proposed in research since the early days of connectionism.[2, 3] There is a history of using the terms symbolic/subsymbolic/connectionist to indicate different formulations. In this chapter, symbolic systems are considered systems based on knowledge encoded in terms of explicit structures, where inferences are based on rules that operate on these structures and require time and processing power to be processed; whereas subsymbolic/connectionist neural networks represent knowledge in terms of correlations,[4] coded in the weights of the network, which result in processes that are opaque, nonconcatenative and efficient.[5, 6] The primary motivation for subsymbolic AI has been to account for high-level cognitive phenomena that are statistical or intuitive in nature.[4] But this is not where high order discernment takes place. The subsumption[1] of subsymbolism to symbolism implies in fact that subsymbolic computing exists as a subordinate relation to symbolic representation. If AI had developed taking the inherent subsumption relation of subsymbolic to symbolic into account, there would have been no need to resort to neuro-symbolism. As AI took a radical connectionist turn, machine learning has disjoined symbolic from subsymbolic representation, leading to a dichotomy that needs to be addressed urgently. Hence the renewed interest in neuro-symbolic approaches. As machine learning algorithms become embedded in general systems at multiple levels with disparate functions, their analysis, understanding, evaluation and management require explicitly stated logical constructs, even when inherently the algorithms are probabilistic in nature. Adopting robust life cycle development in AI, which includes appropriate stakeholder analysis and requirements specification, is necessary to ensure that each resulting AI function, whether implemented via natural language or neural networks, actually corresponds to intended and desirable systems outcomes and that the system benefits and performance can be evaluated based on how well these features satisfy requirements.

The lack of adoption of robust life cycle approaches in AI systems development lies outside the immediate scope of this chapter; however, pointers are provided to structured analysis and integrated model driven methods that were followed to produce the outline of the proposed model card, which is in essence an explicit declarative synthesis of the system complexity and multidimensionality.

The model card aims to pragmatically bridge the fragmentation that confines the different disciplines, such as semantic web (SW), data science (DS), and AI, and aims to increase the interoperability of the respective constructs as well as the understandability of the underlying ML algorithms and their combinatorial design. The proposed structure for the model card contains both symbolic and subsymbolic/connectist elements of the models, as well as SW and DS constructs. This is required for the system level to emerge, whereby intelligent automated systems integrate

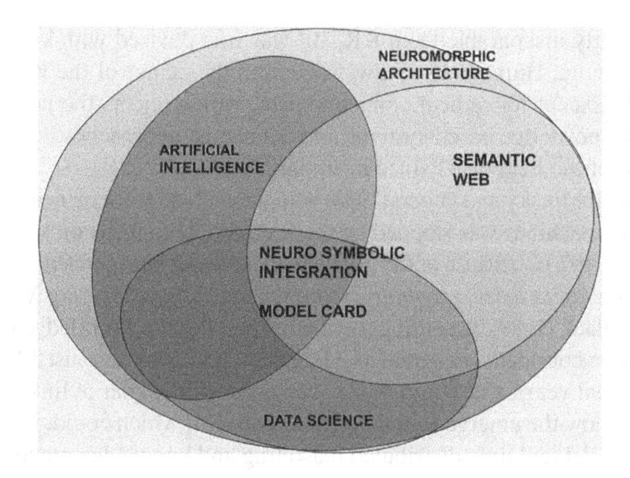

FIGURE 9.1 NSI MC in the context of SW, DS and AI.

and combine multiple techniques and approaches, facilitating the explanation and communication of their composition. The model card design outline will be refined during multiple iterations and responses from public consultation if it is considered as a possible web standard.

Future generations of computers will be based on neuromorphic engineering,[7, 8] a paradigm emerging from the co-evolution of synthetic biology and neuroscience. In this area, advances rely on neuroscience models, knowledge and data. Nonetheless, access to neuroscience research data without explicit data models and without domain specific and platform specific expertise is severely limited due to fragmentation, heterogeneity of data formats and lack of labeling and overall context. This specific problem is better described and partly tackled elsewhere,[2] and in this chapter considered as a use case to support and validate the benefits of the proposed approach.

Neural-symbolic integration (NSI) is known for bridging divergent approaches in AI—symbolism vs connectionism—for at least 20 years.[9] Nonetheless, it has not been leveraged to date solve open problems in AI systems integration nor to facilitate the explainability of complexity or to ensure reliability and enable replicability of algorithmic outcomes of automation. Neuro-symbolism is not even formally taught; in fact, it is not even mentioned in AI textbooks and coursework materials in use as of the time of writing.[3]

This chapter provides an opportunity to leverage NSI to expose the latent connection among AI, data science and the semantic web, and proposes that neuro-symbolic knowledge representation (NSKR) plays a role in supporting these connections. The solution offered is an integrated model card synthesizing and integrating the diverse perspectives. AI and the semantic web have a shared trajectory: both promised great impact on revolutionizing computing but ended up becoming constrained by technicalities and governance problems and became relegated to scholarly niches or used quietly to power backends.

AI, inherently inseparable from KR,[10] was first devised with knowledge based systems at its core. Humans have always been at the center of the loop in the early vision of AI. Except for radical connectionism, which rejects the need for KR, the symbiosis of knowledge based systems and connectist approaches has characterized the evolution of the field of AI since inception.

Starting with Minsky and Papert (1988) who wrote that "work on neural network and other learning machines was stopped by the need for AI to focus on knowledge representation in the 1970s, because of the principle that *no machine can learn to recognize X unless it possesses, at least potentially, some scheme for representing X*"[11] to Hinton (1990) and Pollack (1989), according to whom some level of knowledge representation has always been considered essential to AI, including in connectionist AI.[12, 13]

In subsequent years, the World Wide Web pursued the goal of linking knowledge resources to allow the emergence of a *global brain*,[14] which could, in theory, power massively parallel real time distributed reasoning and knowledge processing capabilities over vast amounts of openly accessible and unstructured data.

A trend toward unsupervised machine learning has become established in recent years,[15] thus contributing to a state of the art where machine learning based AI formed independently from other areas of research more closely related to human cognition and logic as well as human centered AI. Similarly to AI, data science has also been around several decades [16] and has been characterized with techniques to query and manipulate big data, ie. large distributed datasets, and eventually became a discipline of its own. Nonetheless, DS shares concepts, vocabularies and methods with other disciplines—mostly quantitative, structured data—while increasingly becoming qualitative and unstructured corpora. A system level model is required to produce an integrated view capable of reflecting shared aspects of these disciplines and capture their interdependence.

9.2 THE SYSTEM LEVEL

Intelligent systems capable of supporting the complex knowledge and processes that characterize the socio-technical scenarios, mediated and impacted by new technologies, must be capable of capturing and addressing the system level (SL), which can be defined according to multiple dimensions and properties: from the point of view of structure, function and behavior, for example,[17] the SL describes concepts, entities and relations that span the whole system rather than these elements being modeled separately. From the point of view of the life cycle, the SL can be defined based on levels of analysis (requirements, design), design (structure), and implementation (languages, environment, platforms).[18] In relation to the task at hand, i.e. devising a neuro-symbolic integration model card, the system level is used to guide the conceptual design of the target system using a structure function behavior approach as well as a life cycle approach.

9.3 KNOWLEDGE REPRESENTATION

Knowledge representation consists of methods and techniques used in many disciplines to capture domain knowledge.[19]

In AI, KR is used to capture and codify knowledge so that it can be expressed as system logic, using different methods and languages. AI includes a broad spectrum of techniques—from expert systems, decision support systems, robotics and intelligence automation. Machine learning (ML), deep learning (DL) and neural networks (NN) are considered subsets of AI. Similarly, knowledge representation (KR) includes a wide range of artifacts and techniques that support the engineering of autonomous intelligent systems, irrespective of the implementation method of choice.[20] In addition to declaring and formulating logical constructs for the purpose of AI software development, KR can also be useful to capture and convey domain knowledge in complexity science and engineering as well as to model specific data and logical sets for the purpose of algorithmic composition and programming. Knowledge representation roles correspond to distinct knowledge levels, such as:

Implementational: Including data structures such as atoms, pointers, lists and other programming notations.
Logical: Symbolic logic propositions, predicates, variables, quantifiers and Boolean operations.
Epistemic: Concept types with subtypes, inheritance and structuring relations.
Conceptual: The level of semantic relations, linguistic roles, objects and actions.
Linguistic: Deals with arbitrary concepts, words and expressions of natural languages.
Operational: How knowledge is deployed and operationalized.

System level knowledge representation (SLKR) can be used to satisfy the requirement for the modeling of multidimensional complexity in adaptive systems by articulating conceptual artifacts such as knowledge schemas across the dimensions.[21]

Systems are characterized in the first instance by the definition of their boundary.[22] Contemporary system architectures require increasingly dynamic extensible and reconfigurable boundaries. From a systems design and management point of view, functional processes, when decoupled from the context in which they are developed and deployed, can easily dissociate and become scrambled. This can deliver some advantages—such as extreme agility, reconfiguration and remodeling—but also disadvantages, such as fragility and loss of unity of purpose and logical integrity. Dynamic ML algorithms are known to be prone to misalignment, mislabeling, and wrong outcomes even when technically correct.[23] Nonetheless, as long as the intended system logic is preserved throughout the transformations, internal validity can be systematically evaluated and tested at any stage of the processing. By adopting an indicative model for an appropriate subdivision of knowledge levels described in structure modeling approaches, such as IDEF,[24] constructed according to an integrated structure function behavior framework, robust logical knowledge models can be devised that can be described in model cards, in the following sections.

9.4 TRUE AI, DATA AND SEMANTIC WEB

ML is increasingly important for data science [25] in that it provides the computational methods to process and analyze the large datasets. KR is foundational for both AI and SW. As computational power, data storage capabilities and speed soon increase, distributed reasoning capabilities in the open web are still, generally speaking, not a reality. The limits of computation until now have been constrained by Moore's Law (which states that the number of transistors on a microchip doubles about every 2 years), although the cost of computers is halved[4] and soon to be overcome through quantum computing and neuromorphic architectures. "The gradual end of Moore's law will open a new era in information technology as the focus of research and development shifts from miniaturization of long-established technologies to the coordinated introduction of new devices, new integration technologies, and new architectures for computing."[26]

Human cognition, however, the ability of humans to carry out information processing, critical reasoning and decision making, remains constrained by a range of factors including emotional states and even environmental conditions. Humans learn, process data and perform reasoning through a limited ability to concentrate their cognitive apparatus—memory, reasoning and to apply undivided attention on a single discrete issue/problem/event at the time; this is an ability which can be easily disrupted by psychological, physiological or external factors.

Despite great intelligence and creativity and the target model for intelligence automation, humans remain highly fallible.[27]

Connectionism, in a way, seeks to overcome the limitations of human cognition by reproducing biological intelligence with cellular automata [28] and the underlying motivation to the adoption of ML in AI. However, KR is essential to support the balanced integration of diverse approaches.

9.5 OPEN CHALLENGES

ML has strongly dominated AI research in recent years, thanks to the successes of the adoption of NN in statistical analysis.[29] With neural networks, certain types of AI capabilities—mostly those based on large scale data analytics and computation—have increased exponentially. They have also produced new categories of risks [23] and include a lack of explainability, reliability, accountability, verifiability, transparency and reproducibility.

Knowledge analysis and representation has not been considered in AI developments based on ML, at least not in the sense KR is used in symbolic AI (such as declaring, specifying and making explicit terms, concepts, axioms, relations, functions and logical inferences used in the algorithms). The shifting away from explicit and shared KR which resulted in dilution and even distortion of its role has motivated the research and promulgated the notion of system Level and neuro-symbolic knowledge representation which constitutes the background and foundation for the work proposed in this chapter.

Implications are found in research that artificial neural networks (ANN) and statistical analysis including various types of code are considered 'new types of KR' [30] and [31] and this could result in misleading conclusions about knowledge and its representation and handling in systems powered by machine learning, Neural networks cannot substitute entirely the function of symbolic KR.

The risk of misrepresentation—that something (a fact, function, process or a computational outcome, especially in relation to a person or an event or a notion) is represented to be something that it is not, that is, facts are represented as true when in reality are false, is one of the greatest risks in AI. Only robust explicit KR can reduce this risk.

Misrepresentation can cause, trigger and reinforce different kinds of bias in AI.[32] The phenomenon referred to as deepfakes—the manipulation of facts and information using AI-based technology to misrepresent a fact or a person by altering images and videos [33]—is the directly observable result of intentional knowledge misrepresentation.

One of the arguments supporting the lack of explicit KR in ML is based on the fact that human learning at its best, that is as observed in infants, is inherently experiential. Thus, if a child can learn without language and explicitly formulated KR, so can a machine.[34] The idea of intelligence without representation did not gain widespread acceptance because "bottom-up research on mobile robots, although valuable, is neither necessary nor sufficient as a foundation for core AI research".[35] Nonetheless, it is still implicit in much of the notion of embedded ML as applicable today.

Despite currents of radical connectionism, anti-representationalism and non-conceptualism, there is general agreement that at a minimum, from a systems engineering point of view, explicit and shared KR is necessary to facilitate affordable development, maintenance and communication among developers and users of intelligent systems, even when ML is embedded. Furthermore, it is necessary to explain, verify, evaluate and other necessary activities, without which AI risks increase exponentially. Ultimately, 'trustworthy' and ethical AI can only be achieved through shared explicit KR.

9.6 INTEGRATED KNOWLEDGE REPRESENTATION

Symbolic KR consists of explicit (logically and consistently written, structured, codified and labeled) natural language and pseudocode structures that can serve as descriptions and annotation of facts (the knowledge base) and their logical relations. The latter can be non-linear, polyvalent, representing more than one thing, hidden or not formally proven. Symbolic KR artifacts are typically based on natural language and include frames, production rules, semantic nets and Bayesian networks. [36] Non-symbolic KR is sometimes referred to as connectionist KR; knowledge stored as visual imagery with subsymbolic KR generally refers to knowledge used in computational operations based on mathematical functions.[37]

A KR movement referred to as anti-representationalism [38] advocated that KR is really not necessary to intelligence, while an abstractist/non-conceptualist movement proposed that deep in our brains there exist non-conceptual representation.[39] Although these ideas have never become seriously accepted, in practice they have influenced contemporary machine learning theories and research and created new categories of risks. Even the early proponents of connectist AI agreed that "such radical connectionists claim that symbolic processing was a bad guess about how the mind … So radical connectionists would eliminate symbolic processing from cognitive science".[40]

An important link between symbolic and non-symbolic approaches in AI, neuro-symbolism denotes autonomous intelligent architectures that adopt both symbolic and subsymbolic methods in a variety of configurations and following different strategies.

The terms connectionist and subsymbolic and neuro-symbolic have been used interchangeably in AI literature throughout the history of computer science, a source of confusion to this day. In essence, KR consists of determining what knowledge is required by a system or a process to operate, encoding this knowledge and making it available for this purpose. Explicit semantic and contextual representation is limited in ML driven AI, which focuses on quantitative aspects of computation and performance. In sum, it is explicit shared KR that makes systems explainable among humans who communicate using natural language.

9.6.1 Neuro-Symbolic Knowledge Representation (NSKR)

Understanding neuro symbolism is essential in present and future computing, including cyber-physical systems and the Internet of Things. An appreciation of the underlying arguments and constructs in neuro-symbolism is essential to anticipate and manage algorithmic risks and build awareness considering that intelligent systems in use are increasingly complex, distributed and powered by embedded logic. Recent scholarly trends attempt to build the trajectory of neuro-symbolism by tracking its history in AI, without explicit regard to its role and importance in KR, where it really belongs. Neuro-symbolism pertains to the knowledge representation aspect of AI. Neuro-symbolism is related to knowledge representation, as explained in this chapter.

The history of neuro-symbolism can be traced back to the challenges of designing neural networks attributed to McCulloch and Pitts [41] and the subsequent history of AI developed based on the refinement of machine learning artifacts and the need to complement these with cognitive approaches.[42]

The foundations of neuro-symbolism integrate knowledge expressed as rules into neural networks [43] with innumerable examples of combinations and architectures in between, subsequently documented in literature.

Typically, neuro-symbolic integration bridges symbolic and connectionist computing using hybrid or unified approaches [44] by either combining explicit logic and KR into connectionist systems or via the connectionist implementation of symbolic processes, thus following diverse compositional approaches. Knowledge representation can be used to make explicit and replicable the configuration of such composition, thus contributing to its validation. Early examples of neural-symbolic integration explored neural networks as computational mechanisms to power expert systems with strategies described in literature as ranging between unification vs integration.[44]

Neuro-symbolic programming is an emerging field combining deep learning and program synthesis where the goal is to learn functions from data and where functions are represented as programs [45] which leverage the foundational principles of neuro-symbolism. Neuro-symbolic knowledge representation (NSKR) as described in this chapter is concerned primarily with the explicit representation of knowledge using natural language expressions aimed at describing, whereas possible, the hidden computational processes that take place in ML algorithms, such as deep learning and in neural networks, as well as to make explicit compositional strategies.

In addition, NSKR is concerned with making explicit the correspondence between representation of natural cognitive processes and functions, as studied in cognitive and behavioral science, and neuromorphic systems (including neuromorphic architectures, decisions support systems) described, which are currently emerging from research.

Representational integration [46] was initially described as emphasizing either the connectivist of the symbolic alternative or both, the latter pointing to hybrid architectures. Novel NSKR artifacts for example, such as neurules, were described as "a type of hybrid rules that integrate symbolic rules with neurocomputing" [47] and provide an example of hybrid KR artifacts.[48]

NSKR can be achieved by associating to each machine learning algorithm a natural language explanation. The adoption of symbolic representation for non- symbolic computing can be made on at least two grounds:

1. Some level of symbolic KR is necessary to explain and communicate non-symbolic AI (algorithms and ML).
2. Non-symbolic computation should be explainable in symbolic terms providing epistemological and logical evidence as justification and argumentation for the reasoning.

In this research NSKR is applied to help address open challenges in ML, for example, to facilitate the explainability and replicability of ML algorithms, in particular to make explicit the hidden or unpredictable outcomes of processes powered by neural networks. There is no single way to produce exact explainability in AI because what is considered understandable by humans varies greatly among individuals; however it is generally accepted that explainability is related to interpretability, comprehensibility, intelligibility [49] and fidelity.[50] ML is delivering benefits in scaling computational performance and is increasingly adopted in intelligent systems automation at the cost of limited reproducibility, reliability, explainability and trustworthiness, which results in new classes of risks. Explainable AI needs to be understandable to humans, although depending on language, background knowledge and their education, there is no single standard definition. Humans communicate largely using natural language and visual communication, therefore explainability in ML requires annotated elements in natural language supported by visualizations such as diagrams and schematics. Limited or no explicit knowledge modeling and representation is applied in ML, partly due to the probabilistic nature of neural networks and partly because the emphasis in ML is not knowledge modeling but automated function execution.

From the point of view of systems reliability, when systems are powered by ML, the unpredictability of algorithms constitutes an infamous weakness in AI. Although Bayesian networks are considered a form of KR in terms of inference representation, there is consensus that Bayes inference is limited.

To communicate, discuss and evaluate the effectiveness and the performance of automated learning functions, at least some level of symbolic KR is necessary, or at least greatly beneficial, as it makes its understanding and evaluation of embedded functions and their impact clearer and less ambiguous.

KR is necessary in systems design processes and problem solving,[51] especially where diverse stakeholders collaborate to deliver and evaluate solutions based on the convergence of multiple intelligent technologies, which results in some form of AI. Representational schemas used in learning systems play a computational (as well as a semantic) role in determining how knowledge is used.

Other examples of use of symbolic KR in machine learning advocate that rule induction systems use simple propositional logic representations, enabling search to be adequately constrained, and that learning systems can easily augment such rules to include measures of certainty or probability.

9.7 MODEL CARDS

Neuro-symbolic knowledge representation is concerned with the conceptual abstraction level of the system layer and their semantic integration. It can be distinguished from more general neuro-symbolic integration in AI as the latter is focused on the implementation methods.

In NSKR, integration of knowledge and data models can be achieved: model driven approaches, such as, for example, Model-Driven Architecture (MDA) and, more specifically, the Model-Driven Engineering (MDE) class of methods that leverage generative and transformational techniques in software engineering, system engineering or data engineering. A detailed explanation of model driven approaches is outside the scope of this chapter and dealt with elsewhere.[51–56] The model card can also serve as a matrix to create interoperability among different syntaxes and models, and it is constructed using a Lightweight Object-Oriented Structure (LOOS) approach based on Integrated Definition (IDEF) [57], a set of standardized methods and a family of graphical language used for informational modeling in the field of software engineering (SE).

The standard modeling components of IDEF are *functions* (represented in a diagram by boxes) and *data and objects* that interrelate those functions (represented by arrows) resting mainly on a functional model of the system. The proposed adaptation of IDEF incorporates system dimensions and allows for the use of text, numerical values and imaging for each individual entry of a vocabulary. Model decomposition is undertaken to break down the complexity—such as enabling the distinction of the representation of data at structural, functional and behavioral level using different scales.

The Model Card facilitates the representation of three main dimensions (structural, functional and behavioral), each intersected with individual representation (such as scales) and each corresponding to a system function which can then be associated with states and dynamic properties. The *structural model* representing the components; the *functional model* includes processes, decisions, actions and activities of an organization or system and a *behavioral model* represents either biological behaviors (of cells and synaptic networks) or physiological and psycho/cognitive behaviors (of the individual subjects, human or animal). Models essentially are representation of systems as well as of real-world situations and when expressed using code, they can be processed by language specific compilers and interpreters (thus are implementation dependent), and they are intelligible and explainable based on the familiarity with the specific coding language and platforms. By including natural language expressions—for example, English—explainability and intelligibility can be generalized and the algorithms can be understood by humans. Neural-symbolic integration from a systems development perspective is a system modeling task, albeit a complex one, and can be achieved following structured development and model driven engineering approaches. The resulting artifact, equivalent to a system specification, can take the form of data model cards.

It is here proposed that a data model card for neuro-symbolic integration can serve as the basis for a possible open specification[5] and a web standard, currently being discussed in a W3C AI KR community group.[6] The vCard ontology[7] is the proposed blueprint for such a model card because it is both machine and human readable, and it is considered meeting general systems development good practice.

9.7.1 Model Cards

Decades ago, Levesque (1986) suggested simplifying the knowledge representation problem by using "vivid knowledge bases," which are essentially models.[52] Data cards have been in use since the beginning of computing, and long before being digitized they were hand punched or handwritten and serve a variety of uses. In their most recent applications,[53] they are used to encourage such transparent model reporting by disclosing the context in which models are intended to be used, details of the performance evaluation procedures and can be used to document machine learning models to increase transparency and accountability of both machine learning and data science.[58] Model cards can also be used as a vehicle to represent data statements described as design solutions and professional practice for natural language processing in both research and development and as a tool to mitigate algorithmic and data bias.

Inspired by recent efforts to address the ethical concerns and data biases in trained models,[59, 60] data model cards are adopted as templates for recording the details of individual human evaluation experiments in natural language processing (NLP), facilitating the recording of properties in sufficient detail and consistency to support comparability, meta-evaluation, and reproducibility assessments for human evaluations.[61] In diverse forms, they are also widely in use in a variety of software engineering context such as Microsoft cards, Twitter cards,[8] etc.

9.7.2 Model Cards as Declarative Frames

Model cards are being proposed in contemporary AI literature as short pieces of documentation that can be leveraged in ML to summarize and make transparent machine learning models, as well as executable programs to test/evaluate inputted data using a model.

In this sense, declarative model cards are what in early AI were called frames: explicit knowledge structures written in natural language that describe the knowledge (declarative) and processes (procedures) used by the AI. In declarative modeling, models are not a series of assignment and control statements but a set of facts that are true about the model.

Frames are one of the earliest known mechanisms for knowledge representation in AI.[62] They can be designed to integrate diverging modeling approaches and to integrate modeling dichotomies in AI, such as declarative vs procedural knowledge,[63] and to address problems such as default values in updating knowledge and databases.[64] Model cards and data cards are essentially declarative and procedural frames.

They can be used to provide natural language and graphical descriptions of ML algorithms designed into a single document and adhere to one or more template schema, as well as the explicit code structure template for validation and testing of data models, in data science and ML. Many examples exist that can be easily found via web searches.

Given the proliferation of neural-symbolic integration approaches in use, as well as the many model card templates, when discussing NSI, some level of integration and standardization would facilitate sharing and interoperability.

This chapter offers the neuro-symbolic model card approach for standardization.

9.7.3 NSI Model Card (NSI MC)

The need to capture and make explicit the many modalities of neuro-symbolic integration emerging from research and praxis are one of the motivating factors behind the proposal of the development for the model card template for neuro-symbolic integration, provisionally named NSI MC, where the meaning of the neural-symbolic integration is extended to include domain knowledge integration and, possibly, scientific paradigm integration in the context of applied epistemology.

In its current version, the NSI MC is a prototype (class) consisting of a design outline for a declarative knowledge representation frame (object) that captures and describes the key elements, values and variables of the neuro-symbolic model composition intended as the representation of symbolic and connectist processes and their intersection with data science, semantic web and knowledge representation in the context of neuromorphic systems. More detailed prototype implementation will be carried out through a series of refinements and may be extended to include procedural aspects.

NSI MCs replicate the core elements of general model cards already widely in use and include a symbolic representation (in the form of natural language or pseudocode description) of the algorithmic processes in use based on case scenarios in literature. See Figure 9.2 for the NSI MC outline.

It integrates S, B, F [65] together with other system dimensions loosely adopted from IDEF,[57] and the multiple viewpoints correspond, in this instance, to SW, DS, and ML constructs set against the neuromorphic engineering knowledge domain.

NSI ModelCard_Elements

MMCID—Identifier
MCV—Version
NSI-MCO Owner
NSI-MCUse (standalone, integrated)
NSI-MCDefiniton (concept, motivation, theory)

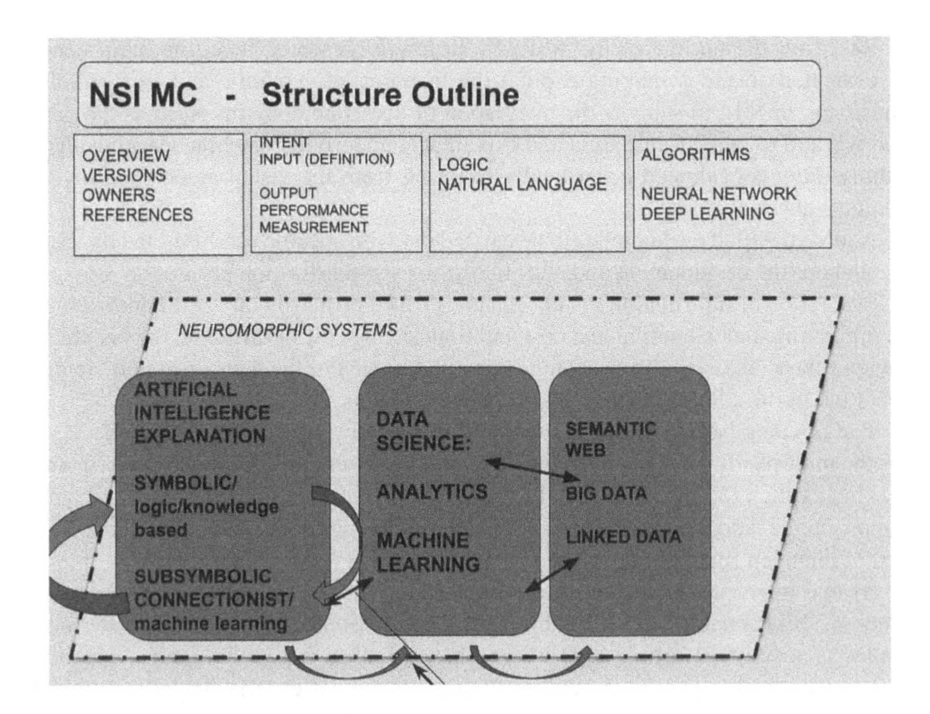

FIGURE 9.2 NSI model card structure outline.

NSI-MC Application Domain Definition
NSI-MCSL System Layer Definition
NSI-MCALApplication Layer Definition
NSI-MCML Machine Layer *embedded
NSI-MCMLD Machine Layer Description *explicit representation of the embedded layer
NSI-MCSM Structure Model (components: neurons, networks, regions)
NSI-MCFM Function Model (processes, dynamics)
NSI-MCBM Behavior Model (synapses. Activation, state changes) (cell, subject)
NSI-MCDM Dynamic (Process)
NSI-MCSM Static (state)
NSI-MCSD Spatial (nm, NSI-M, cm)
NSI-MCTM Temporal (ms, minute, hour, day, month, epoch)
NSI-MCSC Scale (nano, micro, macro, meso)
NSI-MDS Design Specification*context view

9.7.4 Evaluation

The main intended contribution of this chapter is a novel approach to neuro-symbolic integration using model cards. To the best of the author's knowledge, at the time of writing (May 2022), this is the first NSI model card ever invented.

Based on the outline of the NSI MC structure presented here, the main benefit evaluated so far is making explicit the mapping of symbolic and subsymbolic constructs in ML, to support the integration of disparate domains (such as SW, DL and AI) and to make transparent and explainable underlying machine learning algorithms. This is evaluated heuristically resulting from the visual association of the components on the card.

A more detailed evaluation is to be carried out with specific use cases and datasets throughout the development and, eventually, the standardization process.

Evaluation of information system design is based primarily on value benefits such as improving understanding and cost and time savings.[3] In this case, the benefit of saving time in figuring out the composition and impact of the neuro-symbolic representation in the context of diverse knowledge domains.

The obvious benefit is representational immediacy by providing a single view for the multiple dimensions that constitute what is essentially a complex system and making explicit the intended use of connectionist algorithms in the context of symbolic systems, and the corresponding boundaries and analytics methods, in the context of intended function.

By making explicit the conceptual space shared by ML and symbolic based techniques, NSKR enables the verifiability of the consistency and validity of the outcomes of machine learning algorithms—hence makes it possible their evaluation and benchmarking against a number of socio-technical criteria, including ethical criteria and fairness of outcomes, and not benchmarked in terms of computational performance.

NSKR implemented via model cards can also be used to make explicit, and thus publicly scrutable, the design choices and the functioning of hybrid architectures, where machines are built combining features from biological design to power automation, as in neuromorphic engineering, that by default combines both symbolic and subsymbolic approaches.

A full evaluation of the NSI MC using case literature will be reported on in detail in future occasions.

9.8 AN OPEN SPECIFICATION FOR NSI USING MODEL CARDS

Open standards (OS) are shared systems specifications. They are central to a culture of open technology which has emerged in the digital age.[66] They are formats and protocol that typically conform to certain characteristics,[9] and they are subject to full public assessment and use without constraints. They do not use components or extensions that have dependencies on formats or protocols that do not meet the definition of an OS themselves; they are free from legal or technical clauses that limit the use utilization and are managed and further developed independently of any single vendor. Web standards are what open web platforms for knowledge sharing and application development[10] are built on.

The web is the de facto largest openly shared repository for knowledge, research and scientific data, including new research output datasets. As such, it is proposed that NSI MC be published as implementation independent open specification via the W3C.

A system level knowledge model, adherent to a standard template, and the objects derived from it, such as data cards, can be useful general tools to serve as a map to guide learning, training—both human and machine—system development, from backend architectures to systems design and user interfaces.

The model card can also be used as a learning interface supporting the intuitive navigation of specialized knowledge by offering users multiple choices that reflect user tasks removing the inherent complexities (and responsibility) of understanding the intricacies of robust data structures.

These interfaces can be learning objects themselves, and, provided the datasets are labeled using natural language and coupled with shared metadata, the original formats and structure of the data do not need to change from whatever native format they are in, ie. the interfaces serve as a mask.

The NSI MC concept outline is being put forward for consideration throughout W3C AI KR CG and will be undergoing consultation among group members initially, and subsequently through public consultation.

9.9 USE CASES AND APPLICATION SCENARIOS

As AI becomes applied to a wider range of scientific and pragmatic problems, and as it is increasingly modeled using insights from brain data and neuroscience, KR has to fulfill the role to make the constructs from the corresponding domains generally understandable.

In some cases, ML-based AI may not include any form of explicit knowledge representation at all, while others may rely on different kinds of formalism for their communication and dissemination, mostly different types of diagrams and graphical visualizations, sketches and drawings, generally accompanied by narrative annotations.

The nature of intelligence is co-emergent resulting from processing different kinds of information using different senses, following diverse types of reasoning, processed by multiple parts of the brain and leveraging a combination of physical and cognitive elements.

When creating AI, which leverages biologically inspired approaches, the complexities and compositional aspects of intelligent functions reflected in the computational representation should also be made intuitively understandable to humans from diverse disciplinary backgrounds.

In the first instance, evaluation for the usefulness and validity of the model card is carried out using neuro-symbolic integration in neuro-informatics scenarios. For example, to transpose the integrated representation of biological brain functions and features to the information systems engineering domain, data science and machine learning constructs as the computational mechanisms and natural language descriptions to describe and explain the processes.

By using a working convention that provides a structured conceptual schema for the neurons in the physical brain, using a model driven approach, it should be possible to develop equivalent frames (data cards) for ANNs (artificial neural networks). If machine learning is modeled on human or, at a minimum, biological learning, then it should be possible to find some correspondence between the two.

However, representational equivalence between biological neural networks and artificial neural networks is notional, loosely established and not a direct fit.[67]

Using model cards facilitate the mapping and transposition of models from one domain to another.

9.9.1 How to Use NSI MC

An initial evaluation of the proposed NSI MC approach has been carried out heuristically by asking the question, what problems does the NSI MC solve? What benefits does it deliver and at what cost?

The benefits of adopting model cards in general, and in particular a model card that makes explicit how neural symbolic integration is achieved, are manyfold.

By drawing use cases from scenarios in neuro-informatics the MC demonstrates to facilitate the following uses.

1. **To compare, evaluate, benchmark and align different knowledge bases, datasets and models:** There are similarities and overlaps, for example, between Brain Architecture Management System (BAMS),[68] which can be accessed via BrainInfo[11] and contains three knowledge bases: NeuroNames, indexing brain structures; the Template Atlas, containing structures in the primate brain; NeuroMaps, which is not being actively maintained; and Ebrains.[69] Using a system level schema, it is possible to identify gaps and overlaps. Nomenclature for human brain in BAMS XML, for example, cannot be retrieved as a list, and the metadata set for Ebrains is minimal and unstructured. The two cannot be compared using web interfaces; they can, however, be compared using model cards that summarize the respective data models.
2. **To create a consistent logical structure:** Navigation model for interfaces that support conceptual and representational coherence across distinct resources, knowledge bases and datasets.
3. **To facilitate transfer learning**[70]**:** By attaching dimensional labels to data using GUISs.[71]
4. **To create blueprints:** To facilitate the replicability of experimental protocols. Taking, for example, recent experiments harnessing Locus Cereleus data in macaques,[72] a high level schema could map the experimental procedure to make it applicable to humans (using fMRI, ultrasound or other noninvasive sensing technique).
5. **To refine categorization structures**: Brain computer interfaces (BCIs) are currently categorized according to a limited set of criteria. Representing interfaces according to structure function and behavior and analyzing the categories using system level view can lead to the identification of additional categorization criteria [73]. For example, distinguish (by function) BCI that acquires data from the human brain in normal state from BCI that stimulates the human brain to study responses and understand behaviors. This distinction is going to be useful as neuromorphic BCIs become in use. Duly refined, system level models and artifacts can help to design repositories and

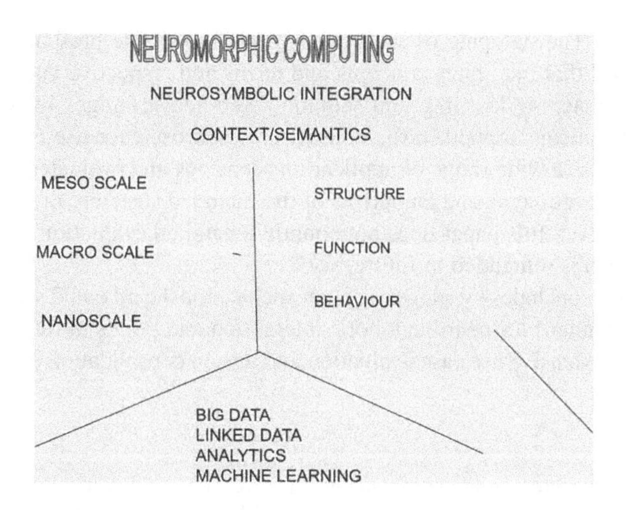

FIGURE 9.3 Systems dimensions for NSI MC.

interface that make explicit at a minimum the level of representation of the resource such as

Scale micro- macro- meso scale, for each structure function behavior

State Static vs dynamic view, for each structure function behavior. (See Figure 9.3.)

6. **To support scale free system wide queries using multiple views:** The diversity of possible application scenarios shows that ordering the system properties using multiple configurations changes their relationship. A certain structural configuration provides a given combination of functions and behaviors, but a different function will impact the definition of structure and behavior, and a given behavior can determine an entirely new structure and function, thus pointing to the requirement for dynamic analysis and visualization of different views of the brain as a system according to different scenarios. New findings suggest that the sample size of the studies may not be sufficient to explain certain correlations. The hypothesis here is that data and models currently in use to search for such correlations, not adequately representing whole brain networks, may result in computational bias.

The data and description about the evaluation of the model in relation to the use cases described is remanded to future work.

9.10 LIMITATIONS AND CONCLUSION

This chapter identifies the need for an integrated approach to model knowledge domains such as DS and SW and discusses how in modern computation these cannot be separated from AI and ML. It provides background and a discussion of open issues in the respective areas. An integrated engineering approach is proposed that leverages neuro-symbolic integration, named integrated neuro-symbolic knowledge representation. The outcome of the system level knowledge presentation approach is a model card that integrates concepts and terms and respective vocabularies from data science, machine learning and semantic web and leverages knowledge representation formalisms applied in the context of a neuroscience use case by pointing to benefits across a wide range of application scenarios and evaluated heuristically.

Supporting arguments and an outline of the intended structure of the NSI MC are provided; however, this paper does not contain a detailed evaluation of the proposed approach which is remanded to future work.

The chapter concludes by suggesting that such a model card could serve as the basis for an open standard for neuro-symbolic integration and points to future iterations to conduct more extensive use case evaluation and design consolidation of the standard.

NOTES

1 www.sciencedirect.com/topics/computer-science/subsumption-relation
2 Di Maio, P. (2021, September). System level knowledge representation for metacognition in neuroscience. In *The international conference on brain informatics*. Cham: Springer, pp. 79–88.
3 http://ceur-ws.org/Vol-2786/Paper32.pdf
4 Source: 1965, Gordon E. Moore, Intel
5 Work agenda proposed or W3C AI KR CG in 2022.
6 www.w3.org/community/aikr/
7 www.w3.org/TR/vcard-rdf/
8 https://developer.twitter.com/en/docs/twitter-for-websites/cards/overview/abouts-cards
9 https://fsfe.org/freesoftware/standards/standards.en.html
10 www.w3.org/standards/
11 http://braininfo.rprc.washington.edu/

REFERENCES

1. Cummins, R., & Schwarz, G. (1988). Radical connectionism. *The Southern Journal of Philosophy*, 26(5).
2. Touretzky, D., & Hinton, G. (1989). Connectionist models summer school. In *Proceedings 1988 connectionist models summer school*. San Mateo, CA: Morgan Kaufmann.
3. Arnovick, G. N., & Gee, L. G. (1978). Design and evaluation of information systems. *Information Processing and Management*, 14, 369–380.
4. Miikkulainen, R. (1994, November). Integrated connectionist models: Building AI systems on sub symbolic foundations. In *Proceedings sixth international conference on tools with artificial intelligence. TAI 94*. IEEE, pp. 231–232.
5. Smolensky, P. (1990). Tensor product variable binding and the representation of symbolic structures in connectionist systems. *Artificial Intelligence*, 46, 159–216.
6. Touretzky, D. S. (1991). Connectionism and compositional semantics. In J. A. Barnden & J. B. Pollack (Eds.), *High-level connectionist models*. Norwood, NJ: Ablex, pp. 17–31.
7. Mead, C. (1990). Neuromorphic electronic systems. *Proceedings of the IEEE*, 78(10), 1629–1636.
8. Furber, S. (2016). Large-scale neuromorphic computing systems. *Journal of Neural Engineering*, 13(5), 051001.
9. Hilario, M., Lallement, Y., Alexandre, F., & Lorraine, C. I. (1995). Neurosymbolic integration: Unified versus hybrid approaches. In *The European symposium on artificial neural networks*.

10. Brachman, R. J. (1990, July). The future of knowledge representation. In *Proceedings of the eighth National conference on Artificial intelligence*, vol. 90. Boston, MA, AAAI Press. pp. 1082–1092.
11. Minsky, M., & Papert, S. (1988). *Perceptrons*. Cambridge, MA: MIT Press.
12. Hinton, G. E. (1990). Preface to the special issue on connectionist symbol processing. *Artificial Intelligence*, 46(1–2), 1–4.
13. Pollack, J. B. (1989). *High-level connectionist models*. Columbus, OH: Department of Computer and Information Science, Ohio State University.
14. Heylighen, F. (2011). Conceptions of a global brain: An historical review. *Evolution: Cosmic, Biological, and Social*, 274–289.
15. Jordan, M. I., & Mitchell, T. M. (2015). Machine learning: Trends, perspectives, and prospects. *Science*, 349(6245), 255–260.
16. Naur, P. (1974). *A basic principle of data science*. Lund: Concise Survey of Computer Methods.
17. Qian, L., & Gero, J. S. (1996). Function–behavior–structure paths and their role in analogy-based design. *AI EDAM*, 10(4), 289–312.
18. Pfeiffer, H. D., & Pfeiffer, J. J. (2007, July). Representation levels within knowledge representation. In *International conference on conceptual structures*. Berlin, Heidelberg: Springer, pp. 484–487.
19. Davis, R., Shrobe, H., & Szolovits, P. (1993). What is a knowledge representation? *AI Magazine*, 14(1), 17–17.
20. Brachman, R., & Levesque, H. (2004). *Knowledge representation and reasoning*. Elsevier, Netherlands.
21. Di Maio, P. (2021). System level knowledge representation for complexity. In *2021 IEEE international systems conference (SysCon)*. USA, IEEE.
22. Bunge, M. (1992). System boundary. *International Journal of General System*, 20(3), 215–219.
23. Müller, V. C. (2016). Editorial: Risks of artificial intelligence. In *Risks of artificial intelligence*. Chapman and Hall/CRC, pp. 12–19. Naur, P. (1974). *Concise survey of computer methods*. New York, USA: Petrocelli Books.
24. Mayer, R. J., Painter, M. K., & de Witte, P. S. (1994). *IDEF family of methods for concurrent engineering and business re-engineering applications*. College Station: Knowledge Based Systems.
25. Loukides, M. (2011). *What is data science?* O'Reilly Media, Inc., California.
26. Theis, T. N., & Wong, H. S. P. (2017). The end of Moore's law: A new beginning for information technology. *Computing in Science & Engineering*, 19(2), 41–50.
27. Reason, J. (2000). Human error: Models and management. *BMJ*, 320(7237), 768–770.
28. Yilmaz, O. (2015). Machine learning using cellular automata based feature expansion and reservoir computing. *Journal of Cellular Automata*, 10.
29. Arel, I., Rose, D. C., & Karnowski, T. P. (2010). Deep machine learning-a new frontier in artificial intelligence research [research frontier]. *IEEE Computational Intelligence Magazine*, 5(4), 13–18.
30. Li, X., Zhang, S., Huang, R., Huang, B., Xu, C., & Zhang, Y. (2018). A survey of knowledge representation methods and applications in machining process planning. *The International Journal of Advanced Manufacturing Technology*, 98(9), 3041–3059.
31. Neelakantan, A. R. (2017). *Knowledge representation and reasoning with deep neural networks*. Doctoral Dissertations, University of Massachusetts Amherst, USA.
32. Markman, A. B. (2013). *Knowledge representation*. London, UK: Psychology Press.
33. Korshunov, P., & Marcel, S. (2018). Deepfakes: A new threat to face recognition? assessment and detection. arXiv preprint arXiv:1812.08685.
34. Brooks, R. A. (2018). Intelligence without reason. In *The artificial life route to artificial intelligence*. Routledge, USA, pp. 25–81.

35. Etzioni, O. (1993, Winter). Intelligence without robots: A reply to brooks. *AI Magazine*, 14(4), Articles. https://doi.org/10.1609/aimag.v14i4.1065
36. Brachman, R. J., Levesque, H. J., & Reiter, R. (Eds.). (1992). *Knowledge representation*, vol. 41. USA, MIT Press.
37. Boden, M. A. (Ed.). (1996). *Artificial intelligence*. Netherlands, Elsevier.
38. Haselager, P., De Groot, A., & Van Rappard, H. (2003). Representationalism vs. anti-representationalism: A debate for the sake of appearance. *Philosophical Psychology*, 16(1), 5–24.
39. Cussins, A. (1990). *The connectionist construction of concepts*. Stanford, CA: CSLI, Center for the Study of Language and Information, pp. 368–440.
40. Garson, J. (1997). *Connectionism*. Stanford Encyclopedia of Philosophy First published.
41. McCulloch, W. S., & Pitts, W. (1943). A logical calculus of the ideas immanent in nervous activity. *Bulletin of Mathematical Biophysics*, 5, 115–133.
42. Mey, M. D. (1982). The development of the cognitive view. In *The cognitive paradigm*. Dordrecht: Springer, pp. 3–18.
43. Boz, O. (1995). *Knowledge integration and rule extraction in neural networks*. EECS Department, Lehigh University. Bethlehem, Pennsylvania, United States.
44. Hilario, M. (1997). An overview of strategies for neurosymbolic integration. In *Connectionist-symbolic integration: From unified to hybrid approaches*, Chapter 2. Lawrence Earlbaum Associates." Inc., Mahwah (1997). pp. 13–36.
45. Chaudhuri, S., Ellis, K., Polozov, O., Singh, R., Solar-Lezama, A., & Yue, Y. (2021). Neurosymbolic programming. *Foundations and Trends® in Programming Languages*, 7(3), 158–243.
46. Hatzilygeroudis, I., & Prentzas, J. (2004). Neuro-symbolic approaches for knowledge representation in expert systems. *International Journal of Hybrid Intelligent Systems*, 1(3–4), 111–126.
47. Corchado, J. M. (1995). Neuro-symbolic reasoning-a solution for complex problems. In *International conference on intelligent systems*. NA, London, UK.
48. Orsier, B. (1995). Etude et application de systèmes hybrides neurosymboliques. Doctoral dissertation, Université Joseph-Fourier-Grenoble I.
49. Clinciu, M. A., & Hastie, H. (2019). A survey of explainable AI terminology. In *Proceedings of the 1st workshop on interactive natural language technology for explainable artificial intelligence (NL4XAI 2019)*. Association for Computational Linguistics, Tokyo, Japan, pp. 8–13.
50. Velmurugan, M., Ouyang, C., Moreira, C., & Sindhgatta, R. (2021). Developing a fidelity evaluation approach for interpretable machine learning. arXiv preprint arXiv:2106.08492.
51. Sattler, U. (1998). *Terminological knowledge representation systems in a process engineering application*. Mainz, p. 221.
52. Levesque, H. J. (1986). Making believers out of computers. *Artificial Intelligence*, 30(1), 81–108.
53. Mitchell, M., Wu, S., Zaldivar, A., Barnes, P., Vasserman, L., Hutchinson, B., . . . Gebru, T. (2019, January). Model cards for model reporting. In *Proceedings of the conference on fairness, accountability, and transparency*, Association for Computing Machinery, New York, NY, United States. pp. 220–229.
54. Soley, R. (November). *The OMG staff, model-driven architecture*, OMG Document. www.omg.org/mda
55. Bézivin, J. (2001, August). From object composition to model transformation with the MDA TOOLS'USA. Santa Barbara, Volume IEEE publications TOOLS'39. www.sciences.univ-nantes.fr/info/lrsg/Recherche/mda/TOOLS.USA.pdf

56. Bézivin, J., & Gerbé, O. (2001, November 26–29). *Towards a precise definition of the OMG/MDA framework ASE'01*. San Diego. www.sciences.univnantes.fr/lina/atl/publications/ASE01.OG.JB.pdf
57. Dašić, V. Š. P., & Labović, R. J. D. (2009). *Functional and information modeling of production using IDEF methods. Journal of Mechanical Engineering*, 55(2009)2, 131–140.
58. Pushkarna, M., Zaldivar, A., & Kjartansson, O. (2022). Data cards: Purposeful and transparent dataset documentation for responsible AI. arXiv preprint arXiv:2204.01075
59. Bender, E. M., & Friedman, B. (2018). Data statements for natural language processing: Toward mitigating system bias and enabling better science. *Transactions of the Association for Computational Linguistics*, 6, 587–604.
60. Gebru, T., Morgenstern, J., Vecchione, B., Vaughan, J. W., Wallach, H., Iii, H. D., & Crawford, K. (2021). Datasheets for datasets. *Communications of the ACM*, 64(12), 86–92.
61. Shimorina, A., & Belz, A. (2021). The human evaluation datasheet 1.0: A template for recording details of human evaluation experiments in nlp. arXiv preprint arXiv:2103.09710.
62. Minsky, M. (1974). *A framework for representing knowledge*. MIT-AI Laboratory Reprinted in The Psychology of Computer Vision, P. Winston (Ed.), McGraw-Hill, 1975. Shorter versions in J. Haugeland, Ed., Mind Design, MIT Press, 1981, and in Cognitive Science, Collins, Allan and Edward E. Smith (eds.) Morgan-Kaufmann, 1992 ISBN 55860-013-2]
63. Winograd, T. (1975). Frame representations and the declarative/procedural controversy. In *Representation and understanding*. Elsevier, Morgan Kaufmann, pp. 185–210.
64. Hayes, P. J. (1981). The logic of frames. In *Readings in artificial intelligence*. Elsevier, Morgan Kaufmann, pp. 451–458.
65. Gero, J. S., & Kannengiesser, U. (2000, June). Towards a situated function-behaviour-structure framework as the basis of a theory of designing. In *Workshop on development and application of design theories in AI in design research, artificial intelligence in design'00*. NA, Worcester, MA, pp. 1–5.
66. Russell, A. L. (2014). *Open standards and the digital age*. Cambridge, England: Cambridge University Press.
67. Hasson, U., Nastase, S. A., & Goldstein, A. (2020). Direct fit to nature: An evolutionary perspective on biological and artificial neural networks. *Neuron*, 105(3), 416–434.
68. Bota, M., Dong, H. W., & Swanson, L. W. (2005). Brain architecture management system. *Neuroinformatics*, 3(1), 15–47.
69. Schirner, M., Domide, L., Perdikis, D., Triebkorn, P., Stefanovski, L., Pai, R., . . . Ritter, P. (2022). Brain simulation as a cloud service: The virtual brain on EBRAINS. *NeuroImage*, 251, 118973.
70. Sharma, A., Jayakumar, J., Sankaran, N., Mitra, P. P., Chakraborti, S., & Sreenivasa Kumar, P. (2021, September). ConnExt-BioBERT: Leveraging transfer learning for brain-connectivity extraction from neuroscience articles. In *International conference on brain informatics*. Cham: Springer, pp. 235–244.
71. Raffel, C., Shazeer, N., Roberts, A., Lee, K., Narang, S., Matena, M., . . . Liu, P. J. (2020). Exploring the limits of transfer learning with a unified text-to-text transformer. *Journal of Machine Learning Research*, 21(140), 1–67.
72. Zhou, X., Li, M., Zhou, H., Li, L., & Cui, J. (2018). Item-wise interindividual brain-behavior correlation in task neuroimaging analysis. *Frontiers in Neuroscience*, 12, 817.
73. Hendler, J. A. (1989). Marker-passing over microfeatures: Towards a hybrid symbolic/connectionist model. *Cognitive Science*, 13(1), 79–106.

10 Semantic Web Technologies

D.V. Chandrashekar, Syed Md Fazal, Y. Suresh Babu and A.V. Senthil Kumar

CONTENTS

DOI: 10.1201/9781003310785-10

10.1 INTRODUCTION

The World Wide Web has been an enormously successful endeavour if judged by the material that is accessible and the increasing rate at which people are using it. The ability to be accessed by anyone, anywhere, is arguably the web's most valuable characteristic. It is a vast data warehouse that is not connected to any other systems. It is made up of information repositories that are not connected to one another but belong to a variety of different fields and sectors. The Internet is used for a variety of purposes, including but not limited to the following: searching for and making contact with other people, browsing the catalogues of online stores, placing orders by filling out forms, and more. Discovering information on the Internet is made easier with the assistance of search engines such as Google and Yahoo, amongst others. However, a significant number of users are disappointed with the results that they obtain because it contains an excessive amount of data that is not pertinent to their needs. This is due, in part, to the manner in which information is presented on the World Wide Web. The vast majority of content published on the web is produced with human readers and not computer programmes primarily in mind.

Computers, in their capacity as tools for posting and presenting information, do not have any access to the content itself. Because of this, the data that it stores isn't particularly rich in terms of semantics. If it were semantically enriched, automatic or semi-automatic retrieval would become significantly more effective. As a consequence of this, computers are only able to offer a limited amount of assistance when it comes to gaining access to and analyzing this data. Improved web services and the semantic web are two fresh ideas that have emerged in recent years as a means of addressing this gap on the World Wide Web. On the other hand, the semantic web is predominately concerned with semantics and not the variety of syntactic structures. In the interest of brevity and clarity throughout this chapter, we shall refer to these two technologies collectively as "semantic web" services. Additionally, description logic and several logic languages like the Web Ontology Language (OWL) and N3Logic are covered in this chapter. In this section, we will also discuss some of the most significant logical approaches that were utilized during the course of this research.

Since service-oriented architecture (SOA) were originally made available, web services have brought about fundamental changes to the manner in which programmers communicate and share data over the Internet. They make it easier to integrate business processes while simultaneously cutting down on the amount of time and money needed for the creation and maintenance of online applications. Web services get rid of the need for businesses to negotiate a communication route or syntactic representation in advance. This makes it possible for businesses to exchange (or trade) capabilities with an unlimited number of business partners.

A web service (WS) is a piece of software that has been given a uniform resource identifier (URI) and can be accessed over the Internet using an open interface. This program also has a URI allocated to it. The interface description details the capabilities of the service, the types of messages that are passed back and forth, as well as the actual locations of the ports used by the service. A binding, which comes next, tells the computer and port numbers on that computer to which messages should be

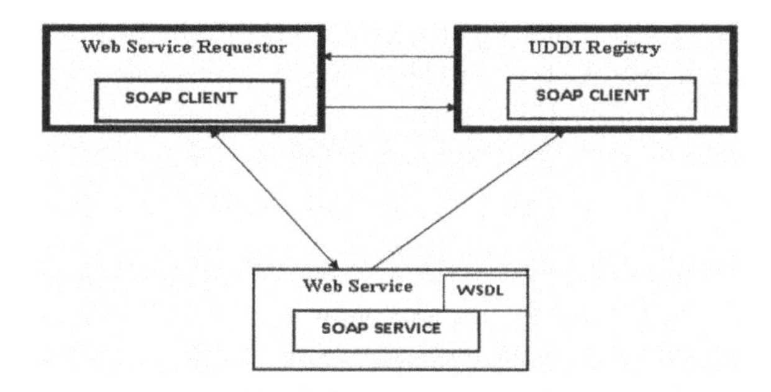

FIGURE 10.1 Fundamental web service architecture.

transmitted. One way to think about the relationship that exists between an application and the Internet is as a web service. According to the tutorial provided by IBM WebService, web services are a new type of web application.[1, 2] These programmes are self-contained, self-descriptive, and modular in nature. They are publishable, locatable, and executable anywhere on the Internet. Web services can be used for a wide variety of purposes, from simple querying to complex corporate processes. Once a web service has been established, it is possible for other apps and web services to discover and utilize it.[3] The web service, according to the W3C definition,[4] is "a software program me that can be identified by a URL and that allows direct interactions with other software applications utilising XML-based communications via internet protocols and that can be defined, described, and discovered."

The following architecture (Figure 10.1) is an excellent example of one common application of web services. By letting developers build web services, the introduction of a set of standards (like SOAP, WSDL, and UDDI) has made web services possible. This is how the great majority of people conduct business. The paradigm consists of three parts: the registry, where services are either advertised or published; the service requester, represented by SOAP clients; and the service provider, represented by SOAP servers. The ad has both a service provider profile (like the name and address of the company) and a service profile (like the name and category of the service; WSDL description, for example).[5]

10.2 A PROTOCOL FOR GAINING ACCESS TO OBJECTS (SOAP)

SOAP (Version 1.2) is a "lightweight protocol" that can be used to communicate structured data in a decentralized and distributed context, per the W3C. An XML-based messaging framework is the key instrument when designing a message structure that can be transmitted over a variety of protocols. The framework does not necessitate the use of a specific programming model or other implementation-specific semantics.[6]

Before being transmitted across a network, the data included within web service request and response messages is encrypted using an XML-based messaging protocol. SOAP messages can be transmitted using SMTP, MIME, HTTP, or any other Internet protocol. Transporting SOAP messages is possible using any operating system or protocol currently in use. Due to the fact that its communication is conducted over XML over HTTP, SOAP is immune to issues caused by firewalls.[7]

10.2.1 The Language for Web Service Description (WSDL)

However, what exactly are web services? What is their purpose there? And how exactly will SOAP-enabled applications gain access to them? WSDL provides answers to these questions. Nonetheless, it is in no way intended to be read by humans. Every standard web service is accompanied by a WSDL [8] document. Per the W3C, "an XML format for representing network services as a succession of endpoints acting on messages containing either document-oriented or procedure-oriented information." As per the W3C, in an XML format to represent network services as a collection of endpoints operating on messages. To design an endpoint, first the operations and messages must be abstracted, and then the actions and messages must be tied to a specific network protocol and message format. In order to reach the desired abstract endpoint, it is necessary to integrate the pertinent concrete endpoints (services). There are no restrictions on the message formats or network protocols that can be used while communicating with WSDL endpoints and the messages they generate.

10.2.2 Description, Discovery, and Integration at the Global Level (UDDI)

The first and most crucial step toward Universal Description, Discovery, and Integration (UDDI) is the construction of an XML-based register that will allow businesses situated anywhere in the world to create online listings for themselves. By taking part in an open industry effort supported by OASIS, companies will be able to name their services, find each other, and say how Internet-based services and software applications talk to each other.

You can register and publish accessible services through the use of a UDDI registry, which then makes it possible for other users, services, and applications to browse and query those services.[9] Businesses can use SOAP programmes [10] to publish a service description, and customers can acquire services of a specific type and bind them together.

The following is a list of some of the advantages that using web services can provide:

1. Interoperability
2. The encapsulating process
3. Availability
4. The modular approach
5. Self-description

The aforementioned advantages will reduce the complexity while maintaining high capacity in development.

10.3 PROBLEMS WITH WEB SERVICES

Despite its many advantages, web services still have some issues that need to be addressed.

1. The resources and services that are provided are not in a form that can be understood by machines.
2. Internet resources and services are presented in an unorganized and unconnected manner.
3. Currently, web searches for resources and services are keyword-based and do not consider the semantics of the resources. In other words, webmasters can make their pages more apparent by employing well-known keywords. Endusers may choose different keywords for the same page, which causespages to be lost and results to be returned that aren't relevant.

10.4 SERVICE AND TOOL INTEROPERABILITY IS EXTREMELY DIFFICULT TO IMPLEMENT

10.4.1 Distributed Information Systems (DIS) and Artificial Intelligence (AI)

Researchers have paid a lot of attention to semantic web technologies. More and more people who study distributed information systems (DIS) want to use semantic web technologies in their software designs.

The goal of the semantic web approach is to create machine-process able languages for conveying information. Tim Berners-Lee, who is known for making the World Wide Web, was the first person to think of the semantic web. Based on data that is machine-readable, this system enables automatic access to information. Using ontology and domain theories to express the semantics of data will allow the web to provide a qualitatively new level of service. There is a reason for the semantic web. [11, 12]

The semantic web is an extension of the traditional web that confers meaning on the data that is stored there in order to make it simpler for humans and machines to collaborate. Web-based data is characterized and linked in order to provide more effective application-wide discovery, automation, integration, and reuse of that data. [13] Internet users have a high level of confidence in Google, Yahoo, and AltaVista, which are examples of keyword-based search engines. On the other hand, using these can be fraught with danger.[14] It is possible to find the core important pages, but doing so will be of little use if hundreds of other documents that are only moderately related or completely irrelevant are also located. Having a strong recall but poor precision in the results will lead to low performance. The opposite of too much is too little, and vice versa. When we search for something, we don't always find the answer we're looking for, nor do we always get the sites that are relevant to our search. Even while newer search engines are less likely to have problems with low recall, it is still possible for those problems to arise. Frequently when the initial keywords do not produce the expected results, the relevant articles have vocabulary that differs from that used in the original query. This is wrong because queries with the same meaning should get the same answers, but they don't do that right now.

In the event that we require information that is dispersed across a number of documents, we are required to make a number of calls related to the data in order to obtain all of the documents that we require. After that, we must manually extract the pertinent information and put it all together. Before the vision of the semantic web can become a reality, the following conditions must be met: Is there anything else on the web that computers can do besides automate tasks? The solution is to make the content of websites more "machine-friendly." Integration of data refers to the capability of combining data from a variety of sources (so big organizations can avoid duplication). The assemblage of web services has been discovered by identifying the required semantic and its duplication method. A significant departure from the tool paradigm that exists now would be to put computers to work for us rather than just using them as tools. This entails eliminating the role of people in the process to the greatest extent possible.

10.4.2 Explanation of the Semantic Web

WordNet's definition of semantics is that it is "of or associated with the study of meaning." When used interchangeably with the word "semantic," "web" refers to a web that is comprehensible to both people and machines. Both people and robots are able to quickly access data on the World Wide Web in order to accomplish objectives that are beneficial to both parties. It is a concept that will be used in the next generation of the web. It will allow online applications to automatically get web documents from a variety of sources, combine and process information, and talk to other applications in order to do complex tasks for people.[15]

The semantic web's primary goals are as follows are to help people learn from raw data and understand what it means.

10.5 REUSE OF INFORMATION

10.5.1 Components of the Semantic Web

In a layered fashion, semantic web technology is constructed, with each stage built on top of the previous one. Taking little measures makes reaching agreement simpler but attempting too much makes it considerably more difficult to get everyone on board.[16]

Because they have comprehensive knowledge of one layer, agents (also known as software elements) that operate independently and proactively [17] ought to be able to comprehend and make use of the data that is provided at lower levels given that they have comprehensive knowledge of one layer. If agents, for instance, are able to comprehend the semantics of the Resource Description Framework (RDF) and RDF Schema data, they will be able to utilize it to its full potential.

If a system is to be successful, it must be able to make use of at least some of the information that is located at a higher level. For example, an agent who only knows

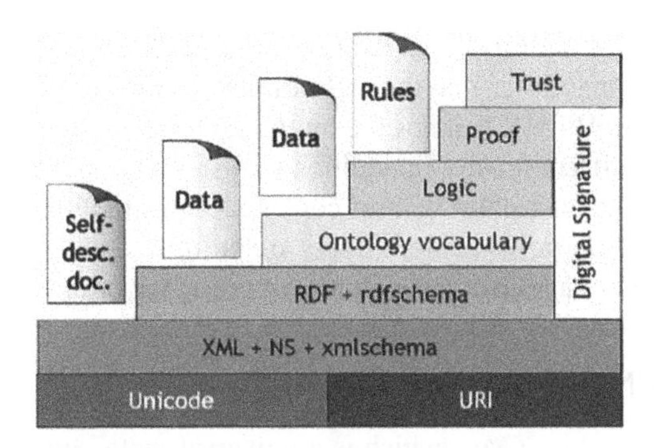

FIGURE 10.2 The semantic web layered approach.

RDF and the semantics of RDF Schema can figure out what a certain piece of OWL knowledge means.

A "layer cake" of semantic web technology may be seen as depicting the fundamental levels of the semantic web architecture and vision in the way that is depicted in Figure 10.2.

10.6 XML AND HTML

XML, also known as Extensible Markup Language and XML Schema are positioned at the very bottom of Figure 10.2, which depicts the semantic web layer. XML can be thought of using the Standard Generalized Markup Language (SGML) subset syntax. Hypertext Markup Language, or HTML, continues to dominate the market. Due to the complexity of SGML, it was determined that it could not be utilized for anything related to the Internet. In addition, the inadequacies of HTML inspired the development of XML.

XML tags, which are effectively labels that are not immediately visible to the user but are used to designate specific elements of page content or specific web pages, are known as "hidden labels." Tags are a sort of annotation. On the other hand, the structures themselves remain shrouded in mystery. The transmission of documents over the Internet is a practical application of XML. XML is loaded with numerous useful features.

XML possesses a number of valuable characteristics, including [18]

1. Extensible tags can be developed and used in a range of applications.

It is possible to produce some content and provide information about the purpose that it serves thanks to a markup language.

2. Information that is accessible to machines can be represented using this method.

It is easier for machines to access XML documents as each piece of information is described in full. The nesting structure in which they are housed also serves to characterize the interactions that exist between them. As an example, the 'author' tag is combined with the 'book' tag to identify the characteristics of a certain book.

Instead of relying on proximity considerations, like HTML does, a machine parsing the XML document may infer that the author refers to the element that encloses the book. In comparison, HTML handles the matter differently. In addition, the content can be exhibited in different ways in addition to just being displayed, which implies that there is no need to maintain several copies of the same information because it can be used for other purposes as well. Users are permitted to specify their own tags in a markup meta-language that does not have a preset set of tags.

10.7 CONCERNS RELATING TO XML

Markup can be defined in XML, which is a universal meta language, and it is possible to do so. A standardized framework and a collection of tools, like parsers, are made available in order to make the transfer of data and information more straightforward. However, there are a number of downsides.

Because XML does not have a standardized vocabulary, it can be read in a variety of different ways depending on the reader. An author and a writer are both terms that can be used interchangeably to refer to the same person or thing. It is evident to humans that both of these things are the same, but it is not known how a machine or computer would come to this conclusion. Because of this, there are often misunderstandings when two machines try to talk to each other by exchanging data. The meaning of tag nesting is not standard; it is up to each application to determine what it implies. David John is an example of a professor who specializes in thermodynamics. There are many different ways that this sentence can be expressed using XML. There are two possibilities: the name of the course is "Thermodynamics," the name of the lecturer is "David John," and the name of the course is "A university-level instructor by the name of David John teaches a course on thermodynamics." The two formalizations that have been presented thus far have essentially the opposite nesting order, despite the fact that they reflect identical information.

First, it has been determined that the name of the "course" is the primary element that nests the element "lecturer." In contrast, "lecturer" is handled as though it were the most important component in the second example. The name of the course should be used as the value for the "teaches" nesting element. There is not one single interpretation that applies to the layering of tags. As a result of the user's ability to provide their numerous domain-specific markup languages have been created. Chemical Markup Language (CML)[19] and MathML [20] are examples of such programming languages. Using domain-specific markup languages increases the likelihood that the markup languages used to describe various types of web resources are incompatible.

This degree of flexibility and extensibility will also result in poor resource descriptions if it is not permitted. This suggests that a standardized model or framework is necessary to connect all of the disparate schemas. Figure 10.2 depicts the current configuration of the semantic web pyramid; the resource description framework (RDF) stands atop the pyramid. RDF is not a programming language, but rather a basic data model. The RDF paradigm describes web content in broad strokes (also known as the rendering of metadata to the documents), which paves the way for a wide range of applications to utilize the metadata. Each three-part triplet, consisting of a resource, a predicate, and an object, conveys the meaning of RDF. This structure can be comprehended by comparing the subject, the verb, and the object of a simple phrase. Using XML components, these triplets may be built.

10.7.1 Triplets Derived from the RDF

These are the components of a fundamental RDF model, which is shown in Figure 10.2. If a resource is required to be described as a resource or topic, it is termed a subject. It may be a "webpage" on the Internet or a "person" in a group setting. The features of a resource, also known as its attributes, can be characterized using the terms properties and predicates, respectively. The "Title" of a website, for example, as well as a person's "Name," for example, are both examples of things that can be identified. As a consequence of this, one can use the other to identify things such as "webpages" and "people," respectively.[20]

An "object," which is the word for the actual value of the property itself, needs to be present for each and every one of the properties. The individual responsible for maintaining the website of the "Documentation Research and Training Center," also known as "DRTC," goes by the moniker "S. R. Ranganathan." The subject, the predicate, and the object make up the three components of a sentence that are referred to as a "statement" or "rule." David John is the author of the webpage that can be found at http://drtc.isibang.acain/David/. In a nutshell, this assertion can be summed up as in Figure 10.3 when represented diagrammatically.

Using XML, the preceding statement is rendered as follows:

The text namespace="www.w3.org/1999/02/22-rdf-syntax-ns" appears in the document's first line. This identifies the file as being an XML file, which we will be working with. Use of the URI reference is by far the most frequent method for referring to the names included within an XML namespace. However, additional methods are available [RFC2396]. The "namespaces" that are used in XML are different from the "namespaces" that have traditionally been used in computing fields because the "namespaces" that are used in XML have an internal structure, but the "namespaces" that are generally used in computing areas are "sets."[21]

When declaring an XML namespace, a particular syntax must be used. name space-prefix="namespace", The rdf: Description element provides details on the resource located at http://drtc.isibang.ac.in/David. In the description, the attribute is represented by a tag, and its value is the content of the tag. RDF was created with the purpose of being domain-independent from its inception. This indicates that it has a very generic nature and imposes no restrictions on any particular domain of application. You can use this word to talk about any kind of knowledge when talking about any subject.

Using XML, the preceding statement is rendered as follows: ?xml version="1.0"? Encoding="UTF-16"? > (srdf: RDF rdf="http://www.w3.org/1999/02/22-rdf-syntax-ns" xmlns: rdf="http://www.w3.org/1999/02/22-rdf-syntax-xmlns: mydomain="http://mydoamin.org/schema/">rdf:description=rdf: description="http://drtc.isibang.ac.in/David">author > David John//rdf:/rdf: mydomain

FIGURE 10.3 RDF statement.

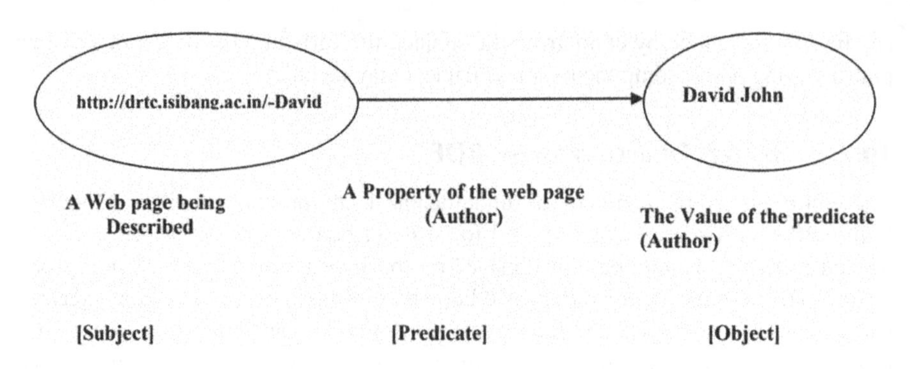

FIGURE 10.4 RDF model.

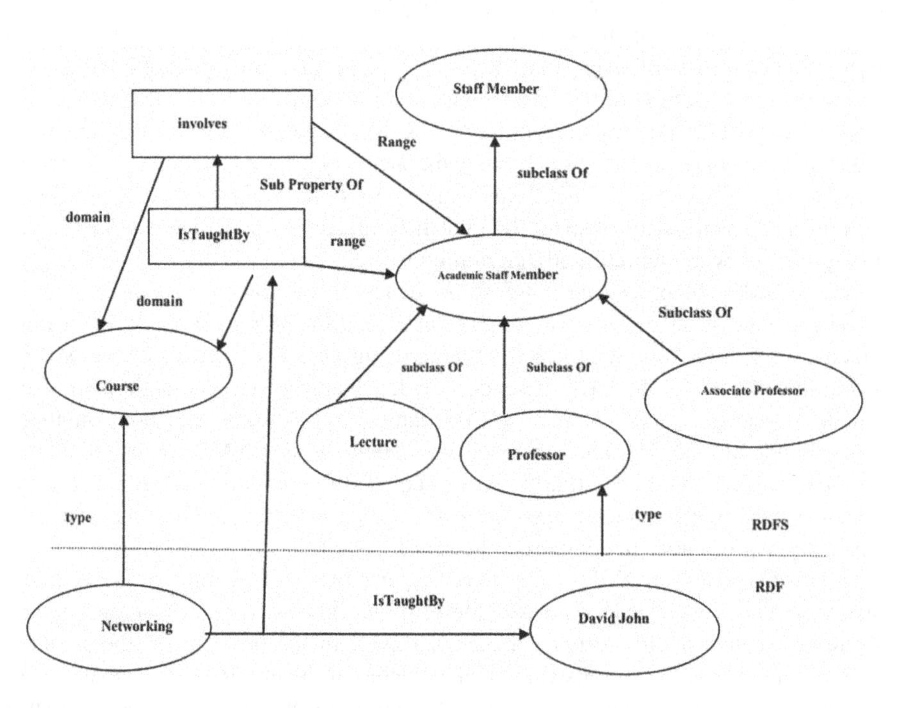

FIGURE 10.5 Layers of RDF and RDFS.

The RDF model serves as an inspiration for the class structure that is used in object-oriented programming The figure 10.4 shows the RDF model (Subject-object-predicate).

In RDF, the term "schema" refers to a collection of classes that have been defined with a particular objective or field of study in mind. The technique known as "sub-class refining"[19] makes it possible to endow these classes with newly developed capabilities. As a consequence of this, the base schema can serve as a foundation for the development of other related schemas. With the assistance of RDF, metadata may be communicated or shared between a variety of schemas, which helps to increase its reusability. In contrast to RDF Schema and RDF Layer, both of these will assert the network layering, as stated by David John. The different layers that are part of RDF and RDFS [22] are shown in Figure 10.5. This statement's schema may include classifications such as lecturers, academic staff, staff members, and courses. Here's an example to show what I mean. Figure 10.5 depicts properties as blocks, classes represented by the ellipses located above the dashed line and instances represented by the ellipses located below the line.

10.8 PROBLEMS WITH THE RDF SCHEMA

RDF and RDFS are capable of expressing certain forms of ontological knowledge. Instances of classes, restrictions on the domain and the range, as well as typed hierarchies and subclasses and subproperty links and subclass and sub property connections are the essential modeling primitives of RDF and RDFS. There are substantial deficiencies which are identified in various number of classes based on their characteristics. The local scope of a property is determined by an RDF range, which in this case is "eats" for all classes. As a direct result, it is impossible to apply range constraints to specific classes within the RDF Schema. We are deceiving ourselves if we claim that cows consume only vegetables, whereas other animals may also ingest meat. Other animals can eat both meat and plants.

In certain circumstances, we will want to refer to classes as being separate from one another. For instance, male and female are not synonymous terms for the same thing. RDF Schema, on the other hand, only lets us show relationships between subclasses, like how "person" has "female" as one of its subclasses.

It is possible to construct new classes by combining existing ones using various Boolean class combinations such as union, intersection, and complement. For example, a class person could be defined as a disjoint union of a man and a female. These sorts of definitions are prohibited under the RDF Schema.

10.8.1 Cardinality Restrictions

There are occasions when we want to limit the number of values a property can have. We may state, for example, that a student has two parents, or that a class has at least one professor. Again, RDF Schema does not allow for such constraints.

When describing attributes, it can be helpful to indicate things like "greater than," "is the mother of," and "is eaten by" are transitive;"is unique," and "is the inverse" of each other.

Therefore, ontology languages that are more comprehensive than RDF Schema and that provide the functionalities described above as well as others are necessary. When developing such a language, it is important to strike a balance between the language's capacity for expressive power and its capacity to effectively support logical argumentation. A good rule of thumb to follow is that the more complicated a language is, the less effective its reasoning support becomes, and it will frequently cross the line into non-computability. A language that is capable of expressing ontology as well as knowledge and that can be supported by reasoners that are both efficient and expressive is therefore necessary.

More than two thousand years ago, Aristotle's concept of categories served as the impetus for the development of this idea. The primary goal was to provide a framework for the categorization of all that was known and available at the time.

Ontologies have since been accepted in a variety of other domains, such as library and information science (LIS), artificial intelligence (AI), and more recently in computer science (CS), as the primary technique for classifying what archivists informally name documents.[22] In the beginning, ontologies were not explicitly referred to as such. There are many different ways in which ontologies have been defined. According to Gruber (1993), "an explicit specification of a conception" is what constitutes ontology.[23] After some time had passed, Studer et al. (1998) described it as "a formal, explicit explanation of a shared idea."[23, 24] Both shared conceptualization and formal links among concepts are included in the definition presented by Studer et al. It is necessary to have knowledge of a particular reality in order for a widespread conceptualization to be able to be conveyed in an explicit and formal manner. Last but not least, the most important part of the mission of ontology is to broadly disseminate the body of knowledge that it represents. Ontology is a way of organizing the formal links that exist between different concepts within a certain domain of knowledge.

Ontology is a set of shared ways of thinking about things that are linked in a formal way through polynomial hierarchies. On the basis of the foregoing arguments, we can derive the following features of ontology:

1. Keyword features have to be selected.
2. Frame the features that derive ontology words.
3. Identify the keyword with the text information available.
4. Extract the appropriate meaning of words.

The idea of ontology is to capture knowledge that is agreed upon by everyone. When something is "conceptualized," it means that it has been put into words in one's mind. Concepts that are linked to that phenomenon are identified and used in its development. In the end, we'll have a common understanding.

To avoid confusion between terms in a single vocabulary and terms in different metadata vocabularies, ontology is the formalization of the relationships between concepts. This makes it possible for computers to analyze semantic links and figure out what they mean.

It takes a polynomial hierarchy approach to development, which is different from a rigid, one-piece hierarchy.

10.9 ADVANTAGES OF ONTOLOGY

The advantages of ontology can be summarized as follows:[25]

First, establish a common vocabulary for the area.
Second, represent a domain knowledge with people and agents.
Third, build a work space of their own.
Fourth, segregate the workspace of people and agent.

It can be used to represent and share domain knowledge by both people and agents that work on their own.

10.10 ORGANIZATIONAL AND NAVIGATIONAL AIDS FOR THE INTERNET

The precision of web searches can be a feature of effective data extraction. Information can be generalized and/or specialized for web searches. Ontology and Logical Logic Languages interacting with logically connected data on the web is made possible by the semantic web. The semantic web provides highly organized data using a rich knowledge representation format, such as RDF. Sharing structured data on the web is now possible for application developers, who can infer information from the various types of data available on the web. RDF is based on the Universal Resource Identifier (URI), which is a simple pointer system. Conventional web usage makes heavy use of uniform resource locators (URLs) to refer to documents and their subsections. By naming everything from abstract ideas like "color" to physical objects like "mountains" and electronic items like "home pages of institutions," the semantic web is showing a new side of the web. Object associations and individual objects can both be named using the RDF naming convention.[26]

Logic plays a significant part in the representation of knowledge. The ontology language is enriched with logic. A known collection of data can be used to deduce conclusions that aren't expressed directly but are required by or consistent with other sets of data already established. The following are some of the most essential logic features:

1. Logic provides a high-level language for the transparent and powerful expression of knowledge.
2. It follows the strict rules of logical reasoning and makes sure that the words it uses are clear.
3. It is possible for automated reasoning systems to derive (infer) conclusions from the input data. For instance,
   Initially, X is a cat.
   Cats are mammals, as are all other animals.
4. A mammal gives birth. The result of this is that X gives birth to new children.

There are proof systems where the logical conclusion in terms of meaning is the same as the derivation in terms of syntax in the proof system itself. The proof that leads to a logical conclusion can be tracked down thanks to proof systems. In this way, logic may explain the answers to questions. Because the web has a number of qualities that can lead us into trouble when we are using current logic, it is important to take care when adding logic to the web. Making inferences, essential actions, and other things that are easier on the web requires using rules. The logic used must be strong enough to describe complicated things, but it can't be so complicated and rigid that it contradicts itself when it tries to figure out what the software agents know.

Ontologies are able to be represented through the utilization of a wide range of knowledge representation paradigms, such as description logics as well as frame logic. The Web Ontology Language (OWL) is an example of a logic-based language that may be utilized to express knowledge.[27]

The Knowledge Interchange Format (KIF) and the Simple Common Logic (SCL) are both part of this category, along with a number of additional logical description languages.

Discussion about OWL and its ancestors follows in further sections. There are many different approaches to defining logic (DL).

First-order logic, often known as FOL, as well as modal logic (ML), are closely connected to description logic (DL). As the reasoning in distinct FOL fragments became more complicated, DL research began to address computing issues. The term "concept languages" was used to underline that the representation language established the basic terminology in the modeled domain, which was followed by "concept languages" in the early stages of DL research.[28] DL has become an important part of the semantic web because it is used to make ontologies. The rise of DL can be attributed to a shift in focus from the underlying logical systems to its qualities. Knowledge representation systems and applications in a variety of domains were studied in depth as part of research on DL, for example, to understand database conceptual models, to represent schemas in an information integration system, or for metadata management, etc.[27, 28]

The term "description logic" refers to formal logics that have clear definitions of their semantics. When working with DL, model-theoretic semantics are used to describe the connections between the syntax of the language and the domain models. The importance of "decidability" as a crucial reasoning problem and the availability of good and thorough reasoning methods are both taken into consideration when developing DL. It is a fundamental quality of DL that it can describe relationships that go beyond the "is-a" ties between concepts.[28] By using a variety of concepts and role constructors, "concept descriptions" are used in DL to explain the most significant aspects of a domain's ideas and roles. Axioms can also be used to limit the ways in which a DL knowledge base can be interpreted by stating facts about the domain.

TBox and ABox are the two primary components of the DL knowledge base and are clearly distinguished. TBox is made up of declarations that explain the generic attributes of concepts and contains knowledge that has been purposefully included in the form of terminology. It consists of statements that describe thought hierarchies.

10.10.1 Family of Web Ontology (OWL) Languages

These use cases were highlighted by W3C's Web Ontology Working Group, which stated that RDF and RDF Schema are not expressive enough for these applications. As discovered by scholars in the United Statesand Europe, ontology construction requires a more powerful language. In Europe, a new ontology language called Ontology Interface Layer (OIL) has been created. A similar initiative called Defense Advanced Research Project Agency (DAML) was launched in the United States.

10.10.2 Distributed Agent Markup Language

This language is a combination of two languages together with DAML+OIL.[29] The W3C Web Ontology Working Group has started developing OWL, the ontology that will use DAML and OIL as the language of the semantic web. The OWL ontology language was constructed with description logic serving as a fundamental component. OWL is built on top of RDF and RDF Schema, which act as its foundations. OWL was developed by the Object Web Working Group. OWL has a significant impact on the expansion of the vocabulary that can be used to describe classes and attributes. In addition to this, it consists of linkages between classes (such as the concept that two classes are distinct from one another), cardinality (such as "exactly one"), equality, more robust property typing, and enumerated classes.[30]

The objective of the OWL language is to give computer-understandable meaning to resources.

In order to improve the realism of machine representations of resources,[31] OWL takes advantage of RDF's web description structure and URI naming to enhance ontologies with the characteristics stated that follow.[30] These are the additional benefits:

1. The ability to run on many platforms at the same time.
2. The ability to adapt to the Internet's demands.
3. Adherence to web accessibility and internationalization standards.
4. Flexibility and openness.

This ontology language is made up of three smaller languages: OWL Full (OWL DL), OWL Lite (OWL), and M-OWL. The ways that these sublanguages can communicate are different from one another. There are pros and cons of each species of OWL. The full OWL uses all of the basic elements of the OWL language. RDF and RDF Schema let you put these primitives together in almost any way. Syntactically and semantically, OWL Full documents are compatible with RDF documents. Any legally valid OWL Full document is identical to any lawful RDF/RDF Schema conclusion. Because it is more expressive, it can't be used for applications that need a lot of help with reasoning. This makes it unusable. The more expressive a person's knowledge base is, the more difficult it is for them to reason. Because time is going faster and faster, software agents will need more time to process a query.

It is meant to bring back computational decidability, which means making sure that all computations can be done in a finite amount of time, while also helping users who want to be as expressive as possible while still keeping computations complete. There are a few ways you can't use OWL DL, which is made up of all the OWL language constructs. For instance, a class can be a subclass of multiple classes, but it cannot be an instance of another class. The reasoning used to describe objects in OWL DL is more similar to that of the less expressive SHOIN (D).

10.11 ONE BENEFIT IS RATIONAL THOUGHT FACILITATION

Before an RDF document can be regarded as a legitimate OWL DL document, it must undergo a variety of restrictions and modifications. So this is a negative event. As stated earlier, all valid OWL DL documents are also valid RDF documents. It's like OWL DL, except with a few more restrictions. Descriptive logic SHIF (D) has a similar structure but is less expressive. When it comes to things like disjointness declarations, OWL Lite does not include them. Start-up and implementation should be as simple as possible. Processors makeit possible for beginners to learn the basics of OWL Lite before moving on to more advanced applications.

Advantage:

Because it's easier to understand and implement, it's better for users (and tool builders).

Disadvantage:

The range of emotions is narrower.

We chose to utilize OWL DL to represent the current work's knowledge base because of the benefits of OWL DL and the tools available for ontology construction.

10.12 TRUST LAYERS

The trust layer at the structure's apex is the most crucial and advanced level of the pyramid. In order for the Internet to realize its full potential, users must believe that the information they discover on the web is accurate and that using the Internet is secure. Digital signatures and other types of information, such as ratings or recommendations from trusted agents, recommendations from certifying agencies, and recommendations from consumer bodies, can be used to build the trust layer. The figure 10.6 explains Structure of the ontology of African fauna, including its classes and subclasses.

Figure 10.8 shows OWL-S framework.

People think of each layer of the semantic web as being built on top of the one before it. As the layers go on, not only does each one become more specialized, but it also gets more complex. It is possible to build the layers and run them in a decentralized way, which is also what you want to do. "SWS" is the short form for "semantic web service." With the help of web standards, a technology called "web services" can make it easy for software components written in different programming languages to be accessed in the same way by everyone. These are reusable software components that have been well-defined and use a standard web-oriented framework to carry out the operations that have been encapsulated. One of the biggest problems with web service technologies is that they can't find and put together web service resources on their own. Because of this, they are not good for use in a business setting that requires a lot of work and input from people. But the technologies that are currently used as the standard for web services, like WSDL, don't make it clear what the syntactic definitions might be.

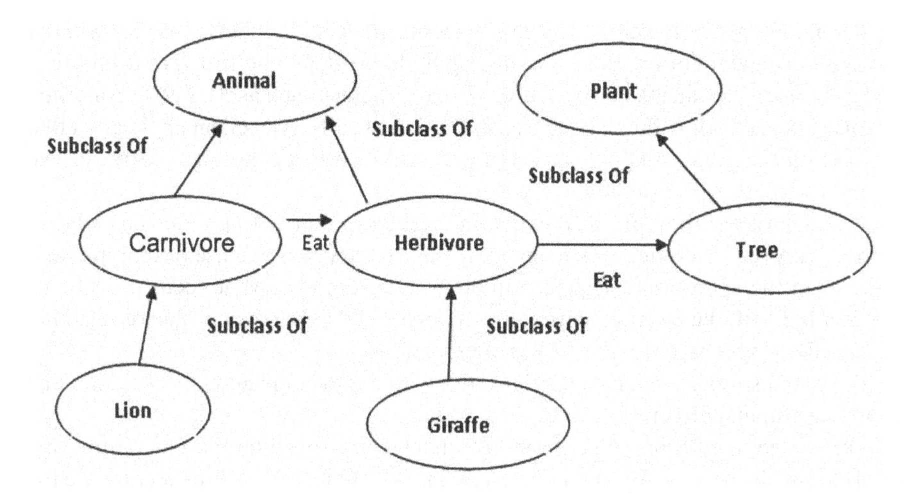

FIGURE 10.6 Structure of the ontology of African fauna, including its classes and subclasses.

Academics who are working on the semantic web have suggested that web services should have semantic descriptions added to them to make them easier to find and use. This is because there is no longer a clear way to describe how online services work. SWS is a technology that brings together web services and semantic web technologies.[32] To put it another way, the SWS is an extension of a web service that adds a representation of semantics. The SWS will automatically find web services, put them together, and run them. This could change the way that businesses and people get and share information and services over the Internet. In her description of the SWS infrastructure, Sabou [31] used three orthogonal dimensions: use activities, architecture, and service ontology. The requirements for SWS can be divided up into the following three categories: operational, physical, and conceptual. It is necessary for a framework for SWS to be capable of meeting the functional criteria determined by the activities associated with using SWS. The architecture of the SWS identifies the numerous components that are necessary to construct a semantic web. It is a service, and the information that explains and supports how to utilize it is known as the knowledge-level model of the data.

10.13 A PERSPECTIVE ON THE SWS

If semantic web services are implemented, developers and users will have greater control over all aspects of web service construction and operation. An agent will be able to automatically locate a web service that meets their requirements, which may also serve as a source of motivation. Currently, a student seeking a class must do so manually byusing a search engine or the UDDI registry to identify potential options until they locate one that meets their requirements. By incorporating semantic markup into web services,[33, 34] it is feasible to define a service in a form that is comprehensible to both machines and humans. This enables us to meet the requirements of both groups. Getting to the same place will take the least time and effort if you use a search engine with ontology or a service registry.

Automatic invocation occurs when an automated agent executes a service without human assistance. For operations requiring multiple steps, the software agent must understand how to interface with the service. Ontologies such as the OWL-S markup language provide an API that enables these function calls. Any client application that employs ontology can comprehend service calls. As a direct consequence, the service can now be performed automatically.

The majority of the time, a single online service is unable to provide an end user with the desired result that was planned. It is necessary to make use of multiple web services in the appropriate sequence in order to accomplish what needs to be done. In order to fulfill the requirements of a user, an agent needs to be capable of selecting and configuring a number of different web services.

The many services involved need to work together to come up with a solution that is both useful and effective.

Automated monitoring is a challenging undertaking that requires a lot of time and effort to finish successfully and effectively. The key objectives of this feature are the monitoring of component tasks and the warning of unanticipated faults. Because of the capability of automatic monitoring to offer feedback to the agents responsible for execution or composition, those agents will be able to react more quickly in the event that any alterations are required. The firm's name is SWS Design Concepts. In this study, Berners-Lee et al. [12] give a discussion of the components of SWS that are organized architecturally and are based on their findings. These are the several parts of the SWS architecture, which the SWS architects consider to be the cornerstone of the many trust and safety systems. The constituent parts area service registration that offers the capability for establishing and modifying service descriptions within a semantic registry, as well as techniques for publishing and finding services. Throughout the whole process, a reasoner is used to help figure out how semantic descriptions and questions should be interpreted. When searching for services, the individual who is in need of them and the individual who is responsible for registering them are referred to, respectively, as "discoverers" and "matchmakers."

A composer (or decomposer) is a piece of software program which is essential to decompose or compose composition model, a composition service, and a mediator between the services. It is also used to manage and correct misalignments between the service and demand offered. They have also brought attention to the fact that these constituents may be known by a variety of names and have their own unique complexity in a variety of ways.

10.14 THE SERVICE PROVISIONING SERVICES ONTOLOGY PROVIDED BY SWS

Ontology is the name of another part of SWS. The service's ontology shows what it can do and what it can't do. At the knowledge level, web service standards like UDDI and WSDL are combined with domain knowledge. Service ontology will usually have the following parts: functional capabilities, like input and output; pre-and post-conditions; non-functional capabilities, like category, cost, and service quality; information about the service provider, like name, address, and cost; information about the service's task or goal; and quality and domain knowledge definitions, like input types.[35, 36] The ontology language that is used to build the semantic web helps the service ontology, which is used to describe SWS, be more expressive and able to make inferences. With a group called SWS, web service description languages, which are compiled at various levels of business process such as BPEL and AWS, suffer from an inability to adequately convey their meaning. Techniques and technologies such as FLOWS, METEOR-S, WSDL-S, OWL Service Ontology, and Web Service Modeling Ontology use first-order logic andhave made semantic web service frameworks viable (OWL-S). The OWL Service Ontology and Web Service Modeling Ontology are two further ontologies (WSMO). Thus far, OWL-S and WSMO have received the greatest attention and appreciation for attempting to describe semantic web services.

10.14.1 METEOR-S

METEOR-S is an application that places a high priority on semantic web service workflow management strategies. As a whole, a web process will be helped by ontologies about data, function, non-functional quality of service (QoS), and execution semantics. In addition to actual execution, annotation, discovery, composition, and optimization are included.[37] This chapter's primary aim is not the creation of a new language like OWL-S but rather the application of ontologies to give WSDL meaning. Both WSDL and UDDI have become more detailed as a result. Using the extensibility of WSDL and the data structures of UDDI, you may create "tModels" by grouping processes with their inputs and outputs (technical models).[38]

METEOR-S is currently developing a framework for self-healing, self-optimizing, and self-configuring online processes by merging multi-paradigm reasoning, semantic modeling, and operations research optimization.[39] With this framework, web processes will be able to self-repair, improve, configure, and self-repair.

10.15 FLOWS

Using the FLOWS model, it is possible to describe the way that services talk to each other in a standard way. This chapterfocuses mostly on the significance of the document's content, not how it is assembled into an XML-based message payload. Using the provided components, it is also feasible to emulate the inner workings of web services. The goal of the FLOWS project is to make it possible to reason about the meanings of web services and other electronic services, as well as their interactions with one another and the "real world." FLOWS is not intended to provide a comprehensive description of online services. Instead, it is intended to provide an abstract model that accurately reflects the semantic aspects of how a service operates. WSDL-S was created with the main goal of making it easier to describe semantic web services. The recent involvement of the business world in web services standards has given a new lease on life to the people who work on business process integration. Because it gives us a standard-based framework for sharing dynamic information on web services, it could be a good way to solve the problem of integrating enterprise applications, especially since the need for interoperability is growing. Since the current version of WSDL does not include semantics, it is not possible to automatically add web services-compliant apps to computer systems that are already in place. WSDL-S is a simple extension that adds meaning to web service requirements and capabilities so that service discovery and execution can be done automatically. This means that this goal can be reached. Web services need a lot of different kinds of semantics, from descriptions to compositions to orchestration and choreography/orchestration, as well as a lot of different kinds of data, functions, behaviors, quality of service, and execution. This range of meanings starts with descriptions and goes all the way through orchestration and choreography/orchestration. OWL-S website offers an ontology structure of the subject related to its kind to be looked at in depth, and it is used by various types of people in both academia and business. It was also the first website of its kind to come out. Because of this, service providers can explain what they offer, and clients can explain what they need. OWL-based web service ontology gives service providers a core set of markup language constructs for describing their web services' attributes and capabilities in a way that a computer can understand. Markup makes it possible to automate a wide range of web service tasks, such as discovery, execution, composition, and communication between services. Using the tiered method for making markup languages (OWL), the current version of OWL-S is built on top of the Ontology Web Language.

Figure 10.7 shows the three different kinds of web services-related information and knowledge. What kind of service does this specific website offer? The results are redirected to check the profile of business information and also advertise the business profile as such.

1. What the service can do?
2. Could you tell me how this web service works? Is it possible to use this web service or get access to it through the service model? Through Service Grounding, you can get this service.

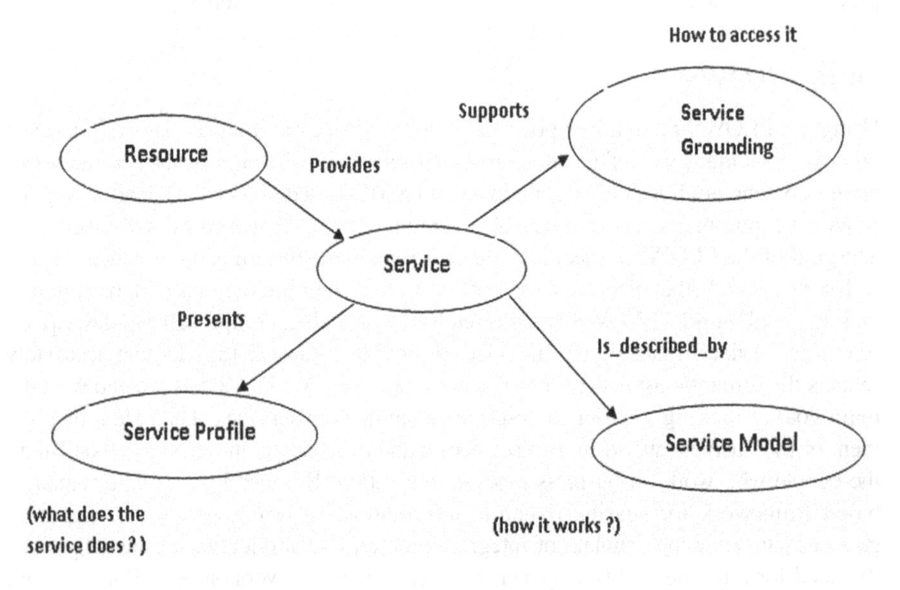

FIGURE 10.7 Web services, upper ontology.

3. Make service descriptions by putting together a lot of basic or atomic services to do a certain task.
4. Be in charge of what the different people who are using an invoked service are doing when the service is being used.

The class service has a list of properties that includes the words "presents," "is described by," and "supports." These needs are met by the service profile, service model, and service grounding classes, in that order. In the paragraphs that follow, we'll talk about the service profile, the service model, and the service foundation in more depth. A description of a service that is included in a service profile in order for it to be discovered. To characterize the service and associated questions, service name, contact information, and an infinite number of additional characteristics are used, eg. an estimate of the maximum response time, the type of concept, the quality of service, and where the service can be used. For example, the profile class can be subclasses and specialized, which lets you make profile taxonomies that define different service classes.[40]

The OWL-S profile is made to model not only the input and output parameters of web services but also the conditions that come before and after them. But so far, there hasn't been a formal method for defining the criteria. Now, only the types of input and output parameters need to be specified to finish a constraint. Since a web service's temporal structure is not limited in any way, users cannot choose between functional and non-functional service properties or set up relationships between inputs and outputs.

After a service has been found, the service profile is no longer needed because it doesn't give any more information that can be used to make calling the service easier. Before a service can be turned on, data must be retrieved from the service model that goes with it. A service model gives an agent a number of different ways to get help when they need it. A thorough analysis has been done to check whether the service standards are met or not.

In this case, the service could be thought of as a task that needs to be done. The OWL-S architecture is made up of processes at the atomic, fundamental, and composite levels. If you want to, you are free to order as many IOPEs as you want. Both the service profile and the service/process model are ways to describe the same service, so they must be brief.[41] Even though the task jobs are different for various cases because of their OPEs elements, one is shown up with the other elements.

With the help of OPEs and third-party service elements, OWL-S is designed to serve the needed and eliminate the elements of unnecessary. In the event that the service profile and the service model do not coincide and an OWL-S expression needs to be verified, the interaction between the services is likely to fail. This is because the OWL-S expression describes the relationship between the service components. Due to these factors, it is abundantly evident that IOPEs must be stated with great care in the service profile to avoid inconsistencies and ensure that the service profile does not contain too many or too few IOPEs.[42, 43]

This is a description of "service grounding," which is the method through which an agent can connect to a service. In this section, we define the protocol, the message format, and other properties such as the host machine's IP address and port number.

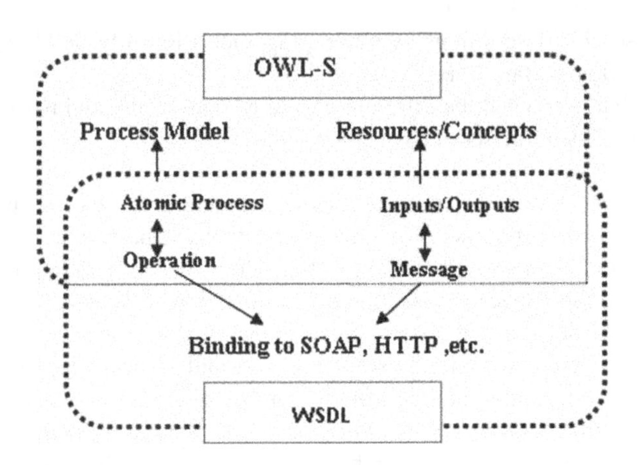

FIGURE 10.8 OWL-S framework.

Unlike service profiles and service models, grounding is viewed as an abstract representation in OWL-S. The grounding process can be viewed as a map from an abstract explanation of the service interaction-required sections of the service description to a more concrete explanation. OWL-methods are WSDL-based.[44]

10.16 TOWARDS MODELING ONTOLOGY FOR WEB SERVICES (WSMO)

The Web Service Modeling Ontology (WSMO), which is a top-level ontology for semantic web services, describes the many parts of these services. The goal of the Web Service Modeling Framework (WSMF) research project is to improve both the idea of web services and the technology that makes them work.

WSMO gives an overview of what semantic web services are and how they work. The WSMO architecture is based on the ideas of strict decoupling, mediation centrality, separation of ontological roles, description versus implementation and execution semantics, and ontology-based implementation.[45, 46] In WSMO, you can use one of four top-level elements to describe the different features of semantic web services (based on Version 1.2). Ontologies, goals, web services, and mediators are the four main ways to group things. Ontology elements are described in an ontology language, which is based on the Meta Object Facility (MOF).[46]

To make sure that item descriptions are complete, properties are used to describe every WSMO element that includes non-functional aspects that are important. The Dublin Core Metadata Set serves as the foundation for this, and other service-specific features, such as information regarding versioning and information, are included, which include quality of service and it is often available to gain stable information to the user about the owner and information related to financial service.

As a side note, it's important to know that the match between purpose and capability can be found right away if the person asking for the service and the person giving the service both use the same ontology. Most of the time, their ontologies are

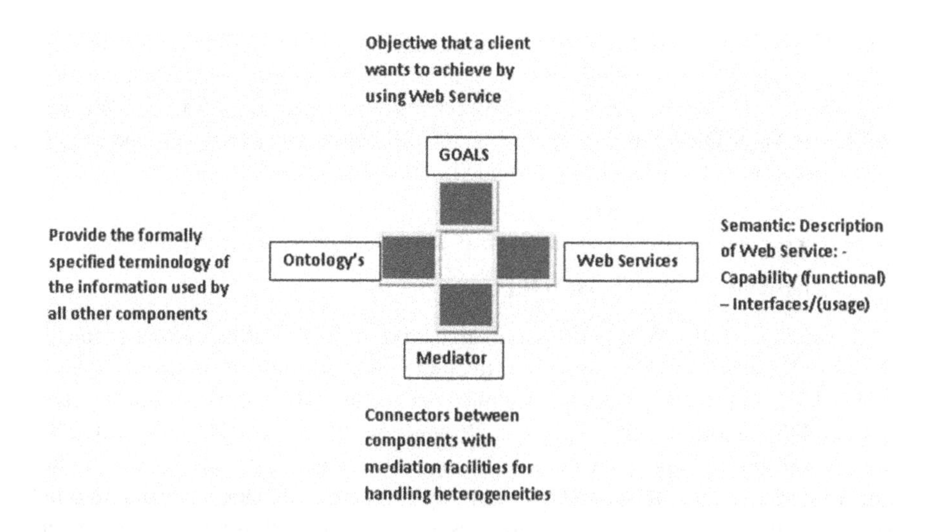

FIGURE 10.9 WSMO is a meta-ontology for SWS.

different from each other. The fact that the requestor and the provider are different makes it hard for them to communicate with each other. WSMO suggested four different kinds of mediators to help solve this conflict. Mediators connect ontologies to each other, web services to each other, web services to goals, and goals to goals. OO mediators are used to link two ontologies together. WW mediators link two different web services together. Both the OWL-S ontology and the WSMO ontology are based on the same basic idea. The most important things about WSMO are that it is easy to use, it is complete, it is more expressive thanOWL-S, and it can be run. OWL-S and WSMO, on the other hand, aren't quite ready for prime time yet.[47] The OWL-S and the WSMO are side by side. Previous research has shown that OWL-S and WSMO are the two most important activities that must be undertaken in order to successfully implement semantic web services. In addition to that, this is something that needs to be taken into consideration.

METEOR-S is different from WSMO in that it puts more emphasis on technology and doesn't have a conceptual model for defining services and the parts that go with them. This is a big change from how things usually are. Domingue et al. (2005) say that the differences between WSMO and OWL-S in terms of epistemology have to do with ontological separation and mediation centrality.[46] The OWL-S and the WSMO are different from each other in a number of ways.

WSMO is an acronym that stands for ontologies, web services, goals, and mediators. These are the most important parts of semantic web services. The OWL-S ontology and language can be used to describe these parts.

However, WSMO distinguishes between the perspectives of the requester and the provider about capability. On the other hand, goals in OWL-S service profiles include both capabilities that are already available and those that are expected to be available (advertisements and requests). In the same way that OWL-S does, WSMO defines a WSDL mapping. However, it places a greater focus on a foundation that is founded on ontologies. In OWL-S, mediation is treated as a byproduct of the orchestration process, whereas, in WSMO, mediators are considered integral components of the whole idea. When it comes to taxonomic knowledge, OWL-S uses OWL and SWRL, but WSMO uses WSML, which is a family of languages that can be used to express taxonomic knowledge and provide inference rules.

10.17 THE FRAMEWORK IS THE E-LEARNING RESOURCE MANAGEMENT MODEL

In this section, we'll talk about the tools that were used to model and build ontologies, as well as the tools that were used to create the learning perspectives.

In order to construct ontologies, we made use of the OWLDL programming language and the ontology editor known as Protégé.[48] OWL's foundations, which are built on description logic, provide it with many advantages, but it also has a number of disadvantages. Even though OWL is highly expressive, its lack of "when"applied to the web, the presence of defined variables and rules, as well as a method for determining "whose" content or person says "what," make it less expressive.[49] Axioms in the description logic must be limited to a tree structure in order to show rule-like axioms, which are frequently required for complicated activities involving data integration, such as queries, views, and transformations. It must to be this way for the axioms to be understood. Rule-based languages and ontology-based languages have been brought together in many different ways. Semantic Web Rule Language (SWRL) [50] shows how rule-based extensions can be used with OWL DL. As we've talked about in the past from the point of view of description logic, the OWL rule language was made so that primary reasoning problems could still be solved. Because of how it is used, SWRL has a bad reputation for being hard to understand. Some people have said that DL-safe is a subset of SWRL,[51] which would be a valid way to extend the rule. In Section 10.4, we talk about Tim Berners-Lee and N3Logic, who both helped us achieve our own goals. Not only is it used to write the rules, but it is also used to ask questions about the knowledge base. Other important reasoning tools were also used, like Euler, who is a backward-chaining reasoner and will be talked about in the next section of thischapter.

The expressive power of N3Logic is comparable to that of description logics (DL), which is a subset of First Order Logic (FOL); it differs from that of the vast majority of other logics. It enables the establishment and transmission of rules through the Internet. N3Logic serializes RDF models in a non-XML format using the Notation 3 (N3) syntax, which is an extension of the predicates vocabulary used in RDF. N3 aims to provide a standardized data model and syntax for logical information. This paves the way for the generation of ontology concepts from the ground up. RDF and rules can be linked together through the usage of N3Logic.

It has a number of built-in features that make it easy to look for and use information you find on the Internet. The main goal of the N3Logic project is to add an extension to the RDF data model that lets rules and data be written in the same language. The fact that both rules and data are written in the same language is what makes N3Logic both easy to use and complete. Using M3Logic, the information should be limited to rules, but it also needs more expressive power than the RDF graph, which should be able to share in the same way that RDF can be shared, as stated by Berners-Lee.[48, 49] When information is shared in this way, it makes it possible for others to use it for a wide range of purposes, some of which the person who first shared it did not intend. We can't make any implicit or closed assumptions about the system's entire body of knowledge because we don't know enough about how it is used. Because of this, N3Logic is known for being repetitive, which ison purpose.[52] In the N3 (disjunction) construction, you can't use either negation or disjunction. It makes things clearer by getting rid of the need to use OR in the reasoning process. If NOT can't be used, "negation as failure" can't be said. No one person could ever hope to understand how big the world is. Even if you don't know anything about a subject, that doesn't mean it doesn't exist. It's the same as saying "does not exist" is false because we don't know anything about anything. If we lived in a closed universe, this would be the case. When talking about something on the Internet, it would be better to say "does not know" instead of "does not exist." "Open world thinking" is the idea that you can't possibly know everything.

N3Logic gives its customers a set of predicates to use. Table 10.1 shows that its vocabulary is made up of both the N3 syntax and a group of URI references. The following namespaces define these URI references, including but not limited to:

@prefix string:<www.w3.org/2000/10/swap/string#> @prefix log: <www.w3.org/2000/10/swap/log#>

In N3Logic, each of these properties is represented by its very own axiom schema. These axioms can be used to do evaluations on formulas as well as variables. These characteristics are referred to as "built-in functions," and they can be used to carry out a variety of activities.

10.17.1 Overview of RDF, RDFS, and OWL

This section gives an overview of how N3Logic has connected to RDF, RDFS, and OWL.

Despite the fact that N3 syntax allows RDF expressions, it does not make full use of the RDF vocabulary. In place of the abbreviated "a" that is used for RDF's "type" symbol, it is possible to use the full URI symbol for rdf:type. The passage [53] presents a lot of illustrations of situations just like this one. The specification of RDFS semantics is made much simpler by the use of N3Logic, which allows for the following set of rules to be used to define rdf:domain and rdf:range.

TABLE 10.1
Partial Set of N3Logic Predicates

@prefix math: www.w3.org/2000/10/swap/math#
@prefix crypto: www.w3.org/2000/10/swap/crypto#
@prefix list: www.w3.org/2000/10/swap/list#
@prefix os: www.w3.org/2000/10/swap/os#
@prefix time: www.w3.org/2000/10/swap/time#

```
string:contains,        string:startsWith,        string:endsWith,
string:matches,       string:greaterThan,        string:lessThan,
string:notGreaterThan,      string:scrape    ...     log:truth,
log:implies,    log:edualTo,    log:conclusion,    log:content,
log:includes,             log:semantics,            log:notIncludes,
log:supports, ... math:function, math:list, math:difference,
math:edualTo, math:greaterThan, math:lessThan, math:cos, ...
crypto:verify,  crypto:md5,  crypto:sign,   ...   list:append,
list:in, list:last, ... os:argv, os:environ ... time:day,
time:hour, time:minute ...
```

{?S [s:domain?C]?O} => {?S a?C}.
{?S [s:range?C]?O} => {?O a?C}.
The expression "?S a [s:subClassOf?C]" means "?S a?C."

"equal to" (=) in N3Logic is synonymous with the "owl: same As" notation found in OWL, which is a form of shorthand notation. This is one way of expressing the notion that a statement's subject and object are on equal footing. All of our attempts to link OWL DL to N3Logic have been fruitless. When there is evidence of using data type functional properties that go against OWL DL, this is a big red flag.[54] We can conclude that writing and reading N3Logic is not hard because it is based on the N3 syntax, which is an RDF syntax that humans can understand. N3 syntax is also easy to learn and use, and it lets you use quotation marks, variables, and a "implication operator." Rules can use the built-in functions of N3Logic to access web resources, figure out what conclusions can be drawn from certain web documents, and do a lot of other useful things like math calculations, cryptographic operations, and manipulating strings. Reasoners (sometimes known as "semantic programs") are computer programmers' that can generate logical conclusions from previously presented facts and/or axioms. Using semantic reasoning, an inference engine can be made more general. The semantic reasoner expands your toolset for discovering new facts from an existing set of facts. Here are the four components of drawing an inference: facts, rules, questions, and responses, respectively.

The definition of inference rules frequently makes use of an ontology language. In addition, descriptions are frequently articulated either in DL or in a language that is very close in nature to DL (e.g., N3Logic). When it comes to getting their reasoning chores done, a good number of reasoners rely on first-order predicate logic.

The reasoning methods that are put to use in the process of carrying out the inference rules can be used to divide the currently available reasoning engines into two primary types. It is true.

When an objective is accomplished, forward chaining begins with the data that is now available and makes use of inference rules to derive additional information (for example, user-provided data). A forward-chaining mechanism is employed when an inference rule has an antecedent that is known to be true. This mechanism searches through the inference rules until the antecedent can be discovered. A reasoner can figure out what comes next, which is called the then clause, and as a result, the reasoner can learn new information (facts).

In automated theorem proofs, proof assistants, and other types of artificial intelligence systems, the process of inference known as backward chaining is utilized. Backward chaining is a technique that begins with a set of objectives (or hypotheses) and works backwards from cause to effect to determine which, if any, of these outputs may be supported by evidence. Backward chaining allows an inference engine to obtain the desired outcome.

It does so by traversing all of the inference rules until it finds one with a sentence that follows the one it is currently examining (then clause). It will be added to the list of objectives if the preceding condition (the "if clause") is false. Sometimes, the term "derivational" is used to explain how backward chaining operates. Also, reasoning engines can use chaining in both directions. The fact that the procedures explained earlier can serve as the foundation for a diverse range of reasoning tools has already been brought to your attention. There are several instances, some of which include Jess, CWM, Euler, F-OWL, Pellet, and Racer, among others. In the following paragraphs, you will get an overview of some of the most effective strategies to assist you in thinking things through.

The lightweight rule engine called Jess was made with the help of a C Language Integrated Production System (CLIPS)-inspired expert system shell (Java Expert System Shell). All of Java's application programming interfaces can be accessed through its powerful scripting language (APIs). Jess handles rules the same way that CLIPS does byusing an updated version of the Rete algorithm, which is a very good way to solve the hard many-to-many matching problem. Some of Jess's special skills are backward chaining, querying the working memory, and directly manipulating and reasoning about Java objects. This is a logic programming system as well as a database that uses deduction. It extends Prolog's semantic and operational capabilities by employing a technique known as tabular logic programming (often referred to as "tabling").

Tabled predicates have the ability to be declared by the system either automatically or manually. This allows for the development of a wide variety of applications in the fields of non-monotonic reasoning and knowledge representation. When asked a question, it will return all of the potential responses by employing a tabling technique known as "local evaluation." This technology is fast and useful, so it can be used for things like programming analysis.

Pellet is the answer you turn to if an OWL DL argument that is robust and comprehensive is required. It satisfies the requirements of the OWL DL standard in every way. Pellet makes use of a variety of different optimization methodologies, some of which include incremental reasoning, nominal reasoning, and conjunctive reasoning. Tableaux methods were utilized throughout the design process for Pellet. Pellet provides support for select subsets of both the SWRL and DL-Safe Rules. Most of Pellet's features and profiles will be added to the next update for OWL 2, which will be out very soon. The Closed World Machine (CVM) is a data processor for the semantic web that can do many different things. One of its constituent parts is called the Semantic Web Application Platform (SWAP). When it comes to text files, it functions fairly similarly to sed, Awk, and other similar programmers. CWM is successful in accomplishing this objective thanks to the implementation of an algorithm for forward-chaining reasoning. RDF Core Rules have been added to the RDF Core, making it possible to query, check, transform, and filter information. RDF/XML or RDF/N3 serializations are used as the RDF Core's principal serialization formats. RDF Core Rules have also been added. Particularly noteworthy is the fact that the layered architectural prototype CWM for the semantic web was developed to evaluate large datasets of various types (as shown in Figure 10.2). As a reasoned, CWM does not appear to worry much about large data sets or rule sets, which is contrary to what you might anticipate in applications that are used in the real world.

The inference engine developed by Euler is able to handle logically sound proofs. The backward-chaining reasoning engine has been updated with the addition of a Euler path detector.[54] The axioms are turned into a logic program me as they are retrieved from the Internet in sequence. The Euler proof engine is able to prevent never-ending deductions because it is based on Euler pathways, which is a concept that was developed by Euler more than a century ago. It is because of this thatyou don't have to think about recursion or merging graphs right now. Jos De Roo created the Euler proof mechanism, which is a proof engine.[55]

It was built inJava. This engine uses Euler routes to figure out what's going on without having to worry about loops that never end. It can understand Notation3, including the rules for N3. N3 can also be used to get in touch with the W3C Community Working Group (CWM). Euler can be written in many different programming languages, such as Java, C #, Python, JavaScript, and Prolog. Table 10.2 lists some of Euler's most important built-in functions.

Classification using FaCT model helps in logical satisfiability in using various semantic terminologies.[56] In the FaCT system, there are two reasoners for SHF logic and SHIQ logic. Both of these reasoners add transitive roles, functional roles, and a role hierarchy to ALC by employing sound and exhaustive tableaux algorithms. The FaCT system contains both of these justifications [57–58].

SHIQ is expressive enough to be used as a DLR logic reasoner, and as a result, it can be used to reason about database schemata. Its effective implementation of

TABLE 10.2
The Built-In Functions of Euler

Length is a built-in function that determines the total length of the subject list. e: max is a built-in function that determines the maximum value of the topic list, whereas e: min determines the minimum value of the subject list.
This already incorporates the e:findall protocol (SCOPE and SPAN). For example, use the findall (?SELECT? WHERE? ANSWER) command. Answer provides a list of all instances of choose that, within the scope of all asserted n3 formulae and their logs, satisfy the where clause.
Similar to the log: conjunction, the inclusive of e: disjunction is used at the end of rules that contain disjunctions. However, rather than representing logical AND, it represents logical OR.
e:distinct is a built-in function that eliminates duplicate list items; e:graphDifference is a built-in function that compares two graphs, for example (a b c. d e f d e f). e:graphDifference "a b c"; the built-in e:trace function, which produces the object.

tableaux, which has become the standard for DL systems across the industry. Its COR. SHIQ is able to reason about database schemata. Its capacity to reason with a variety of different knowledge bases. Racer is a semantic web OWL reasoner and inference server. When it comes to Racer's roots, they can be traced back to the logic of description. Tableau calculus for determining aBox consistency in the description logic. The SHIQ(D) is implemented in the middle of the system by the reasoning kernel (w.r.t. a background tBox) RacerPro.

10.18 CONCLUSION

In this chapter, the technologies associated with the semantic web have discussed. Web services came after the languages and technology associated with the semantic web and semantic web services, which arrived first. In-depth research has conducted on ontology, logic and logic languages, reasoning tools and tactics, and ontology itself. In addition, there is a section on the e-learning resource management model in which we discuss the many methods and technologies that have applied to the completion of this project. The aforementioned methods can be implemented using technical stack, which will be used for e-learning system in the future.

REFERENCES

1. Cabral, L., Domingue, J., Motta, E., Payne, T., & Hakimpour, F. (2004). Approaches to semantic web services: An overview and comparisons. In *Proceedings of the first European semantic web symposium (ESWS2004)*. Heraklion, Crete, Greece: DBLP Computer Science.
2. Gugliotta, Alessio, and Vito Roberto. "Knowledge modelling for service-oriented applications in the e-Government domain." In *AAAI Spring Symposium Series Stanford University*. 2006.
3. Fensel, D., & Bussler, C. The web service modeling framework WSMF. www.swsi.org/resources/wsmf-paper.pdf
4. W3C Web Services Architecture Domain. www.w3.org/2002/ws/
5. Pandis, I. A survey on semantic web services, state-of-the-art, current work, and challenges. www.cs.cmu.edu/~ipandis/resources/IndStudy.pdf
6. Martin Gudgin, M., Hadley, M., Mendelsohn, N., Moreau, Jean-Jacques, Nielsen, H. F., Karmarkar, A., & Lafon, Y. (2007, April 27). SOAP, version 1.2 Part 1: Messaging framework, W3C recommendation. www.w3.org/TR/soap12-part1/
7. SOAP. www.webopedia.com/TERM/S/SOAP.html
8. Benz, B., & Durant, J. R. (2003). *XML 1.1: Programming bible*. New Delhi: Wiley Publishing.
9. Web Services Description Language (WSDL) 1.1, W3C Note. (2001, March 15). www.w3.org/TR/2001/NOTE-wsdl-20010315
10. Universal Description Discovery and Integration. http://en.wikipedia.org/wiki/UDDI
11. Fremantle, P., Weerawarana, S., & Khalaf, R. (2002, October). Enterprise services. *Communications of the ACM*, 45(10), 77–82.
12. Berners-Lee, T., Hendler, J., & Lassila, O. (2001). *The semantic web: A new form of web content that is meaningful to computers will unleash a revolution of new possibilities*. Scientific American. www.scientificamerican.com/article.cfm?id=the-semantic-web
13. Lassila, O. Towards the semantic web. www.w3c.rl.ac.uk/pastevents/TowardsTheSemanticWeb.pdf

14. Antoniou, G., & Harmelen, F. van. (2004). *A semantic web primer*. London: MIT Press.
15. Passin, T. B. (2004). *Explorer's guide to the semantic web*. Greenwich: Manning.
16. Davis, J., Fensel, D., & Harmelen, F. van. (2003). *Towards the semantic web*. West Sussex: John Wiley.
17. MathML. www.w3.org/Math/
18. Chemical Markup Language (CML). http://cml.sourceforge.net/
19. Resource Description Framework (RDF) Model and Syntax Specification: W3C Recommendation. (1999, February 22). www.w3.org/TR/1999/REC-rdf-syntax-19990222/#intro
20. Namespaces in XML. World Wide Web Consortium. (1999, January 14). www.w3.org/TR/REC-xml-names/#sec-intro
21. Giunchiglia, F., Dutta, B., & Maltese, V. (2009). Faceted lightweight ontologies. In A. Borgida, V. Chaudhri, P. Giorgini, & E. Yu (Eds.), *Conceptual modeling: Foundations and applications*, LNCS 5600. Singapore: Springer.
22. Gruber, T. R. (1993). A translation approach to portable ontology specifications. *Knowledge Aquisition*, 5(2), 199–220.
23. Studer, R., Benjamins, V. R., & Fensel, D. (1998). Knowledge engineering: Principles and methods. www.das.ufsc.br/~gb/pg-ia/KnowledgeEngineering-PrinciplesAndMethods.pdf
24. Berners-Lee, T., Connolly, D., Kagal, L., Scharf, Y., & Hendler, J. (2006). N3Logic: A logical framework for the world wide web. www.dig.csail.mit.edu/2006/Papers/TPLP/n3logic-tplp.pdf
25. Genesereth, M. H. (1998). Knowledge interchange format. Draft proposed American National Standard (dpANS). NCITS. T2/98-004. http://logic.stanford.edu/kif/dpans.html
26. Altheim, M., Anderson, B., Hayes, P., Menzel, C., Sowa, J. F., & Tammet, T. SCL: Simple common logic. www.ihmc.us/users/phayes/CL/SCL2004.html
27. Schneider, P. (2003). *Description logic handbook: Theory, implementation and applications*, edited by F. Baader, D. Calvanese, D. L. McGuinness, D. Nardi, & P. F. Patel-Schneider. Cambridge: Cambridge University Press.
28. Agarwal, S. (2007). Formal description of web services for expressive matchmaking. Doctoral thesis. www.digbib.ubka.uni-karlsruhe.de/volltexte/documents/2531
29. Web Ontology Language. www.w3.org/2004/OWL/
30. RDF primer, 2004. www.w3.org/TR/REC-rdf-syntax/#richerschemas
31. Sabou, M. (2006). Building web service ontologies. The Dutch graduate school for information and knowledge systems. http://kmi.open.ac.uk/people/marta/papers/thesis.pdf
32. OWL-S 1.0 Release. www.daml.org/services/owl-s/1.0/
33. Martin, D., et al. (2004, November 22). OWL-S: Semantic markup for web services. W3C member submission. www.w3.org/Submission/OWL-S/
34. Martin, D., et al. (2003, October). Describing web services using OWL-S and WSDL. www.daml.org/services/owls/1.0/owl-s-wsdl.html
35. Gugliotta, A., Tanasescu, V., Domingue, J., Davies, R., Gutierrez-Villarias, L., Rowlatt, M., Richardson, M., & Stincic, S. (2006). *Benefits and challenges of applying semantic web services in the e-governance domain*. Semantics 2006, Vienna (Austria): Austrian Computing Society (OCG). http://kmi.open.ac.uk/projects/dip/resources/Semantics2006/Semantics2006_DIP_Camera_Ready.pdf
36. METEOR-S: Semantic Web Services and Processes. http://lsdis.cs.uga.edu/projects/meteor-s/
37. Semantic Web Services Ontology (SWSO), version 1.0. www.daml.org/services/swsf/1.0/swso/
38. Akkiraju, R., Farell, J., Miller, J. A., Nagarajan, M., Sheth, A., & Verma, K. Web service semantics—WSDL-S. www.w3.org/2005/04/FSWS/Submissions/17/WSDL-S.htm

39. OWL-S 1.1 Release. www.daml.org/services/owl-s/1.1/
40. Martin, D., et al. Bringing semantics to web services: The OWL-S approach. http://www.daml.org/services/pubs/090-093.pdf
41. Agarwal, S., Lamparter, S., & Studer, R. (2009). Making web services tradable: A policy-based approach for specifying preferences on web service properties. In *Web semantics: Science, services and agents on the world wide web*, vol. 7. Italy: OSG Publication, pp. 11–20.
42. Ankolenkar, A., et al. (2002). DAML-S: Web service description for the semantic web. In *First international semantic web conference (ISWC)*. Springer: Sardinia, Italy.
43. Boley, H. (2006). *Semantic web services*. ICEC 2006 Tutorial. www.icec06.net/WorkshopsAndTutorials/SOATutorial/ICEC 06-Tutorial-Semantic-Web-Services.pdf
44. Stollberg, M., & Haller, A. (2005). Semantic web services tutorial. In *3rd international conference on web services (ICWS 2005)*. IEEE SCC: Orlando.
45. Florida. (2005, July). ICWS/SWStutorial-iswc05.ppt
46. Domingue, J., Roman, D., & Stollberg, M. (2005, June). Web services modeling ontology (WSMO)—An ontology for semantic webservices. In *Position paper at the W3C workshop on frameworks for semantics in web services*. Innsbruck, Austria. www.w3.org/2005/04/FSWS/Submissions/1/wsmo_position_paper.html
47. Protégé. http://protege.stanford.edu/Horrocks, I., Patel-Schneider, P. F., Boley, H., Tabet, S., Grosof, B., & Dean, M. (2004). SWRL: A semantic web rule language combining OWL and RuleML. www.w3.org/Submission/SWRL/
48. Berners-Lee, T. (2000). Notation3. www.w3.org/DesignIssues/Notation3.html
49. Berners-Lee, T. (2005). Notation 3 logic. www.w3.org/DesignIssues/N3Logic
50. Jess, a rule engine for the Java platform. www.jessrules.com/
51. XSB. http://xsb.sourceforge.net/index.html
52. CWM. www.w3.org/2000/10/swap/doc/cwm
53. Eulersharp. http://projecteulersharp.codeplex.com/
54. The FaCT system. www.cs.man.ac.uk/~horrocks/FaCT/
55. Racer. www.racer-systems.com/index.phtml
56. DL-safe rules. http://logic.aifb.uni-karlsruhe.de/wiki/DL-Safe_Rules
57. Pellet: The open source OWL reasoner. http://clarkparsia.com/pellet
58. Dutta, B. (2008). Semantic web services: A study of existing technologies, tools and projects. *DESISOC Journal of Library and Information Technology*, 28(3), 47–55.

11 Data Science with Semantic Technologies

Mehmet Milli and Fatmana Şentürk

CONTENTS

11.1 DATA SCIENCE

Many individuals regularly feel the need to use systems or platforms that use data science infrastructure from different channels in their daily lives. Although we use these systems every day, few people probably have any knowledge of how the

DOI: 10.1201/9781003310785-11

mechanism behind them works. In cases where we have difficulty making decisions, it is inevitable to interact with systems using data science infrastructure. Social media sites, navigation applications, video content sites, and online shopping sites are at the top of these systems. In the last few decades, with the diversification of data sources, more active use of social media platforms, the emergence of new work areas that produce data, and the acceleration of digitalization in many fields, especially in finance and commerce, data on the Internet has increased at an unprecedented rate. In many fields (Internet of Things, genetics, commerce, social platform, finance, astronomy, energy, agriculture, education, etc.), the volume and diversity of data have long exceeded traditional database processing capacity and manual analysis capabilities.

In parallel with these developments, to cope with such big data: (i) with the developments in the field of transistors, the processing capacity of microprocessors has been increased;[1] (ii) prescriptive and consistent relational databases have been replaced by more flexible, fast, and functional NoSQL databases;[2] (iii) new protocols have been developed for faster data communication and the bandwidth of networks has been increased;[3] and (iv) algorithms have been developed that can link datasets together to obtain more meaningful information and provide a deeper analysis.[4] The main goal of all these technological developments used to analyze large data sets is to extract meaningful information from complex data sets that are increasing in variety and volume day by day. Today, businesses are increasingly investing in these technologies to increase their competitiveness and market share. In the future, new generation data analysis tools will continue to appear in all situations where traditional methods are insufficient or more costly.

Today, many organizations whose brand value is increasing day by day make effective use of data science to increase their income, reduce costs, and provide a competitive advantage by using their existing resources more efficiently. However, for organizations, the increase in data processing technologies along with data may not always be enough to extract meaningful patterns from data. It is an undeniable fact that it is important for other company employees, as well as data professionals, to understand the technology and principles behind data science at a basic level to make effective use of data science in decision-making processes.[5] The first part of this chapter will explain what data science is for people from all walks of life and how the basic principles behind data science work. In addition, the challenges and limitations of data science will be dealt with. In the second part of the chapter, semantic web technologies and ontologies, which are two of the fields of study that have emerged in recent years, will be discussed and how they will be related to data science will be discussed. It will be demonstrated to what extent semantic web technologies and ontologies will contribute to data science. Finally, possible difficulties that may be encountered during and after the integration of semantic web technologies and ontologies into data science will be shared.

11.1.1 What is Data Science?

Generating such a large amount of data in a short time is a huge challenge for any system or organization. Some very powerful, complex algorithms and technologies are needed to process this data and make it useful for the organization. At this point, data science comes into play when it is necessary to extract useful information from huge data piles. So what is the data science that has entered our lives so much? Data science as a computer science term first appeared in 1974 in the book *Concise Survey of Computer Methods*.[6] In this book, Naur refers to data science as "the science of dealing with data, once they have been established, while the relation of the data to what they represent is delegated to other fields and sciences." Although different definitions of data science have been made in many academic articles since the emergence of data science, it superficially overlaps with the definition made by Naur in 1974 as a basic principle.

In general, data science is the application of quantitative and qualitative methods to solve related problems and predict results.[7] Data science is the systematic study of extracting information from data.[8] Data science is an interdisciplinary science and methodology that represents activities for the analysis, interpretation, and extraction of knowledge and information.[9] In another study, data science is expressed as a set of problem definitions, principles, and processes to extract useful models from large data piles in an incomprehensible way.[10] In other words, the purpose of data science is to reveal patterns that cannot be noticed at a glance by using multidimensional, flexible, and dynamic perspectives apart from conventional theories and methods of the features and hidden structure of complex natural, human, and social phenomena.[11]

Data science brings together different fields of study in statistics and computation under the same roof to interpret data and improve decision-making processes. Data science, which was launched for the first time in the 1970s and experienced its golden age after the 2000s, is an interdisciplinary subject consisting of many fields of study, from basic fields such as mathematics and statistics to general fields such as computer science. Data science has become associated with more specific fields such as artificial intelligence, data mining, and data analytics in recent years.

Undoubtedly, domain knowledge is needed in addition to the basic and specific fields of study to apply data science to a field with a clearly defined framework and to make useful information inferences that affect the decision-making processes related to this field. Today, data science is warmly embraced by more disciplines and fields such as law, history, medicine, and archaeology, apart from its traditionally relevant fields such as mathematics, statistics, and computer science. The number and variety of fields that data science is associated with are increasing day by day.

11.1.2 Data Science Process

In general, data science aims to numerically represent, analyze, and understand real-world situations and phenomena. The processing steps of these real Earth phenomena are complex, multifaceted processes. This complex process begins with the numerical representation of phenomena so that data professionals can more easily understand and process phenomena on their machines. A standard data processing procedure is established by formulating the numerically represented data. However, even at this stage, it is not very possible to draw a clear conclusion and interpretation from the data set obtained. It is inevitable to use convectional mathematical-statistical methods and scientific data processing approaches, as well as machine learning

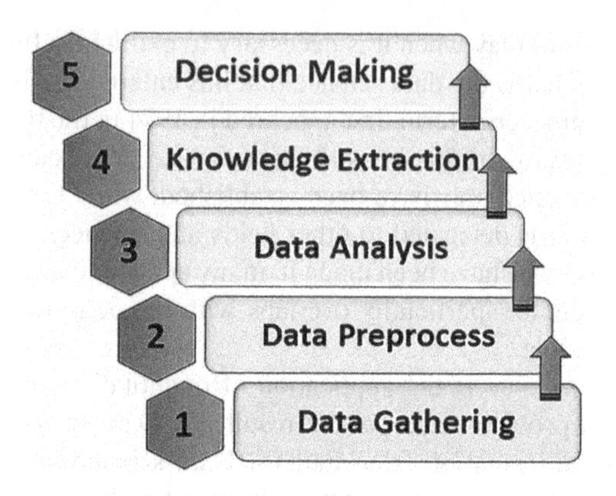

FIGURE 11.1 Data science main operation steps.

and artificial intelligence algorithms capable of in-depth analysis, to obtain predictions from the data that has been structured up to this stage.

The systematic and gradual realization of such a complex process as possible plays a very important role in guaranteeing the possible benefits to be obtained as a result. For this reason, although the process of obtaining insights from data in data science, in general, is carried out in three stages data design, data collection, and data analysis, this process has evolved over time and more specific processing steps have been added. Although it is a widely accepted convention that the Data Science Process Steps model is the typical way to describe data science projects, researchers have yet to reach a consensus on the stages involved in data science processing steps. [12] It will be beneficial for the results to be healthier if the study of extracting useful information from any data is carried out with a data science life cycle defined and accepted in the literature.

Data science processing steps will be examined in five steps within the scope of this chapter. The first of these steps is the data gathering stage, which expresses the collection of data from any source. Other data science stages consist of data preprocess, data analysis, knowledge extraction, and decision making. As data science progresses through the processing steps, the value of raw data collected from real-world applications increases and can be integrated into decision-making mechanisms. The data science process steps that make up the data science processes are shown in detail in Figure 11.1. Although these stages that make up the data science processing processes are independent of each other, they are successive stages to reach a common goal. The output of one stage is an input for the next stage. Therefore, the success of a stage is directly related to how well the product produced by the preceding stage meets the success requirements.

11.1.2.1 Data Gathering

The data collection step is the first of the processing steps shown in Figure 11.1, where useful information will be obtained from the data. Most of the time, at the beginning of the data acquisition phase, it should be done in the definition of the problem. Every successful data science work begins with a clear problem definition for a beneficial outcome of the project. Defining the problem clearly will allow the solution to show itself. Defining the problem often includes a clear statement of where and how the data will be obtained. It is vital for the data scientist to have a better understanding of the collected data and to transform this data into a data science problem.

At the stage of data collection, operations can be performed on the data previously obtained and stored from any source, or the data collection process can be started again according to the definition of the problem. Real-world situations are generally used as the data source in data science projects. Among these resources, there can be a wide variety of resources from the data generated automatically by the systems such as IoT, Sensor, and Log to the data created by users such as social media and enterprise resource planning (ERP) platforms. Also at this stage, a decision should be made on how often and for how long the data should be collected.

11.1.2.2 Data Preprocess

The data obtained in the data collection step, and which has not undergone any preprocessing, is called raw data. The data obtained after the data collection step may be insignificant and mostly worthless information. At this point, data science aims to transform unprocessed data into meaningful data, that is, useful information for businesses and institutions, by applying certain methods and algorithms. To extract more meaningful information from the collected data and to enable machine learning, data mining approaches, and algorithms to process the data more easily, the data needs to undergo some preprocessing. Operations such as data reduction, data selection, data cleaning, data formatting, and data integration are the main data preprocessing steps. These preprocesses may differ depending on the characteristics of the generated data set. There are many methods, algorithms, and approaches in the literature for performing all these preprocessing steps. Which method and approach will be chosen at this stage is very important for the success of the result.

Data formatting is the formatting of the data in such a way that it can be processed into algorithms to be used in the next stages. The collected data set is sometimes too large for the machine learning algorithms to be used. In this case, it may be necessary to reduce or summarize the data set in a way that does not affect the accuracy of the results. Reducing the volume of the data set is called data reduction. Especially in systems that generate data on their own, there may sometimes be data that has nothing to do with the feature. The data set should be purified from these values so that the results are not affected by these irrelevant data. This process is called data cleaning. It may not be necessary to use all the data in the data set to achieve the specified purpose. If some features do not affect the result, these features should be removed from the data set. The process of leaving only the features that change the result and be suitable for the purpose of the data is called data selection. The data selection feature can be on a column basis or a row basis. Data in data science studies do not always come from a single source. If the data comes from different sources, the data must be combined. The process of combining data from different sources is called data integration.

11.1.2.3 Data Analysis

After the data goes through the preprocessing steps, it needs to go through some analysis processes before making an inference from the data, that is, before applying to machine learning and artificial intelligence methods. In the previous stage, selection, cleaning, integration, and reduction operations were important for cleaning the data set. At this stage, the data must change so that the machine learning algorithms do not make biased decisions. For this reason, this stage is also called the data transform stage in many places in the literature. Data transform operations may vary according to the characteristics of the collected data. The most used data transformation operations in the literature are outlier detection, normalization, and imputation.

An outlier (anomaly) can be defined as any observation in the data set that differs from other observations in the data set. It is ensured that these outliers are removed from the data set so that they do not adversely affect the algorithms to be used in the next stages. The data set may sometimes contain gaps due to system or human error. These gaps in the dataset are called missing values. When extracting meaningful information from a data set, large gaps in certain parts of the data set can cause machine learning algorithms to have difficulty processing the data. Therefore, filling the missing values in the data set with a method accepted in the literature will increase the success of the approaches to be applied in the next steps. The process of filling missing values in the data set for any reason is called imputations. Another widely used operation is normalization. If there are obvious differences between the value ranges of the features in the data set, it will be appropriate for the reliability of the result to shift the features in the data set to a certain value. It is inevitable for machine learning approaches to make biased decisions since the feature with a large value in unnormalized data sets is expected to affect the result more.

11.1.2.4 Knowledge Extraction

Up to this point, data collection and transformation into a format that machine learning algorithms can process have been carried out. In the knowledge extraction section, the data that is not worth much at first is ready to be converted into useful information. At this stage where data is transformed into information, machine learning and artificial intelligence methods and algorithms are frequently used. There are many previously accepted approaches to infer information from data. However, no approach or method guarantees that the result will be more successful. Therefore, many methods may have to be tried at this stage to realize a successful data science process and create a utility model. Method and approach which will be chosen at this stage is very important for the success of the result. The model developed as a result of the knowledge extraction stage should provide the data scientist with prediction, association relationships, comments, recommendations, and inferences that can help decision-making processes.

11.1.2.5 Decision Making

The last of the data science processing steps is the integration of the developed model into the firm's existing decision-making systems. At this stage, it is of great importance to discuss the issues of the model developed at which stage of the decision, how effective it will be, and to conclude. In addition to all these, data development processes for a company, institution, and organization are dynamic processes. For this reason, in some studies in the literature, the data science development life cycle is also used instead of data science development steps.

11.1.3 Challenges and Limitations of Data Science

While data science has numerous benefits for businesses, the challenges and risks are too numerous to underestimate. Companies need to avoid risks and overcome challenges as they continue to reap the benefits of data science. Permanent solutions to some challenges and limitations of data science still have not been found. Today, permanent and effective solutions to the risks of data science continue to be sought. The challenges, risks, and shortcomings of data science are listed as follows.

11.1.3.1 Unstructured Data

Most of the data on the Internet is unstructured. Processing unstructured data and extracting useful information is quite difficult.

11.1.3.2 Different Representation of Data

Many systems, institutions, or companies use their infrastructure for data-generating systems. This causes a different representation of data between systems. Therefore, it is not possible to talk about the interoperability of systems in which data are represented differently.

11.1.3.3 Limited Adaptation Ability to Dynamic Processes

Decision-making processes in the real-world order depend on many instantaneously changing parameters. It may not always be easy for the developed model to keep up with such a rapidly changing system.

11.1.3.4 Lack of Intuition and Creativity

Today, decision-making systems based on structured thinking created using data science are successfully implemented in many applications. But since these systems lack human intuition and creativity, they may require human intervention at certain parts of the process.

11.1.3.5 Reusability and Shareability

Today, most of the data is produced by machines and systems for consumption by other machines and systems. This can often cause conflict due to the lack of a common data structure between systems. Therefore, it sometimes causes problems such as not sharing the collected data between systems and not being able to be used again when requested.

11.1.3.6 Deep Domain Expertise Requirement

Data science applications are often specific to a specific domain. To make a successful information extraction based on solid data science foundations, field experts must be included in almost all data science processes. This situation will lead to an increase in the workload of field experts and perhaps the need for more field experts.

11.1.3.7 High Privacy and Security Requirement

One of the biggest problems in every platform where data sharing has been the security and privacy of the data. In the data science life cycle, a data scientist may need to spend a large part of their performance and time maintaining security and privacy.

11.1.3.8 Performance Anxiety

Decisions need to be made quickly so that companies can cope with huge data volumes, demonstrate their competitiveness and meet market challenges. Therefore, the search for better performance in the data science process is expected to be one of the problems that data scientists will address in the future.

Many of the problems briefly mentioned are problems that data professionals have a high potential to encounter during their work. There are a lot of works in progress to address these issues and increase the gains from data science. In addition, many academic communities have stated that some of these problems that data science has to face can be solved with semantic web technologies and ontologies.

The next part of the chapter will discuss what the semantic web and ontologies are and how semantic technologies and ontologies can be integrated into data science. It will be demonstrated to what extent semantic web technologies and ontologies will contribute to data science. In addition, possible difficulties that may be encountered during and after the integration of semantic web technologies and ontologies into data science will be shared. What method should be followed to overcome these difficulties will be included in the discussion.

11.2 SEMANTIC WEB TECHNOLOGIES AND ONTOLOGIES

Most documents in the current web environment are unstructured and designed only in languages that people can understand and interpret. Therefore, machines can't understand and interpret information in current web environments. Another problem arises from the fact that data is only represented specifically to a particular application. The fact that systems produce application-specific data in a format that only their infrastructure can use significantly limits the sharing and reusability of data between systems. In addition, since most of the web content is produced by different channels and sources, it has a heterogeneous structure. This heterogeneity is possible to evaluate in four categories as (i) system (different hardware and operating systems), (ii) syntax (different languages and data representations), (ii) structure (different data models), and (iv) semantics (different meaning of terms in special context).[13]

At the end of the 1990s, the concept of the semantic web was introduced to address these problems inherent in data. The semantic web is not a different technology than the current web but is an infrastructure that enriches the meaning of data by making better definitions and supports the cooperation of machines and humans.[14] In the other words, the semantic web is a technology that allows the use and interpretation of data by machines by providing a common language that aims to understand the contents by machines. Thus, the semantic web provides a framework for sharing and reusing content by creating bridges between similar applications, communities, and companies. For this reason, the semantic web is seen as a way to build bridges

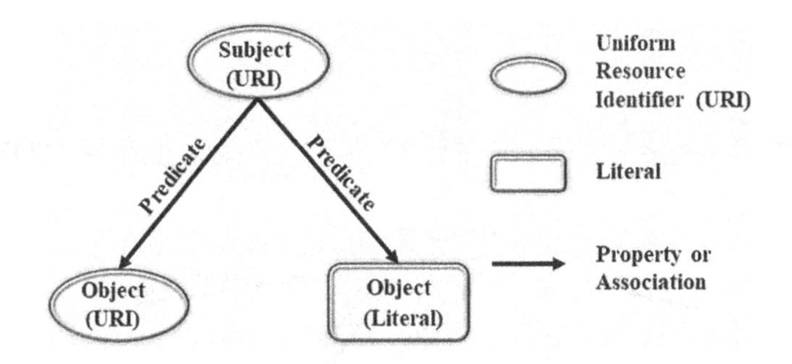

FIGURE 11.2 General structure of RDF triples.

between different content, different systems, and different applications that process information.

Semantic web technology is built on semantic modeling languages such as Resource Definition Framework (RDF) and syntactic modeling languages such as XML to achieve the goal of machines working collaboratively with humans. The RDF metadata model is a flexible W3C specification that represents information about resources on the web. In other words, RDF is a graph-based data model that supports the main component and data representation format for connected data and the semantic web.[15] Figure 11.2 shows how the RDF metadata model represents and presents the data resources.

11.2.1 What is Ontology?

The backbone of the semantic web are ontologies that conceptualize information with metadata models and establish rich relationships between them. According to Gruber's (1995) definition, ontologies are the explicit representation of conceptualization.[16] In other words, shared ontologies that are formally conceptualized in a particular field provide a shared understanding of the potential subject matter that can be collaborated between humans and machines. Putting the information in a certain format through metadata is of vital importance for the machines to understand and interpret the contents. The usage areas of ontologies are increasing day by day. Until now, many ontology examples such as gene ontology, legal ontology, and sensor ontology have been created and successfully executed. Today, ontologies play a key role in technology, integrating data, information, and process with interoperability.

Ontologies represent content in triplets using Ontology Web Language (OWL) built on RDF and XML frameworks. These triples, which ontologies have used to represent knowledge, are similar to simple sentences used in daily life. These simple sentences only consist of subject-verb-object. The ability of ontologies to define concepts using triples enables the creation of metadata in which concepts specific to any field, relations between these concepts, and examples of concepts are defined together. The triples of the ontology, in which some concepts of the mayor and the relationships between them are shown, are given in Figure 11.3.

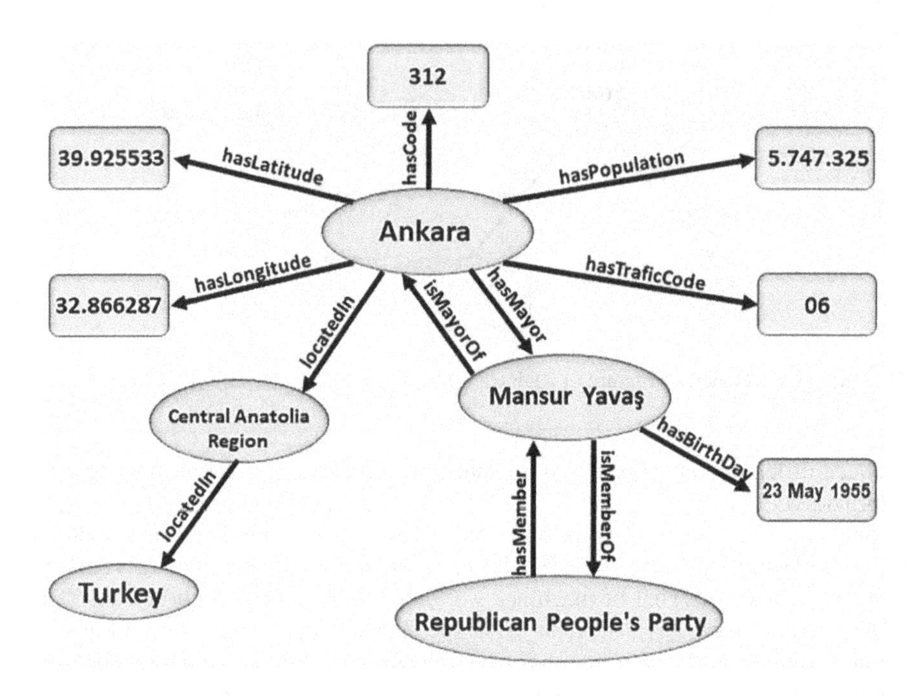

FIGURE 11.3 A simple ontology describing the mayor of Ankara.

11.2.2 Ontology Applications in Different Data Science Fields

Since ontologies emerged, they have become increasingly widespread, and their usage areas have increased rapidly. Especially in fields such as law, finance, and business, where standardization of data and the creation of a common term list are critical, the use of ontologies has become a necessity, not an option. Certainly, the use of ontologies for information representation in areas where data flow is intense will prevent potential misunderstandings arising from the representation of data in different meanings and forms. The differences in the understanding and use of the terminology belonging to the different companies operating in the same field will adversely affect the efficiency of many business processes and cooperation that the companies depend on a common understanding. To ensure that efficiency is not affected by the different terminology used by companies, it is vital to create a concept pool of the field by representing information by ontologies. Thus, a common language is created that allows cooperation not only between humans but also with machines. Finance, healthcare, the Internet of Things, genetics, business, and social networks can be listed among the areas where ontology is used the most. In this section, we will look at examples of ontologies that have been successfully applied in different fields.

11.2.2.1 Financial Ontologies

The financial sector is a part of the economy that consists of firms and organizations that provide financial services on issues that constitute the main pillars of the real sector, such as providing consultancy to commercial and individual customers and providing money flow. When the financial sector is mentioned, it is quite common for many people to think of banks as the first institution. Although banks are the strongest part of the financial sector, the elements of the financial sector are quite high. Apart from banks, it is possible to list investment companies, insurance companies, and real estate companies among the elements of the financial sector. When we include customers (firms, public institutions, individual customers) among all these elements, we realize how wide a stakeholder list the industry can have.

It is vital to use a common language in an area that is so deeply involved with the dynamics of daily life and has a large list of stakeholders. Therefore, a standardization study was required to eliminate possible conflicts of financial concepts that were not officially recognized by millions of financial sectors globally for use in their operational systems. For example, different financial institutions may conceptualize the person or institution to be paid with words such as "payee," "creditor," "encumbrancer," taking into account the systemic, cultural, and handicaps of using different mother tongues. This situation raises the possibility of causing communication disorders during information sharing between organizations. To solve this problem, it is necessary to create a common terminology pool in which the conceptual standards are defined, in which the words "payee," "creditor," "encumbrancer" are clearly stated to correspond to the same concept.

Considering all these requirements in the finance sector, different financial ontologies have been created. The most widely accepted of these is the Financial Industry Business Ontology (FIBO). FIBO was developed by the Object Management Group (OMG) to prevent conflicts arising from the fact that the same word has different meanings in the worldwide financial sector and different words can have the same meaning.[17] FIBO contains semantic connections between financial concepts, explanations of their meanings, and professional use practices. FIBO has been developed for software developers and business analysts as well as the core elements of the financial industry. It is expected that FIBO will become a common language supporting the automation of business processes among many financial companies worldwide soon.

11.2.2.2 Healthcare Ontologies

Data science methods and approaches are frequently used in the field of health, such as diagnosing diseases, improving treatment processes, investigating drug interactions, and directing patients to relevant departments. However, the increase in factors and elements in the field of health along with the developing technology has made IT-based health systems expensive, competitive, and complex. Moreover recently, the development of sensor-based systems in the medical field has led to an increase in data, and developments in the pharmaceutical sector have led to an increase in the number and range of drugs. The development of imaging devices has led to an increase in image data, and over time some other developments in the healthcare field have emerged and different medical fields. As a result of all these developments, the need for consistent definitions of the concepts has increased to establish standards that allow cooperation between different systems and institutions in the health sector.

Just as in other fields, it can use semantic-based technologies to represent and communicate information between different healthcare organizations and their IT-based systems. The treatment process can be evaluated more meaningfully by creating an ontology-based on the patient's complaints and past life history. Again, using such an ontology, the patient's habits, lifestyle, etc. A cause-and-effect relationship can be established by extracting significant patterns between the disease and the disease. Based on this cause-and-effect relationship, suggestions can be made about the patient's future life. Considering similar scenarios in another context, it is thought that semantic-based technologies can provide a basic standardization in the field of health and improve processes.

Today, some ontologies are currently in use in the field of healthcare. Disease ontology (DO) is one of the most well-known ontologies in the field of healthcare, which integrates disease and medical words from different platforms through extensive cross-matching. Gene ontology (GO) was developed to address the need for consistent common definitions of concepts in multiple gene databases. GO can represent information about biological processes, cellular components, and molecular functions. The Foundational Model of Anatomy (FMA) is a domain ontology that can be combined and expanded with different ontologies that represent a coherent body of knowledge about human anatomy. Apart from these, systems such as vaccine ontology (VO), infectious disease ontology (IDO), and Ontology for General Medical Science (OGMS) can be given as examples of ontologies used in this field.

11.2.2.3 Internet of Things (IoT) Ontologies

Recently, the concept of the Internet of Things (IoT) is one of the most frequently mentioned topics both in social life and academically. This newly developing area makes a great contribution to the development of societies in terms of social life and increases the quality of life. Although there is no generally accepted common definition of IoT, it is a technology that enables objects used in daily life to exchange information with other objects over a network and that objects are fully synchronized with each other. IoT platforms enable the collection, sending, and processing of data from the physical environment. Among the basic elements of IoT systems, there are built-in microprocessors, sensors, communication devices, and electrical supply units.

Nowadays, in the age of Industry 4.0, huge data collected by sensors in IoT systems are used to improve decision-making processes by processing with data science technologies so that companies can increase their business values, use resources and assets better, and provide better service to their customers. IoT technology solutions are spreading to more areas day by day. Today, IoT platforms can be used for different purposes in different areas such as smart homes and buildings, healthcare, transportation and logistics, smart cities, and industry. As the numerous benefits of IoT systems and organizations' interest in these systems continue, it is thought that in the future, IoT ecosystems will be warmly welcomed by more disciplines and fields other than the ones mentioned above.

The data that IoT systems obtain from the physical environment are heterogeneous in nature, and there is no semantic closeness between them. In addition, IoT ecosystems are infrastructures that require minimal human-to-human and computer-to-human interaction and aim to communicate with more machines over a common network. Therefore, having a common conceptual language that enables seamless communication between machines is one of the greatest requirements of IoT systems. In other words, to increase the potential benefits of IoT in all areas where it is used, a common language is required in which conceptual standards are defined that can be understood by interrelated devices such as computer-aided devices, digital machines, mechanical systems, and all physical objects that addressable on the network. To meet this need, the semantic sensor network (SSN) approach, which enables us to enrich the meaning of sensor data to provide more advanced access to the data collected by IoT-based systems and add annotations, has been launched by the World Wide Web Consortium.[18] SSN provides a common model by using structural and semantic data representation languages such as XML, RDF, and OWL so that different devices in IoT ecosystems can communicate and work together.

This approach, which includes the ontology of sensor data, will also maximize the potential benefit from data science methods to be used in structured decision mechanisms. Let's assume that there is an IoT ecosystem where many parameters, especially the temperature in the environment, need to be measured. The system will ensure that the air conditioners, windows, and air cleaners are opened according to these values it measures, and the living space will come to normal conditions. For example, temperature data has more than one measurement unit such as Kelvin, Celsius, and Fahrenheit. Therefore, this will lead to conflict between local elements and different measurement systems. Here, the adoption of a common language by all devices in the IoT ecosystem is one of the key points in providing a successful infrastructure.[19] In addition, this sensor ontology developed by W3C will help this IoT system to make a healthy data transfer with other similar IoT systems. Ontologies propose an appropriate approach to generate common concepts, enrich the meanings of data, and ensure interoperability of data, both among local elements and between different ecosystems in IoT ecosystems, which is one of the subject areas of data science.

11.2.2.4 Social Network Ontologies

Undoubtedly, social media is another area where data science is frequently applied and useful information that cannot be obtained by human activities when used in this field is extracted. Social media are digital applications, usually offered as web-based or mobile-based systems, through which individuals or companies can share information over a kind of constantly online network. Social media not only allows individuals to freely communicate with others, but also offers companies the opportunity to reach wider audiences by making use of advertising channels and the potential of social media.[20] Today, the fact that many people can access the Internet through different devices at any time has led to the diversification of social media platforms and the increase in their number. When the current data is analyzed, it is expected that the number of users of many social media platforms will continue to increase day by day.

Considering all these possibilities of social media, it is obvious how much data can be produced in a short time. Thus, from this perspective, social media platforms are an excellent field of study to use data science approaches to extract useful information and patterns that can improve business products and services, political decisions, socio-economic systems, and more. Individual users and companies have the authority to upload daily routine information, exchange messages and even share basic conversations on social media platforms. For this reason, most of the raw data created by users on social media is unstructured. However, each social media system prefers a different way to explain similar concepts that have the same meaning. For example, on different social media platforms, the user can explain the data entered into the system as "message," "tweet," "post," "story," "notification." For social media platforms to work in cooperation with other systems, it is necessary to create a common language.

For data science approaches to reach the desired level of useful information from the heterogeneous and unstructured data produced through numerous social media platforms, the communication between systems must be strong and standard. The lack of a standard and metadata of this raw data obtained from social media platforms has required the creation and use of data models over time. This disadvantage of social media data can be eliminated by using semantic web technologies and ontologies.[21] The creation of a social media ontology that takes into account the similarities and relationships between entities living in the social media ensures interoperability in their systems. In this way, the scope of useful information to be obtained from data science approaches can be expanded by ensuring that heterogeneous and unstructured social media data, most of which are created by different users using different infrastructures, are to a certain model.

Data science approaches are applied at various levels to increase the value of the work done in areas such as finance, the Internet of Things, healthcare, and social media, which are explained in detail earlier. Apart from all these fields, data science is also used effectively in genetics, law, education, business, tourism, municipality, and many other fields. In these areas, there are field-specific problems encountered in integrating data science into structured decision mechanisms. However, apart from the field-specific problems of data science, there are still common problems that await effective solutions. At the top of these common problems is that the data is heterogeneous, dynamic, and unstructured. Recently, the effective use of ontologies in data science processes offers a unique opportunity to solve these obstacles in front of data science.

11.3 FUTURE OF DATA SCIENCE WITH SEMANTIC TECHNOLOGIES

Today the rise of data science in many fields continues rapidly. Data science is no longer a matter of choice for organizations and has become a necessity when it is considered in terms of increasing market shares of companies, providing better service to customers, developing more valuable business processes, increasing employee performance, better use of resources, and better management of human resources. Many studies are showing that the use of structured decision-making mechanisms using data science approaches can improve business performance. It is an undeniable fact that soon investments in data science processes will increase exponentially for companies, organizations, and many other institutions and will benefit more from the blessings of data science.

No matter what field data science is used in, the process of transforming raw data into useful information is somewhat laborious. The principles of data science are idiosyncratic and need to be clearly articulated and implemented for data science to realize its promised potential. To benefit from the blessings of data science, raw data must undergo some changes. The benefit to be gained from data science approaches and models are directly related to the extent to which you provide clean and structured inputs. Even today, some general problems still need effective solutions to get more efficiency from data science. Ontologies can be used to solve some of these problems. In addition, ontologies can be used for systems interoperability, data formatting, and interaction between data. There are some advantages of using ontologies together with data science techniques in data science processes.

- The meaning of the data that is planned to be used in the data science process can be enriched by using ontologies. Through ontologies, additional features can be defined, and concepts can be associated with each other. Thus, the benefits to be extracted from a semantically enriched data set can be maximized by using ontologies.
- One of the most fundamental problems in data science processes is that the data is in unstructured structures due to the lack of standard metadata. The unstructured nature of the data makes it very difficult to process them and convert them into useful information. Common metadata can be created using ontologies and unstructured data can be evolved into more organized data.
- The process that requires the most time, labor, and cost in data science steps in creating a model. Therefore, the fact that a model can be used repeatedly is an important issue for companies that use data science in their decision-making mechanisms. Ontologies can provide a ready and modular structure for building data science models, which enable the model to be used repeatedly.
- The benefit to be gained from data science approaches and models are directly related to the extent to which clean data is provided. Ontologies can be used in the realization of processes that will increase the quality of the data, such as removing outliers from raw data to be used in data science models, filling missing values, and detecting noises.
- In many languages, the language, which is characterized as ambiguity or lack of meaning, may have its definitions. Sometimes different concepts that can mean the same thing, or similar concepts that can mean different things are one of the other important issues that data science should deal with while working on data obtained from different sources. For example, the word "Bass" is both a musical object and a type of fish. By using ontologies, relationships can be established between similar concepts, the meaning of words can be enriched, and rules and special protocols can be created for synonym words and homonym words.

- All languages may have their abstracted conceptual definitions, such as idioms, proverbs, and phrasal verbs, which may cause conflict between the two protocols in which there should no human intervention. Since ontologies are a technology that defines concepts, properties of these concepts, and relationships between concepts in particular fields, metadata can be defined that allows abstract expressions to be understood by systems by enriching their meanings.
- Another big problem in data science processes is the heterogeneity of data, especially from different systems and platforms. The heterogeneous nature of data makes it difficult for data science methods and algorithms to operate on them. By using ontologies, standardization can be achieved for data in different structures. Using this ontological standardization, data from different systems can be enabled to interact with each other.
- By using ontologies, concepts can be defined within the framework of certain rules, their boundaries can be determined, and relations can be established. Using this set of rules, models that can make inferences from statistical data can be created. More patterns can be obtained between concepts by adapting this to association analysis, which is one of the frequently used methods in rule-based data science.
- The ability to make minor changes and reuse a data science model is directly related to how flexible and how modular the model is. Ontologies are flexible and modular structures that can store all kinds of data comfortably. When a change is requested in the model, they can do so comfortably.
- The fact that data is represented in different languages and different formats is another reason for the incompatibility between systems. Ontologies offer different language support and can match concepts that have the same meaning in different languages. Thus, to obtain better results, data sources in different languages are combined, increasing the number of rows (transaction, observation), and columns (feature) diversifying the data set.

It is thought that the benefits to be obtained as a result of including ontologies in data science processes will greatly increase. How ontologies can contribute to data science and which problems they have the potential to solve are explained in items listed previously. Companies need to analyze huge volumes of data and make quick decisions to meet market challenges. Accurate information extraction and visualization from big data can guide decision-makers to make important analyses and accurate predictions. Considering all these benefits of ontologies, the use of ontologies together with data science will become increasingly common soon.

11.4 DISCUSSION AND CONCLUSION

Today, with the developing technology, large data piles have begun to emerge. Therefore, the volume and variety of data obtained over time far exceeded the capacity of manual analysis. Data science processes are successfully applied by many companies and organizations to handle these large data piles and obtain meaningful information from them. In recent years, data science has been frequently used by firms to improve decision-making processes in terms of companies to better use their basic resources, make more profit, meet market challenges, and improve their competitiveness. Decision-making processes, especially in the finance and telecommunication industries, which are the fields where data science is used most frequently, have reached the capacity to be carried out automatically by computer systems.

Detailed analysis of data is indisputably invaluable in establishing a successful infrastructure for decision-making processes. In many applications, decision-making systems based on structured thinking are applied successfully. However, even if new generation algorithms are used, which are based on solid data science, well-defined rules, effective methods, and in-depth analysis, it cannot be guaranteed that the result obtained from the systems based on the structured thinking style will be better than the decision made by traditional methods. Data science, which is shown among the rising values of recent years in the academic and industrial fields, can produce irreparable results unless it is used very carefully. Instead of making a profit, they can make companies lose.

No matter what field data science is applied in, although it has numerous benefits, there are still some common problems waiting to be solved. These common problems include the unstructured nature of the data, the heterogeneity of data from different sources, the only syntax level association of existing data with each other, the constant flow of data, and the inability to define concepts semantically at a depth that machines can understand. All these problems can cause data science approaches and algorithms to have difficulties in processing data and decrease the quality of the information to be obtained. Recently, many researchers have shown semantic web technologies and ontologies as an address for solving these common problems of data science.

Ontologies are metadata that describe entities for a particular field, their properties, instance data, and relationships between these properties. Ontologies can be a great opportunity to bring effective and permanent solutions to the problems that data science has dealt with in different forms and dimensions since its emergence. However, as in every system, semantic web technologies may have some disadvantages depending on the usage area and situation. The ontological representation of data enriches the meaning of the data by adding extra information. However, although adding extra information to the data helps to extract more useful information, they sometimes cause systems to become unwieldy. Therefore, considering the performance of the developed model, the meaning of the data should not be enriched exaggeratedly. In data sets with a lot of observation data, analyzing ontologically represented data with machine learning algorithms may have to face undesirable results in terms of performance, especially in tree structures.

Another benefit of ontologies is that the terminology of a particular field can be determined and used as a common data infrastructure. However, to achieve this fully, ontologies belonging to the field should be aligned with high-level ontologies such as Descriptive Ontology for Linguistic and Cognitive Engineering (DOLCE) and Suggested Upper Merged Ontology (SUMO), which are frequently used in the literature. In this regard, first of all, by whom and how the terminology will be determined and how it will be done according to which criteria, as the biggest problem in alignment, is another problem that awaits the academic community.

Ontologies can offer excellent opportunities to create a common terminology pool and model for the establishment of the infrastructure that will ensure the interoperability of systems. Therefore, for systems to interact with other systems, they must have completed their ontological infrastructure in the systems they work with. However, one of the biggest problems at this stage is that most of the data in the Web 2.0 domain have not yet been converted to RDF format. For the ontologies to be used effectively, the conversion of the existing web content to RDF format is of great importance for the interoperability of the systems.

In summary, in this chapter, we talked about how useful the use of ontologies can be in the field of data science. In addition, it was discussed that ontologies can create an opportunity for data science problems that have been waiting for effective solutions for years and contradict the principle of interoperability. In the field of data science, ontologies can be used effectively in basic subjects such as increasing the quality of data, analyzing data, enriching the data semantically, ensuring the reusability of the designed model, creating a common pool of concepts, and preparing infrastructure for the interoperability of systems. However, in addition to all these benefits, the inclusion of ontologies in data science processes also has some drawbacks such as privacy, security, loss of performance, the extra expense, and labor loss of its alignment. Therefore, the decision of using data science and semantic web technologies together in decision-making systems based on structured thinking should be given by considering its benefits and disadvantages in a delicate balance.

REFERENCES

1. Jorudas, J., Prystawko, P., Simukovic, A., Aleksiejunas, R., Mickevicius, J., Krysko, M., & Kašalynas, I. (2022). Development of quaternary InAlGaN barrier layer for high electron mobility transistor structures. *Materials*, 15(3), 1118.
2. Sahatqija, K., Ajdari, J., Zenuni, X., Raufi, B., & Ismaili, F. (2018, May). Comparison between relational and NOSQL databases. In *2018 41st international convention on information and communication technology, electronics and microelectronics (MIPRO)*. Opatija, Croatia: IEEE, pp. 0216–0221.
3. Rahman, M., Islam, M., Calhoun, J., & Chowdhury, M. (2019). Real-time pedestrian detection approach with an efficient data communication bandwidth strategy. *Transportation Research Record*, 2673(6), 129–139.
4. Chen, Y. H. (2020). Intelligent algorithms for cold chain logistics distribution optimization based on big data cloud computing analysis. *Journal of Cloud Computing*, 9(1), 1–12.
5. Provost, F., & Fawcett, T. (2013). Data science and its relationship to big data and data-driven decision making. *Big Data*, 1(1), 51–59.
6. Naur, P. (1974). *Concise survey of computer methods*. Lund: Studentlitteratur, 397 p.
7. Waller, M. A., & Fawcett, S. E. (2013). Data science, predictive analytics, and big data: A revolution that will transform supply chain design and management. *Journal of Business Logistics*, 34(2), 77–84.
8. Dhar, V. (2013). Data science and prediction. *Communications of the ACM*, 56(12), 64–73.
9. Maslianko, P., & Sielskyi, Y. (2021). Data science—definition and structural representation. *System Research and Information Technologies* (1), 61–78.
10. Kelleher, J. D., & Tierney, B. (2018). *Data science*. Cambridge, MA: MIT Press.

11. Hayashi, C. (1998). What is data science? Fundamental concepts and a heuristic example. In *Data science, classification, and related methods*. Tokyo: Springer, pp. 40–51.
12. Borjigin, C., & Zhang, C. (2021). *Data science: Trends, perspectives, and prospects*.
13. Sheth, A. P. (1999). Changing focus on interoperability in information systems: From system, syntax, structure to semantics. In *Interoperating geographic information systems*. Boston, MA: Springer, pp. 5–29.
14. Berners-Lee, T., Hendler, J., & Lassila, O. (2001). The semantic web. *Scientific American*, 284(5), 34–43.
15. Wylot, M., Hauswirth, M., Cudré-Mauroux, P., & Sakr, S. (2018). RDF data storage and query processing schemes: A survey. *ACM Computing Surveys (CSUR)*, 51(4), 1–36.
16. Gruber, T. R. (1995). Toward principles for the design of ontologies used for knowledge sharing? *International Journal of Human-Computer Studies*, 43(5–6), 907–928.
17. Bennett, M. (2013). The financial industry business ontology: Best practice for big data. *Journal of Banking Regulation*, 14(3), 255–268.
18. Lefort, L., Henson, C., Taylor, K., Barnaghi, P., Compton, M., Corcho, O., . . . Page, K. (2011). *Semantic sensor network xg final report*. Cambridge, MA: W3C Incubator Group.
19. Aktaş, Ö., Milli, M., Lakestani, S., & Milli, M. (2020). Modelling sensor ontology with the SOSA/SSN frameworks: A case study for laboratory parameters. *Turkish Journal of Electrical Engineering and Computer Sciences*, 28(5), 2566–2585.
20. Appel, G., Grewal, L., Hadi, R., & Stephen, A. T. (2020). The future of social media in marketing. *Journal of the Academy of Marketing Science*, 48(1), 79–95.
21. Alamsyah, A., Widiyanesti, S., Putra, R. D., & Sari, P. K. (2020). Personality measurement design for ontology based platform using social media text. *Advances in Science, Technology and Engineering Systems*, 5, 100–107.

12 Ontological Perspective in Cancer Care System

Ujwala Bharambe, Chhaya Narvekar and Prakash Andugula

CONTENTS

12.1 INTRODUCTION

It is estimated that the quantity of cancer cases will rise further in mature populations due to the fact that cancer is the most common cause of death in developed countries. In low-to-middle income countries (LMICs), the incidence of cancer is expected to double by 2035. In the coming decade, cancer research is likely to remain a top priority for saving lives. Figure 12.1 represents World Health Organization's cancer growth projections from 2020 to 2040. Artificial intelligence (AI) has generated a great deal of buzz across healthcare recently, even fuelling a discussion about whether AI will eventually replace doctors. Despite the fact that machines will not replace clinicians in the near future, AI shall help them in making informed decisions and can aid their judgement in some functions in cancer care (e.g., pathology, radiology, etc). A recent successful application of AI in cancer care has been made possible by the increasing accessibility of cancer-related data and the speedy evolution of big data analytic techniques.

In the cancer domain, AI has been successfully applied for tumour image segmentation, mitosis quantification, auto-detection, and classification of benign nuclei from cancer cells;[1] screening mutations;[2] protein alignment; drug discovery;[3] and so on. Low et al. (2018) and Bejnordi et al. (2017) have demonstrated that AI is able to

DOI: 10.1201/9781003310785-12

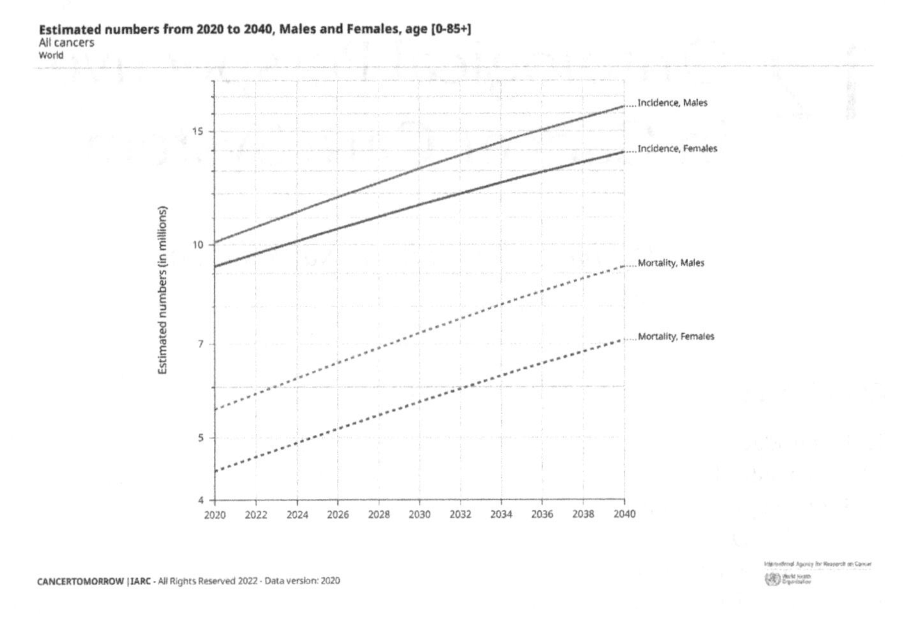

FIGURE 12.1 World Health Organization projection for all cancers, 2020 to 2040. (*Source*: https://gco.iarc.fr/tomorrow/en/dataviz/trends)

diagnose metastatic breast cancer more accurately than pathologists or dermatologists.[4, 5] There are, however, some limitations to machine learning (ML):

- In order for this technique to perform better than others, a large amount of data is required.
- The complex data models make training extremely expensive.
- The selection of the right deep learning tools doesn't have a standard theory because topology, training method, and other parameters must all be known.
- There is a great deal to learn about the output, but it's difficult to understand and explain.
- Domain knowledge cannot be incorporated into the learning process.
- Does not work well with heterogamous and unstructured data.

Semantic network technology is introduced in AI for knowledge representation which supports semantic reasoning. In the next section, we describe semantic technology in detail.

12.2 SEMANTIC TECHNOLOGY

Traditional AI/ML methods (Figure 12.2) fail to address the problem of interoperability, sharing information, and heterogeneity among multiple systems. Semantic technologies, on the other hand, can address each of these concerns. It is a tool that enables semantic links between various underlined data.

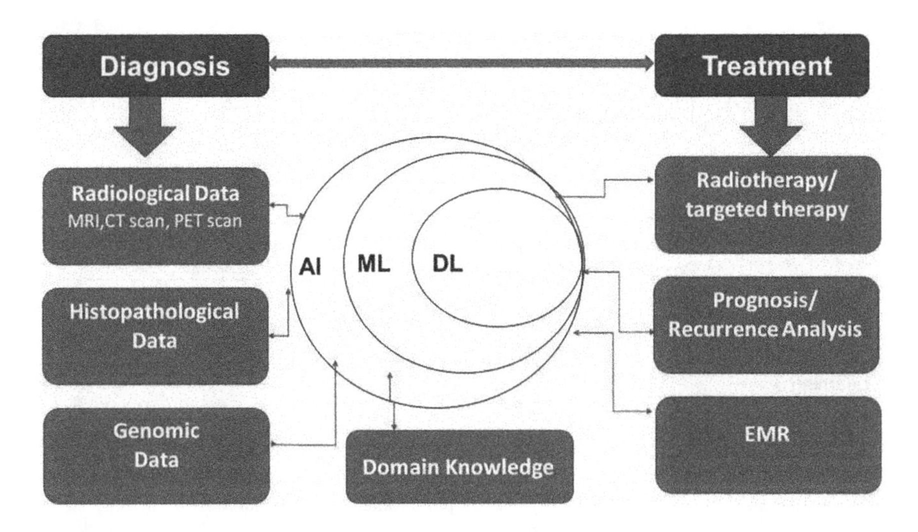

FIGURE 12.2 Application of AI in cancer care domain.

Since the advent of the World Wide Web, semantic web (SW) development has been closely associated with it. A major goal of this technology is to increase the utility of knowledge by making it widely accessible as well as by allowing advanced browsing for evaluation and searching.

A SW vision describes standards in knowledge management as enabling an interoperable system of data so that information can be published, found, and reused beyond its original use.[6] In the era of highly heterogeneous knowledge that grows at an unprecedented rate, semantic technologies offer solutions to accurately publish this diverse expertise. These advantages are: (a) capability to capture knowledge, (b) a platform for publishing and sharing information, (c) a middleware solution for knowledge discovery, and (d) analyze heterogeneous data sources and support reasoning and inference. Figure 12.3 shows AI categorization, a result from which we might be able to (a) uncover new strategies/technologies for cancer characterization, (b) detect cancer earlier, and (c) offer personalized therapy to more patients. The semantic web is shown in Figure 12.4 as a layered approach. XML is a language that allows one to create structured web documents with user-defined vocabulary at the bottom. The XML format is particularly suitable for sending documents across the Internet.

In contrast to entity relationship models, Resource Description Frameworks (RDF) data models do not require XML to write simple statements about web objects. In the triple model, any object with a Uniform Resource Identifier (URI) can be described. Since RDF is based on XML, it is placed on top of XML in the figure. Semantic web components include RDFs that describe triples of subject, object, and their relations.

Web objects (resources) can be organized into hierarchies with RDF schemas. A class or property, a subclass or subproperty relationship, and a domain or range restriction are key primitives. The RDF Schema standard allows data to be represented using RDF. Ontologies can be created with RDF Schemas using its primitives.

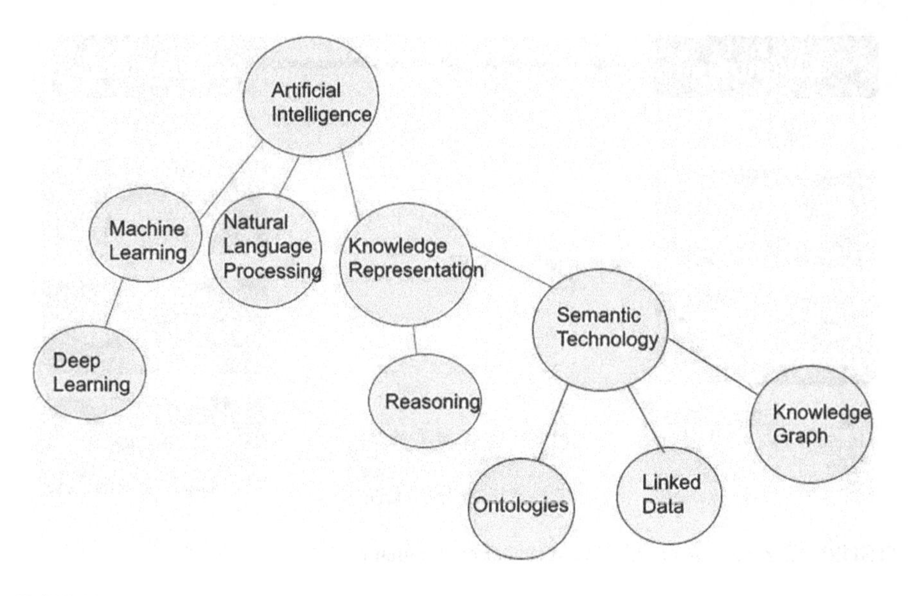

FIGURE 12.3 AI categorization.

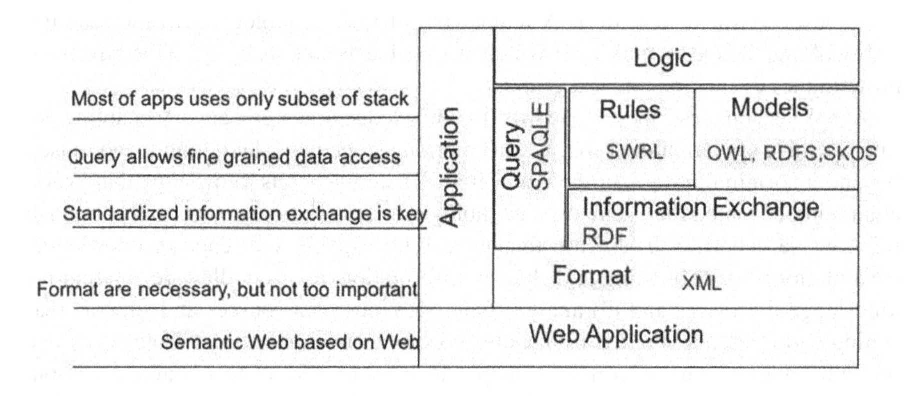

FIGURE 12.4 Semantic web stack.

(Adapted from https://www.ontotext.com/knowledgehub/fundamentals/what-are-ontologies)

RDF Schema needs to be expanded and allowed to represent more complex relationships between objects/entities using more powerful ontology languages.

In Web Ontology Language (OWL), classes, subclasses, and properties are defined above RDF schema in order to create more advanced inter-class relationships, which is useful for forming ontologies. RDF schema is applied to application-specific vocabulary that has classes (concepts), subclasses, and properties, while OWL extends RDF schema in order to create more advanced inter-class relationships. For web ontologies, OWL is being proposed as a standard. By using it, we can describe the semantics of knowledge in a machine-readable manner. With the mapping of OWL on logic, formal semantics and reasoning are supported. This has been accomplished using predicate logic and description logic. The rule system consists of predicate logic with efficient proof systems. A rule axiom is composed of two types of atoms: the antecedent (body) and the subsequent (head). In addition, it is possible to assign a URI reference to a rule axiom in order to identify it. A rule has the form $A_1 \ldots . A_n \rightarrow B$ where A_i antecedent and B antecedent and B subsequent. In the semantic web, this rule can be presented using SWRL. The Semantic Web Rule Language (SWRL) is intended to serve as the semantic web's rule language. Using a formal query language, RDF-based statements are stored in triplestores (like Virtuoso and Stardog). RDF data can be retrieved and manipulated using SPARQL W3C queries. RDF data can be retrieved and manipulated using SPARQL W3C queries. The graph pattern consists of three variables, e.g., subject, property, and object. Several of SPARQL's features are similar to those of SQL. There are three types of SPARQL queries: triple patterns, conjunctions, and disjunctions. A triple pattern is similar to an RDF triple in which the subject, predicate, and object are variable.

12.3 ONTOLOGY ESSENTIALS

During the development of artificial intelligence, ontologies were developed to facilitate knowledge exchange. In the last two decades, there has been an increase in interest in ontologies among artificial intelligence researchers, including work on understanding representation of domains, knowledge engineering, and natural language processing. Of late, the concept of ontology is also entering the fields of intelligence retrieval, digital trade, and knowledge management. The growing recognition of ontologies is a because of what they promise: a shared expertise of a domain that can talk between human beings and alertness systems.

A machine-processable semantic has been developed for information sources in the form of ontologies for the exchange of information between software and humans. A number of ontology definitions have been proposed over the past decade, but the one that best explains how ontologies work is based on a related one:[7] Ontologies define shared conceptualization in a formal, explicit way. Conceptualizations are abstract models of some phenomena which identify relevant concepts for those phenomena. The term "explicit" signifies that the types of concepts and constraints on their use are explicitly explained. An ontology must be formal and machine readable. Different levels of formality are thus possible. The term "shared" refers to the idea that ontology captures consensus knowledge; the knowledge is not restricted to an individual but accepted by the group. Knowledge engineering is primarily concerned with the development of domain models using ontologies. The ontology defines the terms and relationships within the domain.

It is possible to represent knowledge and correlations between different pieces of information using ontologies. The knowledge must be expressed in an appropriate format to be machine-processable. There have been several developed syntactic and semantic languages for this purpose.

Ontologies are conceptual specifications that describe knowledge about a domain in a manner independent of state of knowledge or state of affairs. Furthermore, it constrains the interpretation possibilities of a language's vocabulary in a way that its logical models resemble the structures intended as part of a conceptualization of the domain as closely as possible. In specific formats, such as OWL or RDF, ontologies represent the underlying domain in a semantic way. Classes (or concepts), attributes (or data properties), and relations (also called object properties) provide meaning to an ontology. Ontology definition [8] for general purpose ontologies is represented as four tuple $O=(C, P, A, I)$; where C is the set of concept, P is set of properties, I is the set of instances, and A is set of axioms. It is important to develop an ontology in order to enable domain knowledge to be reused by individuals or software agents, along with sharing a common understanding of the structure of information. Ontology is generally used for

- Information retrieval and extraction
- Knowledge reuse
- Knowledge interchange
- Knowledge integration
- Knowledge representation standard.

12.3.1 Ontological Languages

Even though the ontology in principle is independent of a particular language, it is necessary to choose a language to describe it. This language needs to be formal in order to share, exchange and map ontologies. The use of natural language alone is not sufficient for above as it leaves a significant amount of interpretation for the user. This can lead to potential miss of a significant aspect of the ontology. This issue can be resolved and the knowledge can be made automatically processable by the machine by expressing it as simply as possible in an adequate format.

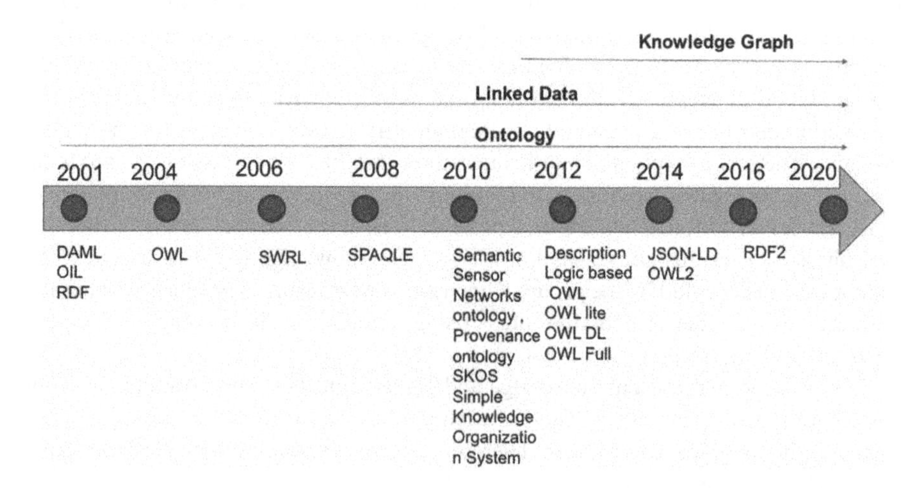

FIGURE 12.5 Timeline of semantic technology.

Several syntactic and semantic languages have been developed and standardized for this purpose (RDF; Resource Description Framework and OWL; Web Ontology Language)

Semantic technology timeline is shown in Fig 12.5. The appearance of semantic web initiative has motivated the development of ontological languages based on XML in the last years, many of them promoted by the World Wide Web Consortium (W3C). W3C published RDF (Resource Description Framework) in 1999 as the basis for describing web resources through triple type < resource, property, value> and RDFS (RDF schema). RDFS permits to build taxonomies that express classes of resource and their subclass relationship, and also define properties and associates with them. RDFS is the base on top of which other ontology languages have been defined, evolving from DAML and OIL to DAML+ OIL and finally to OWL Web ontology language (W3C) [9]. OWL is a standard of W3C and is widely used by the semantic Web community.

The extension of RDFS is OWL which provides an additional vocabulary to describe more semantics, e.g. relations between concepts, characteristics of properties, etc. It is the standard language for defining ontologies. There are three sublanguages of OWL: There are three sublanguages of OWL:1

- OWL Lite: It is the simplest version of OWL which enables the creation of taxonomies and simple axioms.
- OWL DL: It expresses full description logic by extending along with OWL, but has some restrictions to ensure computational completeness and decidability, e.g. a concept cannot be an instance of a concept.
- OWL Full: It is like OWL, but does not have any restrictions

12.3.2 Knowledge Graphs

Big players like Google, IBM, or Microsoft currently use the term "knowledge graph" as part of their marketing campaigns. Using semantic knowledge in web searches is what the term refers to; it is a practice created by Google. The knowledge representation formalism employed by KGs can include abstract modeling languages and probabilistic mechanisms, not merely RDF. As ontologies, data meanings are attached to the graph alongside the data itself. By making knowledge graphs self-descriptive, users can find the data in one place and understand it easily. Inferencing is supported by explicit semantics and formalizations. In KGs, the meaning of main concepts and relationships can be expressed by describing a large volume of items. Their recommendations can help you adjust your data to meet the requirements of a data model. As a result, new insights can be drawn from the available data.

Knowledge graphs should have the following characteristics: (a) describes real-world entities and how they interact in a graph, (b) their focus tends to be on the actual instances (ABox) and a minimal role is played by the schema (TBox), (c) indicates the relations between classes of entities in a schema, (d) incorporates arbitrary entities into an ontology and applies a reasoner to derive new knowledge, and (e) acquires and integrates information into an ontology.

12.4 ONTOLOGIES AND CANCER DOMAIN

There are various sources of cancer domain data: hospitals, insurance, doctor's notes, pharmacies, electronic health records, clinical decision support systems, companies, diagnostic images, electronic medical records, and paper documents. Moreover, cancer-related data have unique characteristics due to the variety of sources and data types. Some of the characteristics of the data are: they are scattered; they are unstructured and structured; they have inconsistent and variable definitions; they are dynamic and large; they are complex. Various ontologies have been developed to address these issues. In terms of the type of knowledge being modelled, these ontologies can be classified into three categories:

Controlled vocabularies: The ontology-based representation of standard cancer terminologies, such as NCI Thesaurus, Foundational Model of Anatomy (FMA),[1] Criteria for Adverse Events (CTCAE), and Systematized Nomenclature of Medicine Clinical Terms (SNOMED CT), which provide a reference terminology for many aspects of cancer management. Common terminology serve the general purpose of data integration and interoperability among disparate information systems.

Declarative knowledge: There are several types of ontologies that represent declarative knowledge, such as Radiation Oncology Structures Ontology,[2] Disease Ontology (DO), Dependency Layered Ontology for Radiation Oncology,[3] and Radiation Oncology Ontology,[4] Cancer staging systems (TMN ontology), Cancer Cell Ontology (CCL), etc. The following is a type of declarative knowledge about

- Types of cancer and their symptoms.
- Aside from radiotherapy, chemotherapy (treatments for disseminated tumours or advanced stages with special approaches to chemotherapy), cancer patients can also undergo surgical treatment (palliative surgery, radical surgery, surgical exploration) and nutrition therapy.
- It also includes cancers and their treatment regimens, including those that involve the head and neck, thoracic, abdominal malignancies, tumours of the bladder, genital, and CNS, as well as cancers of soft tissue, primary malignant bone, and metastatic cancer, mechanisms and classifications of cancer pain, and assessment of cancer pain.
- Pharmacokinetics of drugs and their pharmacological action.
- Drug options, drug doses, relative dose intensity, dose density, intervals, and treatment courses.

Procedural knowledge: There are ontologies for representing procedural knowledge (for example procedures related to cancer treatments, brachytherapy [ENT COBRA ontology]), such as those that represent guideline representations for evidence-based clinical decision support, which describe conditions, decisions, and actions necessary to generate a hierarchical task network model for a dynamic clinical workflow.

12.5 GENERAL ARCHITECTURE FOR CANCER CARE SYSTEM

One of the leading causes of death worldwide is cancer. Cancer mortality rates can be significantly reduced with early detection, accurate diagnosis, and precise individualised treatment. Clinical judgments and decisions can be made more accurately with the help of ontology-driven machine learning. This has included the screening of cancer, the determination of whether a pulmonary nodule is benign or malignant, the classification of histological cancers, the identification of genomics, and judgement of the effectiveness of treatment. Figure 12.6 represent the various cancer challenges and ontological solutions. We have classified ontology driven cancer care system into four categories: (a) ontology driven cancer diagnostic system, (b) ontology driven cancer genomic study, (c) ontology driven personalized medicine in cancer care, and (d) ontology driven cancer decision making and management.

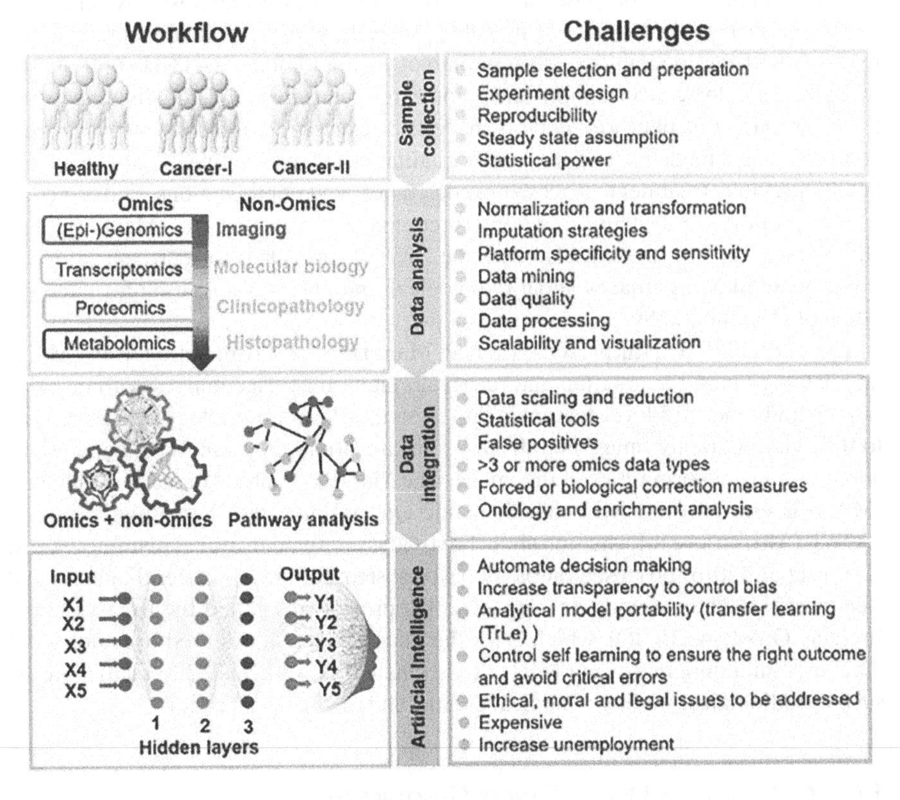

FIGURE 12.6 AI-mediated cancer workflow, challenges, and ontological solutions.

(Adapted from Patel et al. (2020)[10])

12.5.1 Ontology-Driven Cancer Diagnostic System

A main challenge in cancer treatment is finding the cancer before the appearance of its symptoms; the earlier the cancer is detected, better are the chances the patient has for its treatment and cure. Hence, early diagnosis is a very important phase. Most cancer diagnostics systems are based on medical image analysis. The interpretation of a medical image involves the identification of the image area, existence of abnormality or lesions, etc. Each medical image contains extensive information about anatomy and abnormal structures can be detected on examining them. Moreover, the person that examines it must be aware of the features of disease at presentation and the method it can spread. For example, researchers may want to indicate where in certain areas of interest abnormalities are present. This information is often considered "image metadata" and includes image comments, interpretations, and conclusions. Ontologies can help to represent these semantics, which can be easily accessible. Thus ontology based systems help medical professionals to store, manage, and retrieve information, as well as to do interpretation and annotations of images.

Siddiqi et al. (2008) proposed a MATCH system for post-diagnosis investigations to rule out metastatic disease.[11] It enables hospital, pharmaceutical laboratory, and research unit interoperability through ontology. In addition, MATCH supports the treatment of colon cancer tumors and colon carcinomas by automatically diagnosing them. Through this platform, doctors, biologists, cancer researchers, pharmaceutical companies, and medical staff can access qualitative extensive clinical data on patient health, previous treatments, follow-ups, outcomes, and complications. Guliato et al. (2009) proposed a system called Interpretation and Diagnosis of Mammograms (INDIAM).[12] This system also uses the BreastCancerOnto to represent information content in X-ray images about radiological findings. It is also used for interpretation of mammography.

Bulu et al. (2011) [13] presented a new Mammography Annotation Tool (MAT) and ontology based annotation and retrieval system for breast masses with description of high and mid-level features. This system allows a radiologist to get access to the mammography image and, from its visualization, they can make annotations about existing abnormalities in the image. The Mammography Annotation Ontology (MAO) is crafted from RDF and Semantic Query-Enhanced Web Rule Language (SQWRL) to retrieve information about breast masses found in digital mammography.

Isac et al. (2016) proposed ontology based system for breast cancer diagnosis and treatment.[14] The ontology developed in this project was called the Breast Cancer Imaging Ontology (BCIO), which is for the imaging techniques used in radiological diagnosis (mammography and MRI). It also includes terms related to medical tests, such as radiological findings, medical exams, and medical descriptor.

12.5.2 Ontology-Driven Cancer Genomics Study

In addition to providing physicians with more information about cancer's causes, genome sequencing has also changed the way some forms of the disease are treated. On the basis of genome sequencing, AI is transforming our understanding of cancer's origins and dynamics. Genomic sequencing provides detailed information about an individual's cancer development rather than a broad categorization based on location. Cancer genomics data is available from several different databases; for example, Cancer Genome Atlas (TCGA),[15] an International Cancer Genome Consortium (ICGC),[16] the Catalogue of Somatic Mutations in Cancer (COSMIC),[17] and the Cancer Genomic Hub (CGHub).[18] Using cancer genomics and clinical physiology data together could, therefore, help us understand cancer biology and how it responds to treatment.

The combination of clinical physiology data and cancer genomic data could provide a better understanding of cancer biology and the responses to treatment. A combination of histopathological images and genomics can provide more details on cancer tissues architecture. In molecular investigations, cancer tissues' architecture is typically compromised.[19] Using a public dataset (TCGA) and an algorithm,[20] breast cancers are categorized by AI systems. A wide variety of cancers have been successfully predicted using AI algorithms to integrate multimodal biological data with pathology images, including low-grade glioma [21] prostate cancer,[22] non-small cell lung cancer,[23] and renal cell carcinoma.

Althubaiti et al. (2019) have proposed a new technique for locating cancer driver genes.[24] This method makes use of a variety of complimentary information types as characteristics, including cellular phenotypes, cellular locations, functions, and whole-body physiological phenotypes. The identified cancer driver genes are separated into their function in various forms of cancer using the proposed method. Moreover, it also identifies several novel candidate driver genes. In machine learning models for predicting cancer driver genes, this ontology is used as background knowledge. In feature learning, this background knowledge is utilised.

As the name suggests, personalized medicine uses the phenotypes and genotypes of individuals (e.g. medical imaging, lifestyle data, molecular profiling,) to focus therapeutic efforts in a way that is appropriate for each individual at the right time, and/or to determine their risk for disease and/or to prevent disease in a timely and effective manner. It is an advancing field of medicine that draws on each individual's unique clinical, genetic, and genomic information. Figure 12.7 depicts ontology driven personalized medicine in cancer care.

Health outcomes are expected to improve with personalized medicine (PM). With this approach, the right treatment is provided, and side effects are avoided or properly managed. Ontologies are the most effective way of representing actionable knowledge in the cancer domain. This success can be attributed to two things: their capacity to formalize medical information in a straightforward, effective way, and their ability to be applied to reasoning processes performed by clinical decision support systems. In this study, we discuss PM's ontology in detail. Riaño et al. (2012) used ontologies to handle extremely variable patients in order to provide personalization of information regarding both patients' conditions (each patient has unique symptoms, social needs, and signs), as well as action plans for these patients.[25]

Using ontologies, clinical data was personalized and standardized to FAIR standards.[26] Clinical data from relational databases were mapped to radiation oncology using their own ontology. Based on FAIR principles, the ROO consists of 1,183 classes and 211 features among them. SPARQL-based querying of mapped data and ontology using semantic web technologies.

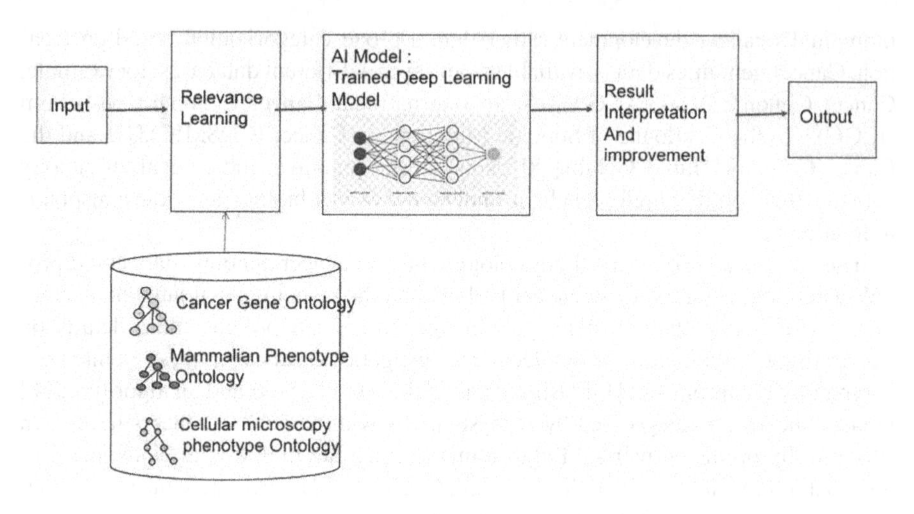

FIGURE 12.7 Ontology-driven personalized medicine in cancer care.

Clinical practice guidelines (CPGs) are textual documents that synthesize knowledge about medical disorders and are meant to provide decision support for clinical situations as a means of promoting evidence-based care. A framework based on ontological modeling of CPG contents was proposed in Galopin et al. (2015).[27] In addition to providing appropriate recommendations at a variety of abstraction levels, the ontology allows patients' data to be adapted and adaptable. To resolve decisional conflicts in the recommendation process, the system proposed a hierarchical graph based on the patient profile.

This multi-institutional and multi-regional clinical pharmacogenomic CDS system is based on the Medicine Safety Code (MSC) system,[28] which provides a web-based, mobile-optimized system dedicated to facilitating personalized medicine. Physicians and patients will be able to present pharmacogenomic data and receive pharmacogenomic guidance at the point of care using the MSC. Using this software, physicians shall provide drug dosage and other treatment based on the genetic profiles of individuals. For formalizing pharmacogenomic knowledge and providing decision support functions in clinical routines, it uses Web Ontology Language 2 (OWL 2).

Luciano et al. (2011) discusses the importance of three factors in designing an experimental personalized medicine system: phenotype categories, population size, and statistical analysis.[29] "Phenotype categories", the first factor, represents knowledge about the phenotypes of human diseases. "Population size" is another important factor to consider because orphan diseases affect a small fraction of the population, which makes it nearly impossible to collect data from a large sample size of patients. A third factor is "statistical analysis", which involves using statistics to classify, predict, or diagnose a patient. As a result of this work, these characterizations are represented by four basic components: (a) knowledge base, (b) ontology, (c) pattern reconstitution, and (d) patient profile. It is the knowledge base that is a source of prior information, as well as the patient profiles; so both of these data are required for the analysis. By combining the two types of data and offering clinically applicable insights, the ontology component adds semantic meaning. Finally, the pattern recognition component deals with uncertainties and errors within both kinds of data.

Characterization consists of ontology components that represent disease categories, while data components, the knowledge base and patient profiles represent the population size dimension, and pattern recognition components represent the statistical analysis dimension. In order to interpret data clinically relevant, the ontology component adds semantic meaning to each other. A component of pattern recognition that manages uncertainties and errors in two different kinds of data. Characterization has an ontology component that corresponds to the phenotype disease categories, whereas the data components are the knowledge base, patient profiles, and pattern recognition components that are defined by the statistical analysis dimension. A direct approach to the practical realization of personalized medicine was gained by relating its characterization with its experimental design. Personalized medicine can thus be implemented on a case-by-case basis, for example, for breast cancer and its subtypes. Over time, an integrated knowledge base and ontology for different disorders are developed.

As an alternative to traditional methods of analyzing gene mutations to discern genetic diseases, Jayaratne [30] develops a novel and generic mathematical model for differential diagnoses of genetic diseases. A genotype-phenotype correlation is represented using an ontology. Human Phenotype Ontology provides a standardized vocabulary to map emerging genetic mutations in a patient, which is then used for differential diagnosis. A differential diagnosis can be made using ontology-based semantic similarity coupled with fuzzy relational theory. There are five complex diseases that can be diagnosed with this system: Cohen Syndrome, Cornelia de Lange Syndrome, Lymphedema-Distichiasis Syndrome, Popliteal Pterygium Syndrome, and Smith-Lemli-Opitz Syndrome.[31]

12.5.3 Ontology-Driven Cancer Decision Making and Management

It has been demonstrated that clinical decision support systems (CDSS) improve practitioner efficiency and workflow. The modern CDSS can store, search, and process heterogeneous datasets, such as medical records, from diverse sources.

Figure 12.8 represents architecture for an ontology-driven decision support system. This technical architecture takes advantage of electronic health record (EHR), data AI/ML/DL model, clinical databases, domain expert knowledge bases in terms of ontologies, and case-based reasoning, and it recommends a provision of decision-making support for healthcare professionals. It is comprised of three-layers:

- **Repository layer:** This layer is composed of various resources such as clinical databases, electronic medical records, data sources from clinical trials, and case studies. Moreover, it contains a knowledge base composed of various ontologies such as disease, medical terminology, cancer gene ontology, etc.
- **Processing layer:** It is responsible for interference and knowledge retrieval from the knowledge base. This layer takes advantage of case-based

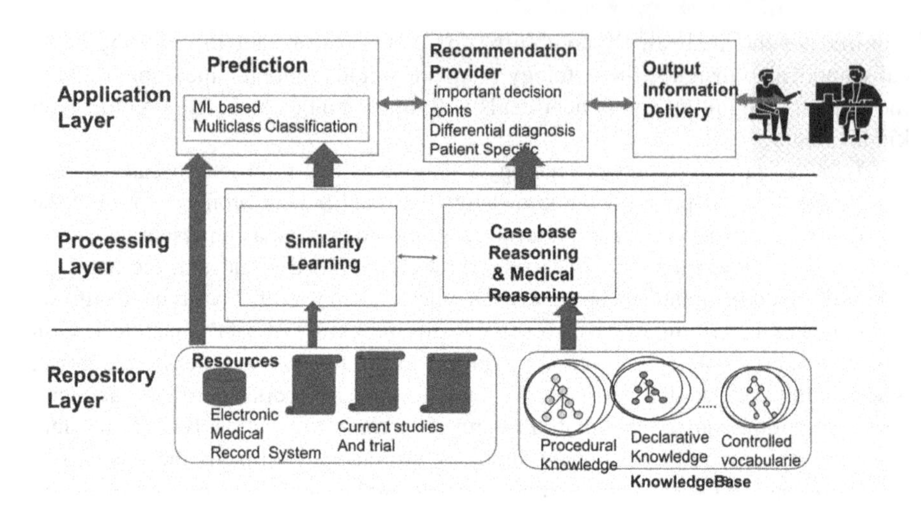

FIGURE 12.8 General architecture of ontology-driven clinical decision support system.

reasoning (CBR). CBR is analogous to problem-solving, which memorizes and restores experience data to solve similar cases.[32] This system generates clinical information like patient conditions, syndromes, and medication characteristics from the repository layer. This clinical information is based on ontological components such as concepts, properties, instances and hierarchy. Patients with similar characteristics are extracted using the reasoning process. Then all these patients are matched to generate similarity in the similarity learning phase to understand the common pattern.

- **Application layer:** It is responsible for communication and output delivery to the user (clinicians and medical professionals). The most crucial part of this layer is to provide a recommendation to clinicians based on differential diagnosis, specific to the patient, and important points to decide treatment for common patterns.

In addition to the capabilities listed above, modern medical decision support systems should support the following capabilities: (a) support both structured and free-form medical data, (b) integration of information from different sources, (c) understanding and translating heterogeneous systems' information, (d) ability to handle sparse patient data, and (e) explain their decisions and recommendations.

In addition to supporting the above capabilities, ontologies will enhance the CDSS in several ways: (a) improving the capability of sharing pathology, radiology, and domain knowledge; (b) retrieving similar cases when organizing and storing information; (c) explicit definitions inherent in the medical field; (d) knowledge analysis; and (e) reuse of data.

The DESIREE project leverages the breast cancer knowledge model (BCKM) to develop a system that offers many modules for decision support in breast cancer patients' care. The BCKM formally expressed as an ontology including two submodels: the data model for representing clinical patient data and the ontological model for representing application domain concepts. Data semantics are described using this ontological model (BCKM), which also enables reasoning at various levels of abstraction. This chapter proposes a decision support module based on guidelines.

Presented a hybrid ontology and learning medical system for supporting treatment decisions. For effective patient treatment, this work combines ontological representation with probabilistic reasoning. The decisions are made using semantic and logic based methods, along with first-order logic explanations for each system that created the results.

CDSS use case-based-reasoning to consider disease manifestations and give clinicians treatment solutions based on similar past cases. Additionally, this system provides treatment recommendations to clinicians using natural language processing (NLP). The NLP processing is enriched with disease ontology to capture the intended meaning of the query. Using DO in this system improves its overall reasoning abilities.

A framework based on ontology is presented in Zhang et al. (2018) for integrating heterogeneous data sources for the study of individual risk factors associated with cancer survival.[33] Through the use of the semantic resource National Cancer Institute (NCI), this work facilitates the integration of data. NCI ontology for cancer research variables (OCRA) links data elements from different sources. With the help of ontologies, a framework has been developed for

- Creating a common vocabulary so both humans and computers can understand the data.
- Inferring semantic relationships from data.
- Linking contextual and environmental factors to patients through geographic variables.
- Being able to document the data manipulation and integration processes clearly in the ontologies. Zheng et al.[34] has shown using an ontology-based data integration approach not only standardizes the definitions of data variables through a common, controlled vocabulary but also makes the semantic relationships among variables from different sources.

A system for diagnosing lung cancer disease and identifying its stages has been developed by GN et al. (2022).[35] The purpose of this work is to utilize ontologies for lung cancer domain knowledge representation for the diagnosis of lung cancer. Ontologies are represented in OWL format and can be queried using SPARQLE.

Cancer care treatment outcome ontology (CCTOO) is provided by Euzenat and Shvaiko [8] for assessing solid tumor treatment outcomes. There are four domains in the CCTOO (health services, physical, cancer treatment, and psychosocial health-related concepts), thirteen subgroups (such as safety, efficacy, and quality of life), and two themes (evaluative and taxonomic) that can be used to organize concepts.

Understanding and making machine-processable information about clinical trials, drug efficacy, safety, and patient-reported outcomes is crucial to assessing treatment outcomes efficiently and effectively. Ontologies will also assist oncologists in justifying and/or prioritizing therapeutic interventions, where holistic decision-making will be further altered by additional outcome variables, such as health-related quality of life and drug cost.

A CDSS for early diagnosis of breast cancer was proposed by Gong et al. (2019). [36] By bridging the gap between multiple heterogeneous data sources, ontologies facilitate interoperability. An ontology that can support breast cancer early diagnosis was created to transform breast cancer clinical knowledge into a computational ontology. An ontology for breast cancer has been created to support the CDSS in this research. The breast cancer ontology (BCO) is primarily based on NCCN clinical practice guidelines for early detection and treatment of breast cancer, as well as Chinese breast cancer diagnosis and treatment guidelines. Breast Cancer Online is a manually curated resource containing concepts and relationships about breast cancer clinical findings, treatments, laboratory tests, pathologies, cancer stages, imaging results, body structure, demographics, and follow-up information. An interoperable standard vocabulary SNOMED CT is used to annotate the curated knowledge. The semantic rules for early breast cancer diagnosis are developed based on this vocabulary.

According to Dissanayake et al. (2020), all CDSS are ontology-based, and the concepts and properties within these ontologies can be identified and classified in order to guide future research. This work demonstrated the diversity of purposes and scope of their ontologies, with a variety of knowledge sources used for ontology development. Medical knowledge concepts accounted for 126, reasoning concepts for 38, and properties for 240 (137 relationships and 103 attributes).

12.6 SUMMARY

Semantic technologies are explored in this chapter to improve the manipulation of cancer care information. Medical professionals are able to share knowledge and retrieve similar cases using semantic technology in this domain area. Ontologies are considered the backbone of semantic technologies that enable the reuse of data, as well as the analysis of shared knowledge and explicit definitions associated with the field.

Early cancer detection is one of the main challenges in cancer treatment; the earlier cancer is detected, the better the chance of treating and curing it. However, human observations are limited, so routine screening often misses a few cases. Ontology-based systems are making it possible for health professionals to diagnose current cases more effectively.

It is possible to separate the domain knowledge of the expert in question for their practical application by using ontology. Due to the fact that medical knowledge is already stored, professionals who develop systems do not have to be experts in the medical field. Ontology-based systems have been a great resource for health professionals in supporting decision making. Organizing and storing information in a coherent and logical manner will maximize its usefulness. In addition, ontologies can also be used in the cancer field to organize and structure medical reports. It facilitates the search for clinical data by standardizing terms and representing the knowledge, improving information transmission and facilitating the exchange of knowledge.

NOTES

1 https://bioportal.bioontology.org/ontologies/FMA
2 https://bioportal.bioontology.org/ontologies/ROS
3 https://bioportal.bioontology.org/ontologies/DLORO
4 https://bioportal. bioontology.org/ontologies/ROO

REFERENCES

1. Sirinukunwattana, K., Raza, S. E. A., Tsang, Y. W., Snead, D. R., Cree, I. A., & Rajpoot, N. M. (2016). Locality sensitive deep learning for detection and classification of nuclei in routine colon cancer histology images. *IEEE Transactions on Medical Imaging*, 35(5), 1196–1206.
2. Coudray, N., Ocampo, P. S., Sakellaropoulos, T., Narula, N., Snuderl, M., Fenyö, D., . . . Tsirigos, A. (2018). Classification and mutation prediction from non-small cell lung cancer histopathology images using deep learning. *Nature Medicine*, 24(10), 1559–1567.
3. Abadi, A., Ben-Azza, H., & Sekkat, S. (2017). An ontology-based support for knowledge modeling and decision-making in collaborative product design. *International Journal of Applied Engineering Research*, 12(16), 5739–5759.
4. Low, S. K., Zembutsu, H., & Nakamura, Y. (2018). Breast cancer: The translation of big genomic data to cancer precision medicine. *Cancer Science*, 109(3), 497–506.
5. Bejnordi, B. E., Veta, M., Van Diest, P. J., Van Ginneken, B., Karssemeijer, N., Litjens, G., . . . CAMELYON16 Consortium. (2017). Diagnostic assessment of deep learning algorithms for detection of lymph node metastases in women with breast cancer. *Jama*, 318(22), 2199–2210.
6. Berners-Lee, T., Hendler, J., & Lassila, O. (2001). The semantic web. A new form of web content that is meaningful to computers will unleash a revolution of new possibilities. *Scientific American*, 284, 1–5.
7. Gruber, T. R. (1993). A translation approach to portable ontology specifications. *Knowledge Acquisition*, 5(2), 199–220.
8. Euzenat, J., & Shvaiko, P. (2007). *Ontology matching*, vol. 18. Heidelberg: Springer.
9. Taye, M. M. (2010). The state of the art: Ontology web-based languages: XML based. arXiv preprint arXiv:1006.4563.
10. Patel, S. K., George, B., & Rai, V. (2020). Artificial intelligence to decode cancer mechanism: Beyond patient stratification for precision oncology. *Frontiers in Pharmacology*, 11, 1177.
11. Siddiqi, J., Akhgar, B., Gruzdz, A., Zaefarian, G., & Ihnatowicz, A. (2008, April). Automated diagnosis system to support colon cancer treatment: MATCH. In *Fifth international conference on information technology: New generations (ITNG 2008)*. pp. 201–205. IEEE.7-8 April 2008, Las Vegas, Nevada, USA.
12. Guliato, D., Bôaventura, R. S., Maia, M. A., Rangayyan, R. M., Simedo, M. S., & Macedo, T. A. (2009). INDIAM—an e-learning system for the interpretation of mammograms. *Journal of Digital Imaging*, 22(4), 405–420.
13. Bulu, H., Alpkocak, A., & Balci, P. (2011, July). Ontology-based mammography annotation and similar mass retrieval with SQWRL. In *International conference on biomedical ontology*. Conference paper. Buffalo, New York, USA.
14. Isac, C., Viterbo, J., & Conci, A. (2016, May). A survey on ontology-based systems to support the prospection, diagnosis and treatment of breast cancer. In *Anais do XII Simpósio Brasileiro de Sistemas de Informação*. SBC, pp. 271–277, Florianópolis, Brazil.
15. Wang, Z., Jensen, M. A., & Zenklusen, J. C. (2016). A practical guide to the cancer genome atlas (TCGA). In *Statistical genomics* (pp. 111–141). New York, NY: Humana Press.

16. Zhang, J., Bajari, R., Andric, D., Gerthoffert, F., Lepsa, A., Nahal-Bose, H., . . . Ferretti, V. (2019). The international cancer genome consortium data portal. *Nature Biotechnology*, 37(4), 367–369.
17. Forbes, S. A., Beare, D., Gunasekaran, P., Leung, K., Bindal, N., Boutselakis, H., . . . Campbell, P. J. (2015). COSMIC: exploring the world's knowledge of somatic mutations in human cancer. *Nucleic Acids Research*, 43(D1), D805–D811.
18. Wilks, C., Cline, M. S., Weiler, E., Diehkans, M., Craft, B., Martin, C., . . . Maltbie, D. (2014). The Cancer Genomics Hub (CGHub): Overcoming cancer through the power of torrential data. *Database*, 2014.
19. López de Maturana, E., Alonso, L., Alarcón, P., Martín-Antoniano, I. A., Pineda, S., Piorno, L., . . . Malats, N. (2019). Challenges in the integration of omics and non-omics data. *Genes*, 10(3), 238.
20. Xu, Y., Zhang, J., Yuan, Y., Mitra, R., Müller, P., & Ji, Y. (2012, December). A Bayesian graphical model for integrative analysis of TCGA data. In *Proceedings 2012 IEEE international workshop on genomic signal processing and statistics (GENSIPS)*. pp. 135–138. IEEE.
21. Nahed, B. V., Redjal, N., Brat, D. J., Chi, A. S., Oh, K., Batchelor, T. T., . . . Olson, J. J. (2015). Management of patients with recurrence of diffuse low grade glioma. *Journal of Neurooncology*, 125(3), 609–630.
22. Robinson, D., Van Allen, E. M., Wu, Y. M., Schultz, N., Lonigro, R. J., Mosquera, J. M., . . . Chinnaiyan, A. M. (2015). Integrative clinical genomics of advanced prostate cancer. *Cell*, 161(5), 1215–1228.
23. Yu, K. H., Zhang, C., Berry, G. J., Altman, R. B., Ré, C., Rubin, D. L., & Snyder, M. (2016). Predicting non-small cell lung cancer prognosis by fully automated microscopic pathology image features. *Nature Communications*, 7(1), 1–10.
24. Althubaiti, S., Karwath, A., Dallol, A., Noor, A., Alkhayyat, S. S., Alwassia, R., . . . Hoehndorf, R. (2019). Ontology-based prediction of cancer driver genes. *Scientific Reports*, 9(1), 1–9.
25. Riaño, D., Real, F., López-Vallverdú, J. A., Campana, F., Ercolani, S., Mecocci, P., . . . Caltagirone, C. (2012). An ontology-based personalization of health-care knowledge to support clinical decisions for chronically ill patients. *Journal of Biomedical Informatics*, 45(3), 429–446.
26. Traverso, A., Wee, L., Dekker, A., & Gillies, R. (2018). Repeatability and reproducibility of radiomic features: A systematic review. *International Journal of Radiation Oncology* Biology* Physics*, 102(4), 1143–1158.
27. Galopin, A., Bouaud, J., Pereira, S., & Seroussi, B. (2015, January). An ontology-based clinical decision support system for the management of patients with multiple chronic disorders. In *MedInfo*, pp. 275–279. São Paulo, Brazil.
28. Samwald, M., Miñarro-Giménez, J. A., Blagec, K., & Adlassnig, K. P. (2014). Towards a global IT system for personalized medicine: The medicine safety code initiative. *Studies in Health Technology and Informatics*, 198, 25–31.
29. Luciano, J. S., Andersson, B., Batchelor, C., Bodenreider, O., Clark, T., Denney, C. K., . . . Dumontier, M. (2011). The translational medicine ontology and knowledge base: Driving personalized medicine by bridging the gap between bench and bedside. *Journal of Biomedical Semantics*, 2(Suppl 2), S1. doi:10.1186/2041-1480-2-S2-S1.
30. Jayaratne, L. (2015). Ontology based approach for diagnosis in personalized medicine. *GSTF Journal on Computing (JoC)*, 4(2), 1.
31. Messaoudi, R., Jaziri, F., Mtibaa, A., Grand-Brochier, M., Ali, H. M., Amouri, A., . . . Vacavant, A. (2019). Ontology-based approach for liver cancer diagnosis and treatment. *Journal of Digital Imaging*, 32(1), 116–130.

32. Shen, Y., Colloc, J., Jacquet-Andrieu, A., Guo, Z., & Liu, Y. (2017, December). Constructing ontology-based cancer treatment decision support system with case-based reasoning. In *International conference on smart computing and communication*. Cham: Springer, pp. 278–288.
33. Zhang, H., Guo, Y., Li, Q., George, T. J., Shenkman, E., Modave, F., & Bian, J. (2018). An ontology-guided semantic data integration framework to support integrative data analysis of cancer survival. *BMC Medical Informatics and Decision Making*, 18(2), 129–147.
34. Zheng, J., Cade, J., Brunk, B. P., Roos, D. S., Stoeckert Jr, C. J., James, S., . . . Vinetz, J. (2016). Malaria study data integration and information retrieval based on OBO foundry ontologies. In *ICBO/BioCreative*, Oregon State University, Corvallis, OR, USA.
35. GN, B. B., Sirisha, N., Rani, B. K., & Bethu, S. (2022). *A frame work design of chicken-sine cosine algorithm-based deep belief network for lung nodule segmentation and cancer detection*. Preprints 2022, 2022060217. https://doi.org/10.20944/preprints202206.0217.v1.
36. Gong, M., Wang, Z., Liu, Y., Zhou, H., Wang, F., Wang, Y., & Hong, N. (2019). Toward early diagnosis decision support for breast cancer: Ontology-based semantic interoperability. *Journal of Clinical Oncology*, 37(15), e18072.
37. Dissanayake, P. I., Colicchio, T. K., & Cimino, J. J. (2020). Using clinical reasoning ontologies to make smarter clinical decision support systems: A systematic review and data synthesis. *Journal of the American Medical Informatics Association*, 27(1), 159–174. 2020. doi: 10.1093/jamia/ocz169.

13 Interoperability Frameworks

Data Fabric and Data Mesh Architectures

Michael DeBellis, Livia Pinera and Christopher Connor

CONTENTS

13.1 DATA FABRICS AND SEMANTIC TECHNOLOGY

In this chapter we will discuss interoperability and semantic technology. This chapter will be divided into the following sections. In Section 13.1 we will discuss the concept of a Data Fabric which is becoming the de facto paradigm for enterprise data interoperability. In Section 13.2 we will discuss the main modules of a Data Fabric and the role that semantic technology can play in each. In Section 13.3 we will discuss the new concept of a data mesh and how semantic technology is essential for a successful data mesh. In Section 13.4 we will provide a brief conclusion.

DOI: 10.1201/9781003310785-13

We will focus on semantic technologies that are W3C standards and assume readers are familiar with them. Specifically, the standards known as the semantic web: IRI, RDF, RDFS, SPARQL, SHACL, OWL, and SWRL. We also expect that readers will be aware of the Protégé ontology editor from Stanford or similar tools. If not, we suggest they read Uschold (2018) [1] or work through the tutorial at DeBellis (2021).[2]

We focus on the W3C standards for two reasons:

1. Standards facilitate reuse and interoperability. Standards are what made the Internet possible. Using standards for ontologies and knowledge graphs provides access to large linked data resources such as DBpedia as well as many reusable vocabularies which will be described in Section 13.2.3.
2. Standards help to future proof information technology. They make it easier to migrate from or integrate with different vendors. Using non-standard technologies increases dependency on a single vendor.

Having said that, we acknowledge that non-standard knowledge graph technologies such as Neo4j property graphs are quite prevalent in industry at this point and there are movements (e.g., the RDF* standards group)[1] to enhance the existing standards to support some of the property graph features. This is a common evolution of standards.

A note about terminology: We will use the term *ontology* when we focus on the T-Box[2] level. We will use the term *knowledge graph* when we focus on the A-Box or both T-Box and A-Box. We will use the term *semantic technology* when talking about knowledge graphs as well as additional technology such as SPARQL, SHACL, and SWRL.

We will reference specific vendors and open source technologies at various points in this chapter as illustrative examples of a certain type of technology or as a case study of certain concepts in action. However, this chapter is in no sense a technology evaluation. The inclusion or exclusion of a vendor or open source product should not be construed as promoting or disparaging that product.

Data Fabric is a concept that was coined by the Gartner Group and has since been embraced by all the major vendors in data science, both vendors of semantic technology (e.g., Franz Inc., Ontotext, Pool Party, Stardog, and Top Quadrant) as well as more traditional vendors such as Oracle and Tibco.

Gartner defines a Data Fabric as:

> A design concept for attaining reusable and augmented data integration services, data pipelines and semantics for flexible and integrated data delivery. It is an optimized combination of data management and integration technology, architecture design and services delivered across multiple deployment and orchestration platforms. This results in faster, informed and, in some cases, completely automated data access and sharing.[3]

The Data Fabric concept is essentially the de facto architecture for data interoperability. However, it is as much a collection of architectural principles as specific modules. To the extent that there are defined modules in a Data Fabric, their specific capabilities and their connection to other modules are loosely defined and vary from vendor to vendor and from one organization to another. For example, a vendor that focuses on data virtualization such as Denodo will include a data catalog as a component of their data virtualization offering or an enterprise data bus vendor such as Tibco will include data virtualization as part of their message bus.

Although Gartner recommends that knowledge graph technology provide the foundation for a modern Data Fabric it is important to recognize that like most architectures, the concepts in a Data Fabric can be implemented in many technologies and do not require knowledge graphs. For example, McComb (2019) describes a case study of a system that in many ways was a Data Fabric developed in 2000 by Standard and Poor's Market Intelligence organization implemented as a database of databases using Microsoft SQL.[4]

However, there are several reasons why semantic technology is the best foundation for a Data Fabric.[5] As described above the collection, analysis, and enrichment of metadata is critical for a Data Fabric. This requires a complex, highly interconnected model that typically spans multiple organizational, geographic, and business boundaries. Relational databases are too rigid to easily store this type of data. Knowledge graphs are perfect to represent such information because they are:

- Flexible. Adding a property or class to a knowledge graph model can be done at run time without bringing down the graph.
- Heterogenous. Modern data resides as much in documents, audio, and video files as in database tables. Since the foundation for knowledge graphs are Internationalized Resource Identifiers (IRIs), it is a natural fit to resources that are unstructured and semi-structured data as well as traditional tabular data.
- Intuitive. Semantic models such as OWL represent knowledge in ways that are highly intuitive such as class and property hierarchies.
- Formal. The formal basis for OWL means that models can be analyzed automatically by reasoners that can determine if a model is consistent. If the model is consistent the reasoner can add a great deal of additional data automatically.
- Fast and Scalable. In the past semantic technology was associated with environments that were significantly slower and unable to scale as well as relational technology. That has changed over the past few years. Several triplestore vendors provide technology that can scale and perform at comparable levels to relational technology. Relational technology still outperforms triplestores on small-grained frequent transactions but for highly connected types of data that would require multiple joins with relational technology, triplestores can perform as well or better than relational technology.

13.1.1 Data Fabric Principles

A Data Fabric is as much a goal as an architecture. The following architectural principles are those that are most often cited in the literature as the essential goals for a Data Fabric.

- Model-driven development. Model-driven development has been the holy grail that developers have searched for since at least the 1980s with CASE tools and the "software through pictures" mantra.[6] More recently this idea was developed by the Object Management Group's (OMG) Model-Driven Architecture (MDA).[7] The OMG is a consortium that defines standards for object-oriented development such as the Unified Modeling Language (UML). Many of the techniques from the OMG's MDA are consistent with the use of semantic technology. The use of semantic technologies has given new life to the concept [4] except rather than using diagrams to generate code as in the CASE paradigm, high-level models such as OWL and SHACL can be used to automate much of the code for functions such as data integrity validation, reasoning about inverse and transitive relations, and automatically classifying instances.

- Work at the logical rather than the physical level.[4] This is essentially a result of model-driven development. Semantic technologies enable developers to eliminate the distinction between logical and physical models. Languages such as OWL and SHACL define the semantics and constraints for data at a logical level. Due to the fact that these models build on lower-level graph languages RDF/RDFS they are already stored as graphs and there is no requirement to transform from logical to physical models.
- Automate the collection and management of metadata. Gartner refers to this as converting passive metadata to active metadata.[3]

13.2 DATA FABRIC COMPONENTS

Figure 13.1 shows a functional architecture diagram for a Data Fabric. There are countless different diagrams of Data Fabrics. This is because Data Fabric is currently a collection of principles rather than a technical architecture. Thus, each

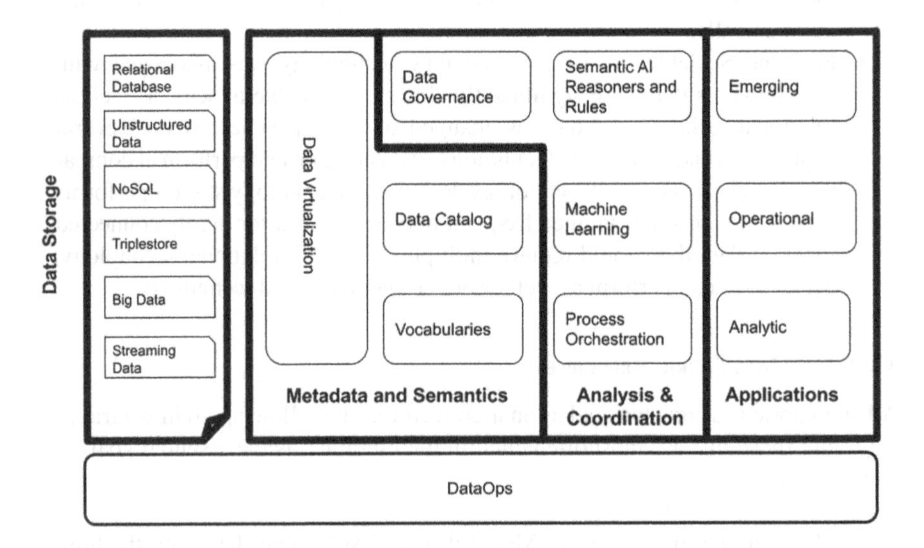

FIGURE 13.1 Data fabric functional architecture.

specific architecture will vary depending on the organization and vendor(s). Data virtualization vendors will place data virtualization at the core of the Data Fabric. Data catalog vendors will place the catalog at the center. EAI middleware vendors will place the message bus at the core. We have attempted to abstract from the various diagrams in order to make something more concrete but not focused on one specific technology.

We have divided the architecture into various layers such as data storage and applications. Within each layer we have placed the most important components. Many organizations will have one or more of these components in place already, and not every organization will need every component.

The layers are:

- Data storage. The Data Fabric ingests data from and stores data to this layer. For example, the Data Fabric transforms data, analyzes it, records metadata, and adds semantics. Two of the most important enabling technologies for this layer are triplestores and distributed file systems. Triplestores such as Stardog, AllegroGraph, and Apache TDB allow storing RDF triples in their native format and are an example of graph databases. While the best triplestores have excellent performance in order to store the petabytes needed for big data a distributed file system is required such as the Apache Hadoop system.
- Metadata and semantics. This is the layer that most distinguishes a Data Fabric from other data architectures. It automates the capture and utilization of metadata and provides one consistent semantic model across the enterprise. This is the layer that is most essential to eliminate data silos.
- Analysis and coordination. This layer utilizes the semantic layer in order to perform sophisticated analysis of data, automated reasoning, and to coordinate processes such as content management workflows and microservices.
- Application layer. These are the business applications. The most important distinction is between applications which analyze historical data and operational applications that run the business.
- DataOps. This layer must span all the other layers. It provides analytics on the performance of the Data Fabric so that data scientists and system administrators can anticipate problems and deal with them before they impact users.

13.2.1 Data Virtualization

The goal of the data virtualization module is to provide a consistent semantic model across the entire enterprise. This goal is typically summarized as "eliminate data silos". This is not a new goal. It can be traced back to the data dictionaries utilized for COBOL systems on mainframes. It is a goal and not something that many large enterprises have completely accomplished, nor is it necessary to completely accomplish the goal in order to achieve benefits. Figure 13.2 illustrates the basic structure of a data virtualization module.

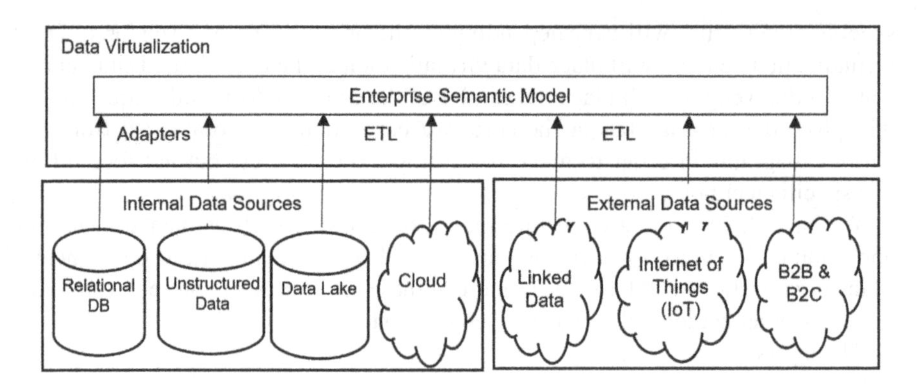

FIGURE 13.2 Data virtualization.

The traditional and still most common way to populate an ontology with data is via extraction, transformation, and loading (ETL). An open source tool that most Protégé users are familiar with is the Cellfie Protégé plugin.[8] Cellfie can import data from CSV files or spreadsheets and define transformation rules to transform rows into instances of the ontology. The most common transformation is to map each row to an instance of a class. Then each column in the table is mapped to an object or data property and converted to the appropriate individual (for object properties) or datatype (for data properties).

While Cellfie works well for importing medium sized data sets (thousands of instances) for enterprise data (hundreds of thousands or more instances) tools that support larger amounts of data are required. All of the major triplestores have similar capabilities as Cellfie that can load very large data sets, for example, the CSV Import Wizard from Stardog.[9]

One recent trend is to reverse the last two steps of the process, i.e., ELT rather than ETL.[10] With ELT the data is loaded as just a Literal and then transformed to the appropriate object(s), datatypes, and properties with SPARQL and/or a programming language. The reason for this is that some strings can map to multiple objects and properties when transformed to a knowledge graph. In addition, it may be required to utilize data already in the knowledge graph to add additional information to the transformation of the property. For example, in the CODO Covid-19 ontology, strings in the Reason column for each patient could have patterns such as "Infected by P346–349". This requires iteration from patient 346 to patient 349. Also, data on those patients (e.g., their social or family relation to the patient for that row) may provide additional information that can be added to the graph.[11]

Another approach that is very common for large enterprise data is to simply use programming languages such as Python and Java to do ETL/ELT. Triplestores have APIs for clients such as Python and Java. Thus, the data can be loaded programmatically and transformed using pattern matching capabilities such as REGEX as well as utilizing existing data in the knowledge graph to appropriately transform new data into the graph.

While transforming and importing data into a knowledge graph is still an important tool, a more ambitious technique is to leave the data in place in a relational database or some other data store and use the OWL model to provide the semantics for the data. In OWL terminology, OWL is used for the T-Box and a relational database is used for the A-Box. Internal data sources such as relational data from applications and ERP systems and external data such as data from sensors on the Internet of Things (IoT) are all mapped to one common enterprise model by various adapters, similar to the adapters that integrate consumers and producers to an Enterprise Message Bus.

Vendors such as Denodo [12] and Tibco [13] have predefined adapters for many of the most common enterprise data sources such as SAP although at the present time, their enterprise models are still primarily relational rather than ontologies. However, the future is to utilize ontologies to provide semantics across disparate relational and other data sources [4] such as data lakes.[14]

Data virtualization has a long history in the OWL community. One of the most important precursors to OWL was the Loom language developed at the Information Sciences Institute (ISI). An important capability of Loom was that it could store A-Box data in relational databases as well as natively in the ontology.[15] A Protégé plugin that provides this type of data virtualization for OWL is the Apache Ontop system.[16]

A new paradigm that is relevant to data virtualization and is being widely embraced by many early adopters is event streaming. Event streaming is in many ways similar to traditional Message Oriented Middleware, but it also has some important differences. One of the advantages of streaming architectures is that the standard ETL transformations can be used but in a pseudo real time mode rather than batch. Event-based architectures combined with the performance of triplestores allows a paradigm shift in the development of systems. Complex machine learning and AI rules can be utilized not just for analysis applications but for operational applications that work in pseudo real time as well.[17]

One criticism of data virtualization comes from those who advocate for a Domain Driven Design approach.[18] Domain Driven Design was first practiced on microservices but is now being advocated by some for data as well as services. In this view, using data virtualization for analytic data is an anti-pattern because operational data is inherently different from analysis data. Operational data requires design for maximum efficiency whereas analysis data should be designed for clarity. See Chapter 2 in Dehghani (2022) for more on this point of view.[18]

13.2.2 Data Catalog

The Data Catalog is the most critical part of the Data Fabric and is all about capturing metadata, both technical metadata and semantic metadata. Examples of such metadata are:[19]

- Identifiers: Keys, codes, synonyms.
- Documentation: Diagrams, manuals.
- Data Format: CSV, JSON, RDF, Avro, relational tables, RDF/RDFS.

- Governance: Who owns the data, who can see it, who can edit it. How it can be used. This includes adherence to relevant regulations described in Section 13.2.4.
- Semantic Metadata: Metadata such as geography, social, and temporal relations. Class and property hierarchies and ad hoc taxonomies.
- Technical Metadata: Data elements, datatypes, min/max cardinality, and other data integrity constraints. Real time data about data usage and performance.
- Business processes: Linking data to business requirements and processes, such as workflows for the life cycle of data and orchestration of microservices.
- Data Quality: Metadata describing issues, Service Level Agreements.
- Data Lineage: How data was derived, e.g., which systems create, read, update, and delete data.

This information is inherently a graph structure. As much as possible, this data is automatically loaded and maintained by the Data Fabric. This can come about in a number of ways. Sample data can be imported and analyzed, high-level languages such as OWL and SHACL that support automated validation and verification can be utilized, etc. The imported data models can also be augmented by owners and subject matter experts with capabilities similar to knowledge management collaboration tools such as threaded discussions.

The steps for creating a data catalog are:[19]

1. Identify the initial data sources. For most data catalogs attempting to catalog all the data in the enterprise at once is too large a problem. It is usually better to pick some business critical domain where a data catalog can deliver measurable value quickly and focus on that first.
2. Decide what metadata is important. This will include technical metadata as well as semantic metadata discussed earlier.
3. Develop/reuse controlled vocabularies for metadata. Vocabularies are ontologies that describe a specific domain as opposed to being designed for a specific application. This is discussed in more detail in the next section.
4. Develop cataloging processes. Defining/importing roles, responsibilities, and workflows. As much as possible strive to automate the collection and maintenance of metadata associated with these processes.
5. Test, redesign, and elaborate. Testing requires business users, but the more business users can be members of the team from the very start (i.e., an Agile approach) the more likely the project is to succeed. After the initial data catalog has shown value, it can be extended to other types of enterprise data.

13.2.3 Vocabularies

Vocabularies are reusable models that describe a specific domain. All of the vocabularies described below are OWL ontologies; however vocabularies can also be defined in other formats such as taxonomies, thesauri, and coding systems. Vocabularies can be vertical (specific to a domain or industry) and horizontal (a technical domain such as metadata). One of the advantages of using OWL is that there are already a large number of open source vocabularies that have been defined or converted to OWL. Examples of some of the most frequently used vertical vocabularies are:

- SNOMED CT.[3] For healthcare clinical terminology. SNOMED dates back to well before OWL but the latest version includes one of the largest and most widely used collections of OWL vocabularies that enable sharing medical records, clinical trials, and other healthcare data.
- FIBO.[4] The Financial Industry Business Ontology is a collection of ontologies developed by the EDM council to facilitate communication and collaboration of corporations and other organizations in Financial Services and related industries.
- Schema.org. A vocabulary for common properties and datatypes used in web systems, email, and social networks. Created by a consortium of leading vendors including Microsoft and Google.

Example of some of the most frequently used horizontal vocabularies include:

- Dublin Core.[5] Dublin Core dates back to well before OWL. It is a vocabulary to describe common technical metadata such as who created a dataset, who last edited it, the dates it was created and last edited.
- Prov-O.[6] Prov-O is a W3C standard for provenance metadata. Provenance is the audit trail of who, when, and how a dataset was created and edited. It utilizes Dublin Core but whereas Dublin Core describes essential metadata, Prov-O defines more sophisticated aspects such as complete audit trails of editors, stakeholders, licensing, and version control.
- SKOS.[7] The Simple Knowledge Organization System (SKOS) is a W3C ontology that defines various techniques to organize, categorize, and index knowledge such as taxonomies, thesauri, synonyms, and classification schemes.

One common point of confusion is the distinction between taxonomies and ontologies. The word *ontology* has a complex history going back to philosophy. In the context of this chapter, an ontology is simply a formal model defined in OWL. The term *taxonomy* simply refers to a tree graph. All ontologies have taxonomies but not all taxonomies are part of an ontology. Before OWL and still with many industry classification schemes, taxonomies were utilized that had no rigorous semantics. The best taxonomies would have a semantics such as kind-of or part-of. However, it is sometimes the case that a taxonomy will merge different semantic meanings into the same tree structure because it seems intuitive. Such ad hoc taxonomies that mix different semantics are an anti-pattern as they have an ambiguous semantics so the map from the taxonomy to other models and code is more prone to error.

13.2.4 Data Governance

Data governance is the definition and enforcement of policies that define what systems and people have access to what data as well as the type of access. This has become much more important for most organizations due to the prevalence of customer data in electronic formats that can be accidentally leaked to the public or accessed by cybercriminals. For corporations that offer services such as online shopping and social networks the assurance that the customer's data is secure is a key differentiator. One data leak can destroy the reputation of a corporation. In addition, recent laws passed in the United States and the European Union mandate that many corporations have data governance processes that are rigidly enforced at the risk of facing significant fines. Some of the most important laws are:

- Sarbanes Oxley. A law passed in the United States to address issues raised by several financial and accounting scandals such as Enron and WorldCom. The act defines requirements for all public companies and accounting firms to maintain audit trails and maintain documents such as financial reports. CEOs and other executives are liable for failure to comply with the regulations.[20]
- Health Insurance Portability and Accountability Act (HIPAA). A law passed in the United States to both enable and secure the adoption of electronic medical records in hospitals and other healthcare organizations rather than traditional paper charts.[21]
- The General Data Protection Regulation (GDPR). A law passed by the European Union that they describe as "the toughest privacy and security law in the world". It not only requires that sites provide rigorous control over user's personal data but defines a "right to be forgotten" that requires a site to delete all of a user's information if they wish. The law "imposes obligations onto organizations anywhere, so long as they target or collect data related to people in the EU".[22] It is also seen by many as a best practice that other nations will either directly utilize or base future legislation on.

The other feature that is provided by data governance is to define what specific data can and *cannot* be used for. An example of the importance of this function is recent news about how Twitter was fined $150M for improper use of personal data.[23] Users were asked for their emails and phone numbers in order to secure their accounts. However, up until 2019, Twitter was also selling this data to advertisers. A data governance tool that modeled that this type of personal data could not be used to generate advertisement revenue would have prevented this.

An example of a Data Fabric solution for data governance is a major international bank that utilized the TopBraid Enterprise Data Governance (EDG) product to define a knowledge graph data catalog that recorded the data lineage for their data.[24] The data was stored in thousands of heterogenous data silos across many different organizations and geographies. In addition, historical data was offloaded from relational databases to a Hadoop data lake in order to minimize the cost of storing large amounts of historical data. This made it difficult for business users to understand where specific data came from. Such knowledge was required to comply with various regulations such as Sarbanes Oxley. The Data Catalog and the EDG product made this knowledge explicit. For example, the data lineage graphs displayed in [24] explicitly show data flows that were buried in code before the Data Catalog.

13.2.5 Machine Learning

As with semantic web technology, machine learning (ML) technology also has its roots in AI research. The two technologies complement rather than compete with each other. There are many examples of the synergy between the two approaches. Some examples are:

- Andrew Ng, one of the leading experts in machine learning, has discussed how improving the quality of data is more important than improving the quantity to provide better results with ML algorithms.[25] Improving data quality is one of the main benefits of a Data Fabric. Languages such as OWL and SHACL provide declarative models for data semantics and data integrity constraints that can automatically be validated by reasoners.
- The emerging field of Graph Neural Networks.[26]
- Using ML to automate enhancement and completion of very large knowledge graphs.[27]

13.2.6 Semantic AI Reasoning

One of the advantages of OWL is that it has a formal semantics that is a subset of First Order Logic (FOL). This enables automated OWL reasoners that can accomplish two goals:

1. Validate the ontology. Guarantee that the model has no contradictions. If there are contradictions, the reasoner can point data scientists to the specific classes, properties, or individuals that cause the inconsistency.
2. Infer additional information from the ontology. Set theoretic concepts such as inverse and transitive properties can be defined in OWL and automatically maintained by the OWL reasoner. In addition, the reasoner can automatically classify objects based on their data values. Thus, logic that would typically be defined in code can instead be defined in declarative models. This is an example of how semantic technology facilitates model-driven rather than code driven development, thus significantly reducing the cost while improving the quality of software.[4]

In addition to the OWL reasoners, there are rule engines that facilitate model-driven development. The Semantic Web Rule Language (SWRL) provides a very powerful forward-chaining language that combined with the OWL reasoner allows many inferences to be defined globally in the model rather than in application code where it often may be duplicated in different applications or even worse have inconsistent definitions across applications. SWRL is so powerful (and hence computationally intensive) that some triplestores don't currently support it. However, it is possible to automatically transform SWRL rules to more efficient formats such as SPARQL SPIN rules.[28] One approach to ontology development is to first use SWRL for rapid prototyping and then utilize SPARQL SPIN or some other more efficient format for production.[11] Many triplestores also provide custom integration with various other rule engines or reasoning tools. For example, AllegroGraph has a powerful Prolog implementation that is well integrated with SPARQL and the AllegroGraph triplestore.[29]

The other semantic modeling language is the Shapes Constraint Language (SHACL). OWL and SHACL overlap significantly in what kind of metadata can be defined. Many of the same kinds of constraints on properties (e.g., domain, range, min/max number of values) can be expressed in both languages. The distinction is not *what* can be modeled but *how* the model is utilized. In OWL, the model is used to reason about the knowledge graph. In SHACL, the model is used to evaluate and report on data integrity constraints.

13.2.7 Process Orchestration

Process orchestration refers to the definition and enforcement of workflows for processes. These can be processes for editing documents and other content, load balancing and coordination tools like Zookeeper,[8] orchestration of container architectures such as Kubernetes,[30] and workflows for batch processes such as Apache Airflow.[9]

13.2.8 DataOps

Gartner defines DataOps as "a collaborative data management practice focused on improving the communication, integration and automation of data flows between data managers and data consumers across an organization".[31] DataOps is as much a process as it is software to monitor the real time status of databases and other components in a Data Fabric. The main idea of both DevOps and DataOps is to apply the same approach used for Agile development to the administration and real time maintenance of software and data.

Specific practices identified by Gartner are:[31]

- Increased deployment frequency. Just as Agile development emphasizes very short development iterations, so does DataOps. Rather than focusing on systems that have new releases a few times a year, the goal is to have weekly or daily updates. The ultimate goal is *continuous delivery* where updates can happen throughout the day.
- Automated testing. In order to achieve continuous delivery and still maintain high levels of quality, automated testing is a must.
- Consistent metadata and version control. Automated creation and maintenance of metadata is the essence of a Data Fabric.
- Monitoring. Monitoring metadata discovers potential problems before they cause disruption to users. One of the key enabling technologies for DataOps are containers such as Docker and technology to distribute and monitor containers in distributed environments such as Kubernetes. They enable components of modern systems to be packaged as virtual machines that can run on any platform and that can make a single server simulate a distributed computing environment. These same technologies provide detailed real time data on the status of each container.

- Collaboration across all stakeholders. Two of the core principles of Agile development are collective ownership and on-site customer[32] (business-people must be dedicated members of the team). This is the combination of these two Agile principles applied to data management.

13.2.9 Applications

Gartner describes three ways that data virtualization can be utilized by an enterprise. [33] These three types of applications also describe the main categories of applications that utilize a Data Fabric.

1. Analytic. A common model for historical data to be used for analysis. A semantic model defined in OWL is especially relevant for the data lake approach where the model provides structure for unstructured data such as web pages, videos, and documents files as well as for structured historical data from relational databases and other structured data stores.
2. Operational. A common semantic model for data utilized to run the business, most commonly in relational databases. Examples include data from ERP systems such as SAP and CRM systems such as Salesforce. One of the most important tools in the data virtualization module for this type of data are intelligent query optimizers that can provide performance similar to directly accessing the data while still providing the benefits of a common semantic model. This is the most challenging type of data virtualization for the semantic approach as in order to achieve acceptable performance it is often still required to either tightly couple applications directly to their relational databases or to utilize a relational rather than an ontology model for the semantic virtualization model. This is an area where there is the greatest need for advances in knowledge graph technology that are comparable to the optimizations that are the result of decades of use of relational databases.
3. Emerging. These are new types of systems built from the ground up to leverage AI technology such as OWL, rules, and machine learning algorithms. These types of systems can be developed using high performance triplestores rather than relational databases for operational data.

Of these three types of applications, the first and the third are where semantic technology is currently being used the most and has the most potential in the near future. Many existing operational systems are so tightly coupled to relational databases that moving to a data virtualization approach with a semantic model is not practical.

Analytic applications are built with tools that utilize machine learning, statistical analysis, semantic analysis, and process orchestration via Dashboard GUIs that allow users to concentrate on the business questions they want to ask rather than on details of mathematical algorithms. Examples include Apache Spark and Databricks.

Data lakes are often stored in cloud environments and utilize built-in cloud tools for analytics. Amazon Web Services (AWS) is a leading cloud provider, and their AWS Lake tool is one of the leading dashboards. Microsoft Azure is also a leading cloud service that provides native analytic and dashboard tools.

Examples of emerging applications are:

- Knowledge graphs for next-gen chatbots. Most chatbots are difficult to use and frustrating. Using knowledge graph technology from AllegroGraph N3 (an automated sales company that is part of Accenture) developed next generation chat bots that can interact with humans in a more realistic way by utilizing domain knowledge and rules that can transform human questions into specific queries that the knowledge graph can answer.[34]
- Edge data access in IoT integration.[13] IoT systems are being developed using the concept of "Digital Twins"—a software object that is a model of a sensor or object (e.g., a self-driving car, a smart appliance). The Edge concept means that processing goes on at the location in the IoT where the object resides (e.g., an autonomous car using AI) as well as having sensor data uploaded for further analysis.
- Systems built using semantic AI reasoning and/or machine learning.[17]

This concludes our discussion of Data Fabrics and how semantic technology is utilized in them. Our next section describes a new alternative approach called the Data Mesh.

13.3 DATA MESH

A new paradigm for data interoperability is the Data Mesh. A Data Mesh applies the same principles used for microservices to the definition and utilization of Data. The four principles of a Data Mesh are:[18]

1. Data ownership by domain. Just as microservices are owned by business units, so is data. Data is designed by domain experts collaborating with data engineers.
2. Data as a product. A business team owns a domain such as sales or marketing and provides its data as internal products to the rest of the enterprise. The various teams that utilize the data are viewed as customers by each domain team. Again, this applies the microservice paradigm to data. Data products have Service Level Objectives (SLO) with their internal customers. One of the most important kinds of metadata for data as a product are metrics that analyze data usage and quality such as adherence to SLOs, number of users, time to change, and change fail ratio.
3. Self-service data. All data is discoverable and can be utilized via loose coupling. Thus, a Data Mesh requires some type of bus that enables a *publish* and *subscribe* paradigm where producers can publish data with no coupling to the consumers that subscribe to it. Producers and consumers can be added without disturbing existing services.
4. Federated governance. Data has integrity and governance constraints specific to each domain, but these constraints must be enforced throughout the organization. Also, the organization will have global constraints such as requirements to adhere to data regulations discussed in

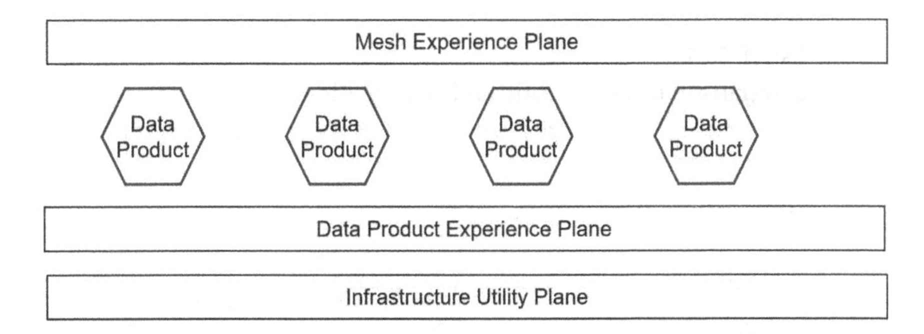

FIGURE 13.3 Data Mesh logical architecture.

Section 13.2.4. Some type of federated governance structure must be defined within the organization with responsibilities explicitly demarcated between the centralized data infrastructure team and the business domain teams. The goal of a Data Mesh is to push as much responsibility and governance to local domain teams rather than to leave them with the centralized infrastructure team.

13.3.1 Data Mesh Architecture

Figure 13.3 shows a high level view for a Data Mesh architecture defined in Dehghani (2022).[18] It consists of three planes: mesh experience, data product experience, and infrastructure utility, as well as the actual data products that are defined, used, and maintained via the three planes. The term plane is used because while there does tend to be a stronger dependency between each plane and the planes above and below it, there is not a rigid hierarchical structure. In other words, the data product experience plane relies heavily on the infrastructure utility plane, but components in any plane can communicate with any other plane.

The functions of each component in Figure 13.3 are as follows:

- Infrastructure utility plane. This plane provides the foundational functions required to power the mesh. For example, data storage such as relational and graph databases, security services, streaming services, and network services.
- Data product experience plane. This is the plane used to create, maintain, subscribe to, and retire data products.
- Data products. A data product is the data equivalent of a microservice. A component defined for a specific domain with interfaces to access, update, and maintain the product. Examples of data products are a lab report or electronic medical record in healthcare and an invoice or order in online shopping. The goal is to push as many capabilities as possible that would typically be defined and enforced at an enterprise level to specific domain products. One of the most radical differentiators of the Data Mesh approach is that the majority of governance is defined in the code of the data product rather than in a centralized data catalog.

TABLE 13.1
Comparison of Data Fabric and Data Mesh

Data Fabric	Data Mesh
Data Sources	Infrastructure Utility Plane
Data Virtualization	Antipattern
Data Catalog	Mesh Experience Plane
Vocabularies	Data Products
Data Governance	Data Products, Product Experience, Mesh Experience
Semantic AI Reasoning	Infrastructure Utility Plane
Machine Learning	Infrastructure Utility Plane
Process Orchestration	Infrastructure Utility Plane
DataOps	Ubiquitous throughout the mesh
Application	Analytic

- Mesh experience plane. This plane is used to navigate and monitor the mesh. For example, to find/monitor a data product, view the lineage of changes to a data product, and to define and enforce global governance policies.

The capabilities of a Data Fabric are all present in a Data Mesh except for data virtualization, which is considered an anti-pattern. The mapping of the components in a Data Fabric and those in a Data Mesh are shown in Table 13.1.

The focus of a Data Mesh is on analytic applications. However, the kind of infrastructure required to create a Data Mesh implies that the organization has a microservices architecture in place with managed components such as Kubernetes as well as an event bus such as Kafka or an enterprise bus such as Tibco. Although in practice, to the best of our knowledge, all the early adopters who are currently building Data Meshes are using Kafka.

Semantic technology is a critical enabling technology for a Data Mesh. The Mesh Experience Plane is described as a graph in Dehghani (2022).[18] Also, it is strongly encouraged that IRIs are used to provide unique addresses for all data products.

Semantic technology allows the data products to be defined and implemented at the logical rather than the physical level.[4] This is critical to enable greater participation by business subject matter experts in the design of data objects. SHACL provides an excellent tool to declaratively define various data integrity constraints on each data product. SPARQL and the OWL and SHACL reasoners provide powerful tools to enable validating data products and searching the mesh.

The most significant difference between a Data Mesh and a traditional Data Fabric is on where and how the data is designed and managed. A Data Fabric focuses on a central organization to define and manage data as has been typical in the past. This is at odds with the paradigm of microservices. Authors describing microservices talk about the shared database anti-pattern.[30] The Data Mesh takes the next logical step and makes data an internal product just as microservices are for business logic and processing.

Another way of saying this is that various approaches in the past (e.g., data dictionaries, data warehouses) have tried and failed to eliminate data silos and provide a "single source of truth" in one enterprise data model. A Data Fabric promises to actually be able to provide this. The Data Mesh approach says that the goal itself is flawed, that data is inherently specialized depending on the domain and that the best architecture is one that recognizes this and gives more responsibility to distributed domain teams rather than a centralized data team.

13.3.2 Criticism of the Data Mesh Approach

The concept of a Data Mesh is new and unproven. The most significant criticisms of the Data Mesh approach are:[10]

- Data Mesh is inconsistent with the goal of model-based development. Rather than putting logic regarding data use in declarative models it pushes this logic into code for the various data products. This makes the logic less visible, less maintainable, and more prone to errors and inconsistencies than a model-based approach.
- Data Mesh is a set of goals with little technology to support it and the few groups that claim to have implemented it are simply rebranding their architecture as a Data Mesh.[35]
- Data Mesh requires business analysts to think like data engineers and vice versa which is not viable.
- Data Mesh is not really a new idea. It is a rebranding of the Data Mart or Information Mart architectures that date back to the 1990s. The same issues that made these architectures impractical will be issues for a Data Mesh. Specifically, that "global governance [of a federated Data Mesh] . . . won't prevent problems with duplications of the datasets and efforts related to their support".[36]

There is currently not enough evidence to evaluate the Data Mesh approach. It shows significant potential but also some serious unresolved questions. However, semantic technology will be at least as important for an enterprise Data Mesh as for a Data Fabric.

13.4 CONCLUSION: INFLECTION POINTS AND PARADIGM SHIFTS

At first glance, the Data Mesh and Data Fabric architectures seem to be polar opposites. In reality, they both share many goals:

- An emphasis on the importance of data for an enterprise and on making data an asset or product supported by a common infrastructure.
- Continuous integration/continuous deployment.

- Explicitly defining and enforcing the ownership, governance, and utilization of data rather than having such processes be implicit in documentation.
- Providing semantic models that are intuitive for businesspeople to understand and to edit.
- Moving much of the code that in the past was associated with each specific system or microservice into the data infrastructure.
- The automated extraction and use of metadata to utilize and govern use of data more effectively, especially to eliminate data silos.
- An Agile approach that has users and developers share responsibility for data domains.

One possible result of the Fabric vs. the Mesh dichotomy is that in reality neither approach will be completely embraced. Rather, many organizations may combine the approaches, utilizing both global data governance as well as data products for certain core data such as customers, electronic medical records, and products. This approach is really consistent with both. The Mesh approach emphasizes data products, and the Fabric approach emphasizes centralized governance, but both agree that some combination of central and local control is required.

Data Fabrics and Data Mesh with semantic technology are a paradigm shift just as the Internet and microservices were a paradigm shift. What drives paradigm shifts are inflection points,[18] changes in technology, and societies that create the opportunities for radical new ways of doing business. The synergy provided by big data, linked data, machine learning, high performance triplestores, and semantic AI provide an inflection point for new ways to develop software[4, 17] that are more affordable, maintainable, and offer radically new functionality.

At the beginning of the new millennium, Watts Humphrey, developer of the CMI Capability Maturity Model said "Every business is a software business".[37] This is truer now than when he said it. Software is no longer one tool among many that helps run a corporation. Corporations *are* their software. It is this trend that will drive semantic technology and the new approaches to data interoperability described in this chapter.

NOTES

1. https://w3c.github.io/rdf-star/cg-spec/editors_draft.html
2. T-Box refers to the schema definitions, i.e., the table definitions in a relational database and the class and property definitions in an ontology. A-Box refers to the data, i.e., the rows in a relational database and the individuals in a knowledge graph. For more on this see Uschold (2018)[1] or Chapter 8 in DeBellis (2021)[2].
3. https://www.snomed.org/
4. http://www.fibo.org
5. https://www.dublincore.org/
6. https://www.w3.org/TR/prov-o/
7. https://www.w3.org/2004/02/skos/
8. https://zookeeper.apache.org/
9. https://airflow.apache.org/
10. Some of these criticisms are based on discussions with practitioners. Any point in the list that does not have a reference is based on such discussions.

REFERENCES

1. Uschold, M. (2018, May 29). *Demystifying OWL for the Enterprise* (First Edition). San Rafael, CA: Morgan & Claypool Publishers.
2. DeBellis, M. (2021). *A practical guide to building OWL ontologies using Protégé 5.5 and plugins edition 3.2*. San Francisco, CA: Michaeldebellis.com, pp. 1–90. Retrieved May 4, 2022, from https://www.michaeldebellis.com/post/new-protege-pizza-tutorial.
3. Zaidi, E., et al. (2019, December 17). *Data fabrics add augmented intelligence to modernize your data integration*. Stamford, CT: Gartner Group, pp. 1–23. ID G00450706.
4. McComb, D. (2019). *The data-centric revolution*. Basking Ridge, NJ: Technics Publications.
5. Polikov, I. (2022, February 14). *Knowledge graphs are key to data fabrics*. Raleigh, NC: Top Quadrant Inc. Retrieved May 4, 2022, from https://www.topquadrant.com/knowledge-graphs-are-key-to-data-fabric/
6. Oman, P. W., et al. (1990, May). CASE: Analysis and design tools. *IEEE Software*, 7(3), 37–43. doi:10.1109/52.55226
7. Object Management Group. *Model Driven Architecture (MDA) MDA Guide rev. 2.0. OMG Document ormsc/2014-06-01*. Needham, MA: The Object Management Group, pp. 1–15. Retrieved May 4, 2022, from https://www.omg.org/cgi-bin/doc?ormsc/14-06-01
8. O'Connor, M. J., Halaschek-Wiener, C., & Musen, M. A. (2010). Mapping master: A flexible approach for mapping spreadsheets to OWL. In *9th international semantic web conference (ISWC)*. Shanghai, China: Stardog.com.
9. Rogers, J. (2020, March 11). *Importing, exploring, and exporting your data with Stardog studio*. Arlington, VA: Stardog.com. Retrieved May 13, 2022, from https://www.stardog.com/blog/importing-exploring-and-exporting-your-data-with-stardog-studio/
10. Bartley, K. (2020, December 9). *ETL vs. ELT what's the difference?* New York: Rivery.com. Retrieved May 13, 2022, from https://rivery.io/blog/etl-vs-elt/.
11. DeBellis, M., & Biswanath, D. (2021, October 8). *The covid-19 CODO development process: An agile approach to knowledge graph development*. New York: Springer Publishing, pp. 1–12. KGSWC 2021 Conference.
12. Denodo Whitepaper. *Logical data fabric*. Palo Alto, CA: Denodo Inc. Retrieved May 9, 2022, from https://www.denodo.com/en/document/whitepaper/logical-data-fabric-whitepaper.
13. Tibco White Paper. *Applying data virtualization: 13 use cases that matter*. Palo Alto, CA: Tibco. Retrieved May 9, 2022, from https://www.tibco.com/resources/overview-brief/data-virtualization
14. Dibowski, H., & Schmid, S. (2021). Using knowledge graphs to manage a data lake. In R. H. Reussner, A. Koziolek, & R. Heinrich (Hrsg.), *INFORMATIK 2020*. Bonn, Germany: Gesellschaft für Informatik, S. 41–50. doi:10.18420/inf2020_02
15. McKay, D. P., et al. (1996). *An architecture for information agents. Advanced Planning Technology*. Cambridge, MA: The AAAI Press, pp. 187–194.
16. Xiao, G., et al. (2019). Virtual knowledge graphs: An overview of systems and use cases. *Data Intelligence*, 1(3), 201–223. Cambridge, MA: MIT Press.
17. Aasman, J. *Entity event knowledge graphs for data centric organizations*. Lafayette, CA: Franz Inc, White Paper, pp. 1–7. Retrieved May 4, 2022, from https://allegrograph.com/wp-content/uploads/2020/06/Entity-Event-Knowledge-Graphs-White-Paper-v692020.pdf
18. Dehghani, Z. (2022, April 12). *Data mesh: Delivering data-driven value at scale*. Sebastopol, CA: O'Reilly Media. ISBN-13 9781492092391.
19. Top Quadrant Video Presentation. *Data cataloging with knowledge graphs*. Raleigh, NC: Top Quadrant Inc. Retrieved May 3, 2022, from https://www.topquadrant.com/project/automating-the-mapping-of-data-elements-to-business-terms-2/

20. Jain, P. K., & Rezaee, Z. (2006). The Sarbanes-Oxley act of 2002 and capital-market behavior: Early evidence. *Contemporary Accounting Research*, 23, 629–654. https://doi.org/10.1506/2GWA-MBPJ-L35D-C4K6
21. Cohen, I. G., & Mello, M. M. (2018). HIPAA and protecting health information in the 21st century. *JAMA*, 320(3), 231–232. Chicago, IL: JAMA. doi:10.1001/jama.2018.5630
22. Wolford, B. (2016, May). *What is GDPR, the EU's new data protection law?* Brussels, Belgium: European Union, pp. 1–5. Retrieved May 4, 2022, from https://gdpr.eu/what-is-gdpr/
23. Allyn, B. (2022, May 25). *Twitter will pay a $150 million fine over accusations it improperly sold user data.* Washington, DC: NPR. Retrieved May 28, 2022, from https://www.npr.org/2022/05/25/1101275323/twitter-privacy-settlement-doj-ftc
24. Coyne, R. (2019, November 13). Finance: A case study for data catalogs and lineage tracking for compliance reporting. In *Case study*. Raleigh, NC: Top Quadrant Inc. Retrieved May 4, 2022, from https://www.topquadrant.com/finance-case-study-blog/
25. Ng, A. (2021, March 24). *From model-centric to data-centric AI.* Alto, CA: DeepLearning.ai Palo. YouTube lecture. Retrieved May 23, 2022, from https://www.youtube.com/watch?v=06-AZXmwHjo
26. Zhou, J., Cui, G., Hu, S., Zhang, Z., Yang, C., Liu, Z., Wang, L., Li, C., & Sun, M. (2020). Graph neural networks: A review of methods and applications. *AI Open*, 1, 57–81. San Francisco, CA: AI Open.
27. Ji, S., Pan, S., Cambria, E., Marttinen, P., & Philip, S. Y. (2021). A survey on knowledge graphs: Representation, acquisition, and applications. In *IEEE transactions on neural networks and learning systems*. Piscataway, NJ: IEEE.
28. Bassiliades, N. (2020, August 5). A tool for transforming semantic web rule language to SPARQL infererencing notation. *International Journal on Semantic Web and Information Systems*, 16(1). Hershey, PA: IGI Global.
29. Franz Inc. *Using prolog with AllegroGraph 7.3*. Lafayette, CA: Franz, Inc. Retrieved May 16, 2022, from https://franz.com/agraph/support/documentation/current/prolog-tutorial.html
30. Newman, S. *Building microservices*. Sebastopol, CA: O'Reilly Media. ISBN-13 9781492034025.2021.
31. Friedman, T., & Thanaraj, R. (2021, February 16). *Introducing dataops into your data management discipline*. Stamford, CT: Gartner Group, pp. 1–9. ID G00376495
32. Beck, K. (2000). *Extreme programming explained*. Boston, MA: Addison Wesley.
33. Ziadi, E., Menon, S., Beyer, M. A., & Jain, A. (2018, November 16). *Gartner market guide on data virtualization*. Stamford, CT: Gartner Group.
34. Aasman, J. (2022, January 26). *Linguistic reduction and knowledge graphs for next-gen chatbots*. Lafayette, CA: Franz, Inc. Retrieved May 17, 2022, from https://aithority.com/machine-learning/neural-networks/linguistic-reduction-and-knowledge-graphs-for-next-gen-chatbots/
35. Oakland Group. (2021, November 18). *Is data mesh a potential data mess?* Leeds: The Oakland Group. Retrieved May 28, 2022, from https://www.theoaklandgroup.co.uk/is-data-mesh-a-potential-data-mess/
36. Zabavskyy, A. (2020, October 14). *Data mesh pain points: Why to think twice before implementing data mesh*. Toronto, Ontario, Canada: Towards Data Science. Retrieved May 28, 2022, from https://towardsdatascience.com/data-mesh-pain-points-b4bebca37357
37. Humphrey, W. S. (2001). *Winning with software*. New York: Pearson Education.

14 Recommender System for E-Commerce
How Ontologies Support Recommendations

Houda El Bouhissi, Archana Patel and Narayan C. Debnath

CONTENTS

14.1 INTRODUCTION AND MOTIVATION

With the rapid growth of the Internet and emerging technologies, the world has moved to an electronic world where most things are digitized and available at the click of a mouse. Most business transactions are performed on the Internet using online shopping, which makes e-commerce more popular. Recently, e-commerce has become more popular; customers are buying more and more products on the Internet, and companies are selling more and more products on the Internet. When a user wants to buy a product on the Internet, they visit an online store and searches for the

DOI: 10.1201/9781003310785-14

item that interests them. There are many popular e-commerce sites, such as eBay and Amazon. These online stores sell many items with many makes and models available for a single item. The ability for the customer to choose from a large number of products increases the information-processing burden before deciding which products meet their needs. If the customer is unsure about choosing a product, they may be faced with the problem of information overload. They may encounter a situation where they cannot decide which product to buy.

Recently, recommender systems (RSs) appeared, which are classes of algorithms that provide personalized suggestions to users based on their preferences. RSs have a particularly important place in online marketing.[1] Thanks to them, e-commerce companies have been able to differentiate themselves from the competition, make life easier for current customers, and reach their potential customers. Whenever a user visits a site and selects a product to buy, the sites recommend other products to buy. E-commerce RS tries to predict, based on product information and the user's profile, which products may be of interest to the user. According to the company's strategy, many recommendation techniques are combined to meet the users' needs.[2] These methods have different advantages and disadvantages and cannot solve all problems. Typically, companies combine multiple approaches to provide the best recommendations. Many approaches have been proposed for building accurate and efficient e-commerce RSs. However, to the best of our knowledge at this time, little effort has been spent on providing more accurate e-commerce RSs.

The aim of this chapter is to review the most important works related to e-commerce RSs and propose a new ontology based-approach to increase the efficiency and accuracy of existing e-commerce RSs. The chapter can be summarized as follows: Section 14.2 presents a summary of preliminaries about the RSs algorithms and domain ontologies. In Section 14.3, we overview the most important related works. Section 14.4 describes the proposed approach. Section 14.5 reports the results of the proposed methodology. Finally, Section 14.6 concludes the chapter and outlines future projects.

14.2 PRELIMINARIES

For a better understanding of the concepts, in this section, we present the theoretical background of the chapter and our contribution. This includes brief overviews of the topic and the ontology used.

14.2.1 Recommender Systems

A RS is a collection of algorithms with a particular type of information filtering that can offer tailored suggestions within a data space. The purpose of a RS is to provide to user pertinent items based on their choices by filtering contents and to guide them to interesting and helpful resources based on their needs. Movies, music, news, websites, books, videos, photos, etc. are all examples of items. Overall, RS is an intelligent software tool that recommends items to a user that will meet their needs or simply stimulate their attention. Recommendation examples include suggestions music to listen or book to buy regarding the individual's tastes.[3]

The three main types of strategies used by recommendation engines are collaborative filtering (CF), content-based (CB), and hybrid recommender systems, which combine the two techniques.[4] In comparison to content-based suggestions, collaborative filtering is currently one of the most widely utilized methods (Spotify, Netflix, and YouTube). CF recommender systems are independent of other users but based on a user's own past records to infer the usefulness of an item based on other users' evaluations. The goal of CB recommender systems is to provide users with recommendations for products that are similar to those that have previously piqued their interest.

Despite their popularity and success, CF methods are still facing a number of serious limitations. These limitations include:[5]

- Cold start: The system cannot recommend an item to someone until a number of users purchase that item. This problem is known as the cold start. The user's cold start refers to a user who has recently joined the recommender system and there is not enough information about him or her.
- Scalability: Due to the sheer volume of customer data and product data, the system has to spend a lot of money and time for extracting the similarities among all the customers, leading to its poor productivity.
- Sparsity of the user-item matrix: Due to the huge number of users, sufficient scores or purchases for an item are not usually available, leading to sparse user-item matrices.
- Change in user preferences: The tastes and preferences of users may change over time. For instance, a mother who has recently had a baby may be interested in buying baby clothes and her recent purchases may reflect this new attention. After a while, she may no longer have any interests in such items.

14.2.2 Ontologies

Ontology is a of the branches of philosophy that deals with the nature and organization of reality, and it can be defined as "a conceptualization of explicit description".[6] One of an ontology's primary goals is to express knowledge about a topic in a machine-readable and concise way. Ontologies' utility therefore depends on data integration, search efficiency, and data organization.[7] For knowledge representation in the e-commerce domain, various ontologies are designed as a result, and the semantic search of applications was enhanced because to these ontologies. Nowadays, ontologies have become a common means for sharing and reusing information across the organization because of their ability to add new knowledge and introduce sharable and reusable knowledge about the desired domain. This scenario creates a platform for developing independent ontologies in the same or different languages (monolingual and multilingual ontologies) within the similar or different domains having some common information, which results in a heterogeneity problem. Data heterogeneity produces variation in meaning or ambiguity in the interpretation of entity, which prevents information sharing between systems. Therefore, without identifying the semantic mappings between entities, we cannot communicate, interact, collaborate, or share information across applications.

14.3 RELATED WORKS

Many approaches have been proposed for building accurate and efficient e-commerce RSs that use a wide variety of techniques.

This section describes the main RSs according to the e-commerce domain and summarizes their main features, advantages, and disadvantages. Proposals are based on different techniques, including machine learning (ML) and ontologies.

Overall, RSs use different datasets and, therefore, different results are gained in the same e-commerce domain. Most of these approaches use the collaborative filtering (CF) techniques; this attempts to find a group of users who have the same preferences. Some systems are based on the content-based (CB) techniques, which are based on profiles. In fact, we build profiles for users and as well as for products. Other proposals are based on ontologies as a semantic model.

Zhang et al. (2019) introduced an improved CF recommendation algorithm for e-commerce called "NewRec" and verified its efficiency through experiment simulation.[8] To better remedying the problems of data sparsity and rating time factor, the authors adopted level filling method to predict the non-rated items and finally combine time weights in the recommendation prediction stage to improve the recommendation accuracy of the algorithm. They used the hierarchical filtering method and improvement of recommendation timeliness. They indicated that their algorithm is more efficient and accurate than the traditional CF recommendation algorithm.

Khodabandehlou (2019) [9] proposed an e-commerce RS based on CF and using a data mining approach. The objective of the study was to resolve RSs limitations involving scalability issue and cold start. To accomplish this, the clients were first categorized into groups according to various features such as frequency and cost, and each group's shopping cart was evaluated using association rules based. Product classes anticipated for each of the target consumers because of segmentation and association rules were used as the CF's input. The author asserted that the results showed that of the performance of proposed system are better than legacy RSs and the accuracy of the new RS has increased significantly. Jiang et al. (2019) [10] presented their proposal, which aimed to remedy to the problem of low accuracy of the traditional and the untrusted ratings in RSs. Moreover, they proposed an algorithm based on the fusion of trusted data and user similarity that can be applied in many applications. They implemented their experiment on Amazon's items rating dataset. The authors concluded that their algorithm performs more accurately than the traditional algorithm. Guia et al. (2019) [11] presented a new hybrid approach that combined the simplicity of collaborative filtering with the efficiency of the ontology-based recommenders. According to the authors, the experimental evaluation showed that the proposed approach presents higher quality recommendations. They concluded that it is simpler with ontology-based models to finding users that have similar preferences with the active user. Nevertheless, the proposal was time consuming since it used the KNN algorithm to find the k-nearest products.

Wójcik and Górnik (2020) [12] presented a proposition to make use of a neural network architecture called Deep Hybrid CF with content as a product suggestion engine. The algorithm was tested on the Amazon Reviews Dataset using repeated pass validation, and it compared with different approaches such as CF and deep collaborative filtering in terms of mean squared error, mean absolute error, and suggested absolute proportion of error. The authors attested the effectiveness of the results.

Liu et al. (2020) [13] studied the issues and challenges of standard online purchasing behavior prediction methods, and they proposed a community purchasing behavior analysis and prediction system. Regarding the client data, the system obtains the customer and purchase conduct rules covered in the customer and stores the located rule expertise in the database. The authors used different models such as the e-commerce consumer behavior prediction model, which is based totally on the decision tree algorithm. Khatter et al. (2021) [14] proposed a new hybrid approach to product recommendations that involved clustering and connections between clusters using CF. The goal was to develop a hybrid model that provide good product recommendations from product information and behavioral data by grouping this data based on their similarity. The algorithm was tested on real products and purchased data from two different companies, a large online bookstore and a small online clothing store. According to the authors, the results were encouraging.

ML techniques like the KNN algorithm, K-means, and naive Bayes are used in the majority of the systems mentioned. These algorithms determine the similarity probability by evaluating how a customer will like a particular product.

Additionally, the clustering technique is commonly applied to produce the cluster since it is an effective unsupervised learning technique for accurately evaluating the substantial amount of data produced by applications. However, only strong clusters will result in an improved suggestion.

This chapter compares all of the selected proposals based on a number of critical performance indicators, including security, response time, accuracy, processing cost, and implicit/explicit data source. However, these indicators are not often considered in many publications; most studies aim to increase the accuracy of recommendations. For this purpose, we wanted to select the top ML algorithms that can manage the RS properties. Additionally, a viable strategy to increase recommendation accuracy is the combination of the ontologies with ML algorithms. Indeed, to our view, combining these two technologies will produce efficient RSs. However, to the best of our knowledge, little research proposes the combination of ontologies and ML algorithms. For this purpose, we propose a new approach based on the use of ontologies and ML algorithms to identify user needs and provide precise and effective recommendations.

Inspired from the existing approaches, our method is based on ontologies and combines the K-means algorithm to provide recommendations that are more diverse. In addition, we aim to provide recommendations for similar products and rank them in order of growth by price.

14.4 PROPOSED APPROACH

RSs play a vital role in dealing with the growth of information on the Internet; they analyze the vast data and produce personalized responses for users that meet their needs. RSs are popular both commercially and among researchers. By providing a curated list of suggested items, they help users with selecting the appropriate items concurrently and in the future. In particular, RSs have been widely used on large e-commerce sites such as Amazon and eBay where millions of products are supplied and making recommendations to thousands of users in real time.

Besides their popularity, the existing e-commerce RSs have limitations, mainly in two aspects:[15] First, the data sparseness problem is the evaluation of the user when there are fewer products to buy. As a result, the quality of recommendation will be reduced. Second, the cold start problem, which refers to the "new product problem" and the "new user problem"; performance will be reduced when both adding a new product without any score data, as well as a new user with no products to score. The research community deployed RSs employing ontologies as a knowledge base across a variety of disciplines, including product recommendation in e-commerce systems, recommendations in tourist applications, and course recommendations in e-learning systems.[16] The goal is to help clients with locating the products and goods that best match their user profiles.

The vector-based method of representing profiles proposed by traditional recommendation techniques, which is based on keywords, has a major drawback: it does not capture the semantics of the profile, as it is mainly dictated by alignment operations between strings. To remedy this, ontologies have been used in some recommendation systems to support the representation of user profiles (respective item profiles).[17] Technologies based on ontologies as a semantic model and e-commerce recommendations can meet products for the users' needs. By using ontologies, RSs can be improved, and recommendations will be faster since we reduce workspace by removing all data that does not match.

The proposed approach addresses the problems of the previous methods to efficiently evaluate user preference on products and balance feature analysis. Our RS is inspired by the state-of-art approaches based on the hybrid recommendation technique which combines CF and CB, since limitations of one technique can be overcome by other techniques, and use a combination of ML algorithms and ontologies. Our system is able to analyze products and similarity with the active user to better generate product recommendations.

In this proposal, different steps are involved online and offline, and new tools are produced to help the recommendation process. An experimental test of the proposed techniques was implemented to show the impact of the proposed algorithms on decreasing the time and the effort of the recommendation process and limiting shortcomings of cold start issues and data sparsity. In addition, using ontologies provide faster and more accurate recommendations.

Figure 14.1 presents the architecture of our proposal, which is a semi-automatic approach and mainly involves two main steps. The offline step involves the preparation of the dataset and the clustering of the products into categories. This step is performed once a new product is added to the dataset.

The online step concerns the recommendation process overall and involves different phases from the introduction of the user information to the return of the top N recommendations.

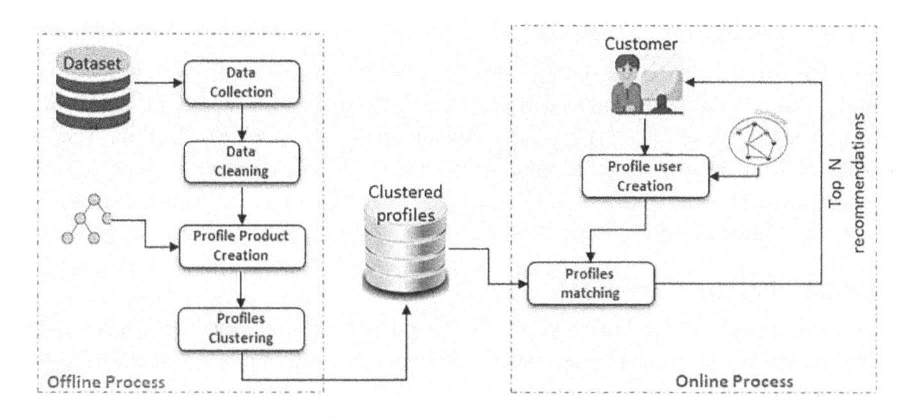

FIGURE 14.1 System architecture.

TABLE 14.1
Dataset Description

	Data type	Description
ID	Integer	Product identifier
Product name	String	Name of the product
Category	String	Category of the product
Selling price	Float	Product price
Product URL	Varchar	Product link

14.4.1 Offline Step

This step is performed casually whenever a new product is introduced to the dataset. It aims at preparing the dataset for the recommendation process. This step comprises different phases.

14.4.1.1 Data Collection

The first phase of the proposal is to import useful data from Amazon's dataset. Therefore, this step is very important for the recommendation process. The dataset includes (Table 14.1): "product name", "category", "price", and other features.

14.4.1.2 Data Cleaning

Retrieved information mostly involved numerous words due to conventional names, thus a preprocessing phase was needed. In this step, we cleaned the collected data to filter it and kept only useful data. We removed all useless characters such as ":", "&", "URL" and corrupted, improperly formatted, duplicated, or incomplete data.

14.4.1.3 Product Profile Creation

This phase involved the product profile model creation based on the earlier information. The profile skeleton comprised the information presented in Table 14.1. According to each product, a profile was assigned and matched to a domain ontology. The aim here was to replace the profile information by ontology instances. For this purpose, we used a semantic similarity measure,[18] which quantified how two entities are corresponding.

14.4.1.4 Profiles Clustering

In this phase, we gathered product profiles based on their common characteristics by using the clustering technique K-means. The purpose of this phase was to make the profile's space smaller.

K-means [19] is a popular ML algorithm which finds similar groups in the feature space in input data. The major focus of this phase was the clustering of N numbers of data into K numbers of clusters, with each data point belonging to the cluster with the smallest mean distance.

Usually, the process of clustering comprises three steps: first, handling the data, then selecting the best distance function to verify how related the items are, and finally obtaining the groups.

We selected K initial centroids in the K-means clustering algorithm, where K is the desired number of clusters. Each point was subsequently assigned to the cluster's centroid, which had the closest mean.

Notice that the ML algorithm cannot understand the clustering rules from textual data. K-means, like any ML algorithm, needs numerical features to understand classification rules. For this purpose, we performed the Doc2vec feature engineering technique to produce a numeric vector representation of each product.

14.4.2 Online Step

This step is the core of the recommendation process. It is conducted in real time whenever a customer accesses the system to purchase products.

14.4.2.1 Product Profile Creation

This phase is similar to the offline phase. We built the user profile model based on the introduced information, i.e. when the customer introduces their preferences (products to purchase). The user profile skeleton was comprised of the same information presented in Table 14.1 because we were concerned with products to be purchased or viewed. Similarly, according to each customer, a profile was assigned and matched to a domain ontology. The aim was to replace the profile information by ontology instances. For this purpose, we used a semantic similarity measure, which quantifies how two entities are corresponding.

14.4.2.2 Profiles Matching

This phase is the most important phase of the recommendation process. We performed a matching process between the user profile and the clustered profiles using a semantic similarity measure, and we kept only the product that matched higher with the user profile.

14.5 EXPERIMENT AND EVALUATION

To verify the efficiency of our approach, we performed experiments in Python language because it is the language that is most popular for implementing ML codes. The experiments included whether or not ontologies were used. The quality of a recommendation algorithm can be evaluated using different metrics. The type of metrics used depends on the type of recommendation technique.

To evaluate our proposal, we used precision and recall metrics since they are widely used in data science to evaluate recommendation and prediction systems.[20]

Precision (Equation 1) measures the accuracy of a classifier. Higher precision means fewer false positives, while lower precision means more false positives. This is often at variance with recall, as an easy way to improve precision is to decrease recall.

$$Precision = \frac{Correctly\,recommender\,products}{Total\,recommended\,products} \tag{14.1}$$

Recall (Equation 2) measures the completeness, or sensitivity, of a classifier. Higher recall means fewer false negatives, while lower recall means more false negatives. Improving recall can often decrease precision because it becomes increasingly difficult to be precise as the sample space increases.

$$Recall = \frac{Correctly\,recommender\,products}{Total\,useful\,recommended\,products} \tag{14.2}$$

According to the obtained results of Table 14.2, we concluded that for the given dataset, using ontologies with ML produces a rather better result.

14.6 CONCLUSION AND FUTURE WORK

To reduce the issue of information overload, which has produced potential issues for many Internet users, it is necessary to classify, filter, and efficiently transmit suitable information on the Internet, where the amount of possibilities is excessive. RSs may resolve this issue by exploring a large volume of dynamically generated information

TABLE 14.2
Presents Precision and Recall for All the Experiments

Proposal	Precision (%)	Recall (%)
Using ontologies	77	68
Using ML algorithm	82	80
Using both ontologies and ML algorithm	93	94

to provide customers with personalized content and services that meet their needs. Depending on the domain for which they are submitted, a variety of RS techniques and algorithms can be employed for various types of recommendations. Usually, the required suggestion will determine the specific technique to be employed in a RS. The main purpose of this chapter is the development of an e-commerce RS that is able to produce personalized products or goods recommendations that meet the needs of users. Using ontologies of user-profiles and product descriptions will help enrich information as well as semantic relationships between customers and products. Scalability and sparsity are two main issues in the design of recommender systems. Accordingly, in this chapter, attempts have been made to solve these issues to improve the performance of RSs. The proposed approach uses ontologies and ontology matching concepts with the aid of clustering techniques. As future work, we would like to improve our proposal to use deep learning algorithms and to predict that RS will be used in the future to expect request for products to enable earlier communication back the supply chain.

REFERENCES

1. El Bouhissi, H., Adel, M., Ketam, A., & Salem, A. B. M. (2021, March). Towards an efficient knowledge-based recommendation system. In *2nd international workshop on intelligent information technologies and systems of information security (IntelITSIS)*. Khmelnytskyi, Ukraine: CEUR Workshop Proceedings, pp. 38–49.
2. Aggarwal, C. (2016). Recommender systems, springer international publishing. https://doi.org/10.1007/978-3-319-29659-3.
3. Kotkov, D., Wang, S., & Veijalainen, J. (2016). A survey of serendipity in recommender systems. *Knowledge Based System*, 111, 180–192.
4. Zhang, Q., Lu, J., & Jin, Y. (2021). Artificial intelligence in recommender systems. *Complex & Intelligent Systems*, 7, 439–457. https://doi.org/10.1007/s40747-020-00212-w.
5. Francesco, R., Lior, R., Shapira, R., & Bracha, S. (2015). Recommender systems: Introduction and challenges. doi:10.1007/978-1-4899-7637-6_1
6. El Bouhissi, H., Patel, A., & Debnath, N. C. (2022). Towards data integration in the era of big data: Role of ontologies. In *Semantic web technologies: Research and applications*. CRC Press, pp. 359–380.
7. El Bouhissi, H., Salem, A.-B. M., & Tari, A. (2019). Semantic enrichment of web services using linked open data. *International Journal of Web Engineering and Technology*, 14(4), 383–416.
8. Zhang, Z., Gongwen, X., & Pengfei, Z. (2019). Research on E-commerce platform-based personalized recommendation algorithm. *Applied Computational Intelligence and Soft Computing*, 1–7. doi:10.1155/2016/5160460
9. Khodabandehlou, S. (2019). Designing an e-commerce recommender system based on collaborative filtering using a data mining approach. *International Journal of Business Information Systems*, 31, 455–478. doi:10.1504/IJBIS.2019.101582
10. Jiang, L., Cheng, Y., Yang, L., et al. (2019). A trust-based collaborative filtering algorithm for E-commerce recommendation system. *Journal of Ambient Intelligence and Humanized Computing*, 10, 3023–3034. https://doi.org/10.1007/s12652-018-0928-7
11. Guia, M., Silva, R., & Bernardino, J. (2019). A hybrid ontology-based recommendation system in e-Commerce. *Algorithms*, 12, 239. doi:10.3390/a12110239
12. Wójcik, F., & Górnik, M. (2020). Improvement of e-commerce recommendation systems with deep hybrid collaborative filtering with content: A case study, 24, 37–50. doi:10.15611/eada.2020.3.03

13. Lu, Y., Liu, C., Kevin, I., Wang, K., Huang, H., & Xu, X. (2020). Digital twin-driven smart manufacturing: Connotation, reference model, applications and research issues. *Robotics and Computer-Integrated Manufacturing*, 61, 101837.
14. Khatter, H., Arif, S., Singh, U., Mathur, S., & Jain, S. (2021). Product recommendation system for e-Commerce using collaborative filtering and textual clustering. In *The third international conference on inventive research in computing applications (ICIRCA)*. New York: IEEE, pp. 612–618.
15. Wang, T. The application of ontology technology in E-commerce recommendation system. In D. Jin & S. Lin (Eds.), *Advances in future computer and control systems. Advances in intelligent and soft computing*, vol. 159. Berlin, Heidelberg: Springer. https://doi.org/10.1007/978-3-642-29387-0_12
16. Le Ngoc, L., Abel, M. H., & Gouspillou, P. (2021). Towards an ontology-based recommender system for the vehicle domain. In *3rd international conference on deep learning, artificial intelligence and robotics (ICDLAIR)*. Salerno, Italy: HAL Open Science.
17. Du, Y., Ranwez, S., Sutton-Charani, N., & Ranwez, V. (2019). Apports des ontologies aux systèmes de recommendation: état de l'art et perspectives. In *30es Journées Francophones d'Ingénierie des Connaissances, IC 2019*. Toulouse, France: AFIA.
18. EL Bouhissi, H., Malki, M., & Sidi Ali Cherif, M. A. (2014, July–September). From user's goal to semantic web services discovery: Approach based on traceability. *International Journal of Information Technology and Web Engineering (IJITWE)*, 9(3), 15–39.
19. Rabi, B., & Kajaree, D. (2017). A survey on machine learning: Concept, algorithms and applications. *International Journal of Innovative Research in Computer and Communication Engineering*, 2.
20. Bekka, R., Kherbouche, S., & El Bouhissi, H. (2022). Distraction detection to predict vehicle crashes: A deep learning approach. *Computación y Sistemas*, 26(1).

Index

E

F

G

H

I

K

L